Current Trends in Human Ecology

Current Trends in Human Ecology

Edited by

Priscila Lopes and Alpina Begossi

CAMBRIDGE
SCHOLARS
PUBLISHING

Current Trends in Human Ecology, Edited by Priscila Lopes and Alpina Begossi

This book first published 2009. The present binding first published 2011.

Cambridge Scholars Publishing

12 Back Chapman Street, Newcastle upon Tyne, NE6 2XX, UK

British Library Cataloguing in Publication Data
A catalogue record for this book is available from the British Library

ISBN (10): 1-4438-3000-3, ISBN (13): 978-1-4438-3000-3

Contents

Contents

List of Tables

List of Boxes

List of Figures

Preface

EMÍLIO F. MORAN
INDIANA UNIVERSITY

The volume Trends in Human Ecology provides a welcomed update on recent trends and issues in human ecology, derived from a recent international symposium of the Society for Human Ecology held in Rio de Janeiro, Brazil, in October 2007. Organized by Priscila Lopes and Alpina Begossi, the volume provides overviews of very recent new trends in the field—such as interest in climate change, vulnerability, biodiversity conservation, fisheries management—as well as excellent updates on more traditional topics such as slash-and-burn cultivation, indigenous knowledge, and the development of human ecology across the world.

It is no easy task to produce any volume on human ecology that intends to be encompassing of this transdisciplinary subject, particularly within the context of SHE, the Society for Human Ecology. SHE has always been a very open society, which began with a focus on environmental education, but quickly encorporated the majority of disciplines regarding the relationship between humans and the environment into its agenda. This has always made the meetings interesting but lacking in some coherence and "take home" messages. The meeting in Rio was different. The organizers kept the door open but they managed to give this one meeting a wonderful coherence around a focus on fisheries and fisheries management around the world. Thus the tradition of SHE was maintained while also having an exceptional focus on a topic very appropriate to its location—since Rio was a fishing community before it became one of the most beautiful tourist destinations.

This volume has several papers on fisheries management, including traditional knowledge possessed by fishers. These are among the strongest papers in the volume, as might be expected given the expertise of the organizers, and the strong focus of the meeting on this topic. This will help readers of the volume to come up to speed on a topic which is less well known to most of the community of human ecologists who are more terrestrial by inclination. This is an important corrective since most of our planet is water and most people live within close proximity to the coast, yet so much of our research in various disciplines (including human ecology) tends to occur far from the coastlines and overlooks the "fishing mode of production". We think about hunting/gathering, agriculture, pastoralism, and industry but tend to overlook fishing (except as an overlooked appendage to hunting). This volume goes a long way to restore some balance to our view of the various activities that make up our systems of production and reproduction.

There is very little in the volume on hunting/gathering, perhaps reflecting the much reduced attention to these populations in recent years, and their ever diminishing numbers worldwide. The chapter on slash-and-burn is a superb update on a persistent material mode of production, and one that always seems to change and adapt to changing conditions. It still constitutes a basis for the living of many millions of people, who now combine it with more intensive activities as well. The meticulous literature review here is a major service to the human ecology community.

The chapters dealing with climate, culture, and vulnerability highlight the role of social scientists in addressing the climate change dynamics around us. It also shows that there is still a major divide between what social scientists are doing and saying about climate change, and the work of atmospheric and climate scientists in the natural sciences. The discussion here is still more from the view of social science, and does not link to the modelling activities and the uncertainties that are at the center of climate models and the changing scales of these studies. By having this discussion here, we can hope that readers and other scientists will join the climate scientists in having a more integrative approach to climate change that takes into account the human dimensions which are surely as important as the natural processes.

This is consistent with the direction of the field of human ecology, as it moves and joins other scientific tendencies towards Integrative Science, wherein natural and social sciences formulate questions jointly, engage in research in a joint enterprise, and provide integrative analyses with real policy implications that do justice to the human and the natural processes. This volume begins to move us in this welcomed direction, and can be seen as representing the trends in research that must characterize work in the 21st century.

Introduction

ALPINA BEGOSSI AND PRISCILA LOPES

The Society for Human Ecology and its contribution to the development of the field

Human Ecology is the study of all relationships that take place between human beings and nature. Under this discipline, human culture and behaviour are not seen as a sole product of society, but also as the result of the influence of and interaction with physical and biological variables. As such, several ecological concepts ranging in scale from population to evolutionary ecology are commonly applied and associated to economic, anthropological and sociological theories and applications.

Writing about Current Trends in Human Ecology is a dynamic task, due to a diverse domain of research and multiple scales that have resulted in the accelerated development and rapid production of evolving approaches. This can be certainly observed within the meetings of the Society for Human Ecology (SHE) and its themes to date.

This book catches a glimpse of today's intellectual diversity and recent trends in the field of Human Ecology. Based on the XVth SHE International Meeting held in Rio de Janeiro (October 2007), this book shows the varied research lines that were presented at that time. However, this volume goes beyond the meeting as it lays out a detailed cumulative summary of the current trends in human ecology.

SHE has been an institutional foundation that collects up-to-date developments in human ecology topics, facilitating critical analysis and making them available through conferences, peer-reviewed journals and scientific gatherings. As an interdisciplinary field, Human Ecology is a novelty and only recently has it started to be included in the academic curricula, despite its critical relevance, given emerging social and environmental challenges. Borden (2008) wrote a brief history of the Society for Human Ecology, highlighting some of these important aspects inherent to an interdisciplinary field:

> Each discipline has a home territory, architecturally walled off as its own domain. Some of the fiercest barriers divide adjacent fields. It is little wonder that interdisciplinary pursuits have struggled to secure a place in modern universities. These are the obstacles the founders of the Society for Human Ecology confronted. They still remain. It is unlikely they will ever disappear.

The challenge of interdisciplinary pursuit has contributed to maintain the vigour of such a self-motivated field. We agree with Borden (2008) that Human Ecology makes no claim that pursuing specialized knowledge is misdirected. Specialized knowledge leads the mind to specific complex questions. Nevertheless, the open minds of human ecologists bring up heretic questions, linking areas that are traditionally separated, contributing with new methods, and then making human ecology continue to develop within an authentic scientific domain.

Box I-1 – Themes of the Meetings of the Society for Human Ecology.

1. Human Ecology: a gathering of perspectives, College Park, Maryland, 1985.
2. Human Ecology: research and applications, Bar Harbor, Maine, 1986.
3. Human Ecology: steps to the future, San Francisco, California, 1988.
4. Human Ecology: strategies for the future, East Lansing, Michigan, 1990.
5. Human Responsibility and Global Change, Gothenburg, Sweden, 1991.
6. Human Ecology: crossing boundaries, Snowbird, Utah, 1992.
7. Human Ecology: progress through integration of perspectives, East Lansing, Michigan, 1994.
8. Human Ecology: livelihood and liveability, Lake Tahoe, California, 1995.
9. Local and global communities: complexity and responsibility, Bar Harbor, Maine, 1997.
10. Living with the land: interdisciplinary research and adaptive decision making, Montreal, Canada, 1999.
11. Democracy and sustainability: adaptive planning and management, Jackson Hole, Wyoming, 2000.
12. Tourism, travel and transport: a human ecological perspective on human mobility, Cozumel, Mexico, 2004.
13. Human Ecology: research and practice, Salt Lake City, Utah, 2005.
14. Interdisciplinary integration and practice: reconciling humans and nature, Bar Harbor, Maine, 2006.
15. Local populations and diversity in a changing world, Rio de Janeiro, Brazil, 2007.
16. Integrative thinking for complex futures: creating resilience in human-nature systems, Bellingham, Washington, 2008.
17. Human Ecology for an urbanising world, Manchester, United Kingdom, 2009.

The annual meetings of SHE creates a forum that transmits developments in the area of human ecology, reflecting the local and regional issues where the meeting was held. Beginning with a Directory of Human Ecologists (Borden and Jacobs 1989), the society has organized meetings since 1985. Meetings typically have integrative perspectives, as it can be seen by some of the conference themes: "Livelihood and Liveability", in Lake Tahoe, USA, 1995; "Living with the Land", in Montreal, Canada, 1997; "Adaptive Planning and Management", in Jackson Hole, USA, 2000; and "Tourism, Travel and Transport", in Cozumel, Mexico, in 2004 (see Borden 2008, for a brief history of SHE and Box I-1 for a complete list of conference themes).

Human Ecology in Rio: Past and Current Trends

The development of human ecology covers an array of different areas integrating humans and their interaction with the environment. Among some of the authors that have contributed to the development of theories and methodologies it is worth mentioning Richerson (1977), who compared theories between biological and social sciences and then went further with studies on the interactions of gene and culture (Boyd and Richerson 1985). Morán (2007), who has been studying human adaptability, especially in the Amazonian frontiers; and Berkes (1989, 2008), whose contributions are directed towards the theory of the commons and the intersection between local knowledge and adaptation.

Under the domain of Human Ecology, there are different fields approaching the relationships of human populations and nature. After the 1980's some areas progressed according to their own dynamics, as they changed and adapted to new information from the biological and social sciences. Examples of this include the case of Cultural Ecology (Steward 1955; White 1943; Harris 1977, 1979; Orlove 1980), Ethnobiology (e.g.: Berlin 1992) and applications of modelling behaviour, such as optimal foraging theory and life history strategies (Winterhalder and Smith 1981; Smith and Winterhalder 1992; Borgenhoff Mulder 1992). Sociobiology must also be mentioned, as it covers many different aspects of human behaviour, and currently addresses other social polemic issues, such as religion (Dawkins 2006). We should not forget about the branches of Cultural Transmission models, such as Richerson and Boyd's (2005), and Lumsden and Wilson's (1981). Ethology had its own development side by side to human ecology, and Bekoff (2002) blends the interface between humans-animals and the environment.

Box I-2 – Topics approached in the XVth Society for Human Ecology Conference, Rio de Janeiro, Brazil.

1. Adaptive Co-Management: Building Resilience in Local Management Systems
2. Advances in Ethnoecology and Ethnobotany
3. Agroecology and Sustainable Rural Development
4. At-Distance Education in Human Ecology: Opportunities and Challenges
5. Behavior and Ecology: Psychology Looks at Mind and Nature
6. Climate and Culture
7. Co-managing Common-Pool Resources: Challenges and Advances
8. Community-Based Faunal Management
9. Contributions of Fishers' Local Ecological Knowledge and Scientific Research to Marine Mammals' Conservation
10. Contributions of Human Ecology to Understand and to Manage Artisanal Fisheries
11. Culture Nurturing Nature: A Chance for (Environmental) Change
12. Diversity and Management: from Extractive to Farming systems
13. Environment and Agriculture
14. Environmental Degradation
15. Environmental Pollution and Cultural Pollution
16. Environmentally Significant Consumption (ESC): Emphasis on Food
17. Folk and Indigenous Transformations in the Amazon region: Implications of Contact, Land-Use Schemes and Market Economies
18. Geographic Information Systems and Remote Sensing in Studies of Population and Environment
19. Human Ecology and Health
20. Integrated Analysis of Local Socioecological Systems: Combining Agent-Based and Stock-Flow Modeling Approaches
21. Local Knowledge (LEK/FEK) and the History of Aquatic Ecosystems
22. New Directions in Human Ecology - Higher Education for Sustainable Development
23. Searching the Real Values of the Sea: the ECOST Perspective
24. Shifting Cultivation and Tropical Forests in a Changing World
25. Social-Ecological Systems Analysis: Understanding Complex Social-Ecological Dynamics: Integrating different geographical scales in coastal systems
26. Sustainability and the Impact of Aid on Indigenous People
27. Temporal and Spatial Cross-Scale Approaches in Human Ecology: Implications for Resource Use from Cultural Evolution and Behavioral Ecology
28. The role of Indigenous knowledge in the Adaptive Process
29. Traditional People, Biodiversity and Cultural Diversity
30. Using and Perceiving a Changing Environment

Aspects related to human cognitive features have also played a role within human ecology. Linguistic and evolutionary psychology have emerged with new questions, methods, and a critical view of previous evolutionary approaches (Hauser 2000; Gottschall and Wilson 2005; Pinker

1994). Education has always been part of each of these branches in human ecology. Reviews covering different areas and authors in human ecology are found in Begossi (1993, 2004), Borden (2008), Laland and Brown (2002), among other researchers.

Recently, the need and urgency for natural resource conservation has been globally recognized. Applications, such as ecological economics, local knowledge and management, collective action, socio-ecological modelling, and adaptive management, have addressed the pressing needs of conservation (Berkes 2008; Ostrom 2002). This aspect is recurrently emphasized in all case studies presented in this book.

The XVth Meeting in Rio de Janeiro, Brazil in 2007, which this book is based on, focused on the theme "Local Populations and Diversity in a Changing World". However, given the broad spectrum of the field, the meeting encompassed a sub-set of the above topics, as shown below. Thirty topics were approached by about 200 people who attended the meeting, representing the developments of Human Ecology in 27 countries from all inhabited continents (Box 2). While integrating this meeting, not all the individual topics were included in the organization of this book, in order to provide a synopsis of the main ideas. However, two of these topics also resulted in special volumes of two peer-reviewed journals:

a) *Environment, Development and Sustainability*, organized by Hens and Begossi (volume 10, 2008), based on the session "Diversity and Management: from Extractive to Farming systems" (session 12);

b) *Boletim do Museu Paraense Emilio Goeldi*, organized by Adams and Murrieta (volume 3, 2008), based on the session "Shifting Cultivation and Tropical Forests in a Changing World" (session 24).

In such regard, we minimized redundancy and looked for more general contributions that could bring a state-of-the-art approach on main fields. We must acknowledge that we had pleasure in putting together the pieces of work that will follow in this book, revealing how human ecology enriches our minds, builds tools to adapt to a changing environment, and shows potential mechanisms and policies to sustain it.

This book is divided into three sections. **Section I,** Human Ecology and the Environment, deals with three aspects of environmental concerns and uses of it: slash-and-burn agriculture and its impacts and trade-offs regarding tropical forests; how local cultures perceive and suffer the consequences of ongoing changes in climate; environmental degradation and the cultural context of decisions concerning the environment. **Section II,** Knowledge and Management, explores how knowledge is used to correct previous unsustainable resource exploitation, using theories and methods from human ecology. Examples are drawn from tropical and coastal marine fisheries. **Section III,** Integrating Human Ecology, brings together areas that have been tangential to human ecology yet inherent to human systems. Specifically, this section focuses on health as a function of environmental conditions, the link between economic valuation and environmental and social costs, and institutional education of human ecology, topics that are typically outside the agenda of human ecology.

Section I, Human Ecology and the Environment, has Chapter One by Pedroso-Junior et al., on slash-and-burn agriculture and tropical forests. It is a much needed review of tropical agriculture, describing impacts of such management, including removal of primary forests and more permanent alterations of the biogeochemical environment. It shows how such changes can affect the soil, for example when the fallow periods of slash-and-burn shrinks, and how subsequent deforestation contributes to greenhouse effects and climate change. Land use changes and alternatives to slash-and-burn are provided through a detailed analysis. The chapter on Climate Change and Culture, by Crate follows, beginning with key concepts from ecology. It describes and compares habitats under transformation using examples of current cultures from high to low latitudes, as well as different altitudes: a review from the current anthropological literature is provided throughout the chapter. Burn's study in Chapter Three draws on culture and its association with environmental degradation. It defines culture in a temporal context, along with its collective meaning and perspective. The ethical, institutional and political foundation of culture is analyzed in the light of environmental changes.

Section II, Knowledge and Management, begins with Chapter Four by Silvano et al. drawing attention to the management of tropical small-scale fisheries. Here, the lack of information in tropical fisheries and the urgency of managing declining and threatened fisheries are described. Paralleling such aspects, an analysis of concepts and methods of local ecological knowledge (LEK) are shown, integrating scientific and local knowledge. A case study illustrating the applications of LEK and its methods with fishermen and cetaceans are shown in the second part of this chapter. In Chapter Five Hanazaki et al. bring a novelty in terms of ethnobotanical applications, describing how fishers use forest resources to improve fishing output, linking biodiversity to local knowledge. The authors show that knowledge about plants can persist among fishers despite current urbanization processes. Morrison and Singh bring a chapter on adaptation and indigenous knowledge. It begins with sustainability concepts, methods and qualitative measures, and links ecological concepts and methods to indigenous knowledge in a historical context. The importance of understanding figurative thought and substantive knowledge (skills and practises), among other aspects of indigenous knowledge, might bridge authentic dialogue between dominant global culture and indigenous culture. Seixas et al. brings an analysis of co-management in Brazil, using a framework of common property theory. After providing the local context, five case studies are shown, covering a wide diversity in aquatic and coastal habitats. Chapter Eight, the last of Section II, focuses on human ecology of coastal and marine systems. The chapter by Glaeser et al. embodies a conceptual-methodological analysis of integrated coastal zone management (ICZM) using social-ecological system analysis. These authors present an historical background of coastal management along with an interdisciplinary and epistemological analysis. Future prospects such as management, integrative methods, and the difficulties of dealing with uncertainty and complexity in human-nature systems are highlighted.

The book ends with **Section III** closing connections that encompass all previous chapters, conceptually, historically and methodologically. Chapter Nine, by Avila-Pires, Human Ecology and Health, emphasizes mainly the human dimension of human ecology. It focuses on the health of rural and urban populations, and how agriculture and its infrastructure affect the epidemiological processes in human populations. Chapter Ten, by Failler et al., emphasizes the importance of linking management of marine resources to societal benefits, using examples from the international ECOST project. Opening with an evaluation of traditional bio-economic models and their limitations, they analyze valuing processes of fishing activity, considering ecological, economic and social costs. Non market values, such as marine heritage and its cultural symbols are also considered. The 11th and final chapter, by Dyball et al., begins with the future of institutional education of human ecology, approaching current technology, communication processes and online resources. Putting human ecology in an historical and interdisciplinary perspective, the authors contextualize educational networks that came out with the development of the area. The use of at-distance education offers promise for academic curricula in human ecology. Examples of recent institutional developments are provided from various universities around the world, along with incentives for future progress.

Acknowledgments

We wish to acknowledge all the organizers of the event and sessions of the XVth International Meeting of the Society for Human Ecology, held in Rio de Janeiro, Brazil, October 4-7, 2007, organized by Alpina Begossi and Priscila F. Lopes. The contribution of researchers, by helping at the organization and giving feedback on the conference is of great value, and is a key part for the development of this book. We thank, in particular, Mohamed Habib, Provost of the University of Campinas (UNICAMP - Brazil), a supporter of Human Ecology in Brazil, and Rich Borden, Sean Berg and Barbara Carter, from the College of the Atlantic, Maine, USA. We would like to thank all the local supporters who worked actively before and during the conference: Claudia Martinelli, Cristiana S. Seixas, Eduardo Camargo, Luciana Araújo, Luziana S. Silva, Luiz Eduardo C. de Oliveira, and Shirley Pacheco (Fisheries and Food Institute [FIFO] and UNICAMP), Benedito C. Lopes, Natalia Hanazaki, Nivaldo Peroni, Tânia T. Castellani (Universidade Federal de Santa Catarina), Célia Futemma (Universidade Federal de São Carlos, São Paulo), José Geraldo W. Marques (Universidade Estadual de Feira de Santana (Bahia), Juarez C. B. Pezzuti (Universidade Federal do Pará), and Renato A. M. Silvano (Universidade Federal do Rio Grande do Sul).

We are also very grateful to all participants who helped at the session organizations or who contributed as a keynote speaker: Almo Farina (Universitá Institute of Biomathematics, Urbino University, Italy); Álvaro D'Antona (Indiana University and Universidade Estadual de Campinas, USA and Brazil); Antonio Carlos Diegues (Universidade de São Paulo, Brazil); Bernhard Glaeser (Integrated Coastal Zone Management, WZB, Germany); Cristina Adams (Universidade de São Paulo, Brazil); David C. Natcher (University of Saskatchewan, Canada); Eleanor Shoreman (Boston University, USA); Emílio Morán (University of Indiana, USA); Fernando Ávila-Pires (Universidade Federal de Santa Catarina, Brazil), Gene Myers (Western Washington University, USA), Heidi Adensam (IFF Social Ecology, Austria), Katie Demps (University of California, Davis, USA), Karl Reinhard (Nebraska Institute of Forensic Science), Katherine S. McCarter (The Ecological Society of America [ESA] Executive Director), Keith Morrison (Lincoln University, New Zealand), Kathy E. Halvorsen (Michigan Technological University, USA), Louis Forline (University of Nevada-Reno, USA), Luc Hens (Free University of Brussels, Belgium), Marion Glaser (Center for Tropical Marine Ecology, Germany), Markus Vinnari (Turku School of Economics, Finland), Mary Gasalla Oceanographic Institute, Universidade de São Paulo, Brazil), Michael K. Orbach (Marine Laboratory, Duke University, USA), Michael R. Edelstein (Ramapo College of New Jersey, USA), Miguel Petrere Jr. (UNESP, Brazil), R. Matthews (IFF & Macaulay I., Austria and United Kingdom), Peter J. Richerson (University of California, Davis, USA), Rebecca S. Purdom (Green Mountain College, USA), Renzo Taddei (Columbia University and Int. Res. Institute for Climate and Society, U.S.A.), Robert

Dyball (Australian National University, Australia), Salvatore Siciliano (Fiocruz, Brazil), Sean Todd (College of the Atlantic, Maine, USA), Simron Singh (IFF-Social Ecology, Austria), Susan A. Crate (George Mason University, USA), Thomas J. Burns (University of Oklahoma, USA), Veronika Gaube (IFF Social Ecology, Austria), and Wolfgang Serbser (German Society for Human Ecology).

Section I

Human Ecology and the Environment

CHAPTER ONE

Slash-and-Burn Agriculture: A System in Transformation

NELSON N. PEDROSO-JUNIOR, CRISTINA ADAMS AND RUI S. S. MURRIETA

Slash-and-burn agriculture has been practised for thousands of years in the forests around the world, especially in the tropics, where it provides for the livelihood of countless poor rural populations. Characterized by an array of techniques based on crop diversification and shifting land use, this cultivation system uses the energy generated from the decomposition of forest vegetation as its main asset. Many studies claim that slash-and-burn agriculture is sustainable when performed under conditions of low human demographic density, maintaining or even increasing local biodiversity. However, it is growing in the academic literature, as well as in development debates, the concern regarding the role that this system has been playing in the deforestation of the planet's tropical forests. This process appears to be closely linked to changes in land use patterns (agricultural intensification) and urban and rural demographic growth. On the thread of these concerns, this article presents a critical review of the international and national academic literature on slash-and-burn agriculture. Thus, this review intend to draw a broad scenario of the current academic debate on this issue, as well as to identify the main alternatives strategies proposed to maintain or replace this cultivation system.

Introduction

Slash-and-burn agriculture has been practised in the world's tropical regions for thousands of years, where it is the cornerstone of the subsistence systems of poor rural populations. Many studies have attempted to understand the dynamic of these systems and their environmental and socio-economic effects. Nevertheless, no hegemonic view emerges from the literature on the issue, where different and often conflicting approaches have had varying degrees of impact depending on the political and academic backdrop of the day. The literature is full of studies that treat slash-and-burn agriculture, to a greater or lesser extent, as destructive and exploitive of the natural resources of the humid tropics, very often proposing alternatives to replace it. One can also find numerous studies that argue for the system's sustainability and draw up proposals for the maintenance of this traditional mode of agriculture.

In this light, the aim of this chapter is to review the literature on slash-and-burn agriculture in order to understand the different ways in which the theme has been approached, as well as to analyse the alternatives that have been proposed to replace it and what would be the benefits of maintaining the system. Firstly, the text will look at studies that describe the system as a whole, as well as its components separately, in an attempt to characterize it and identify the practices and techniques that promote its sustainability. Secondly, the analysis will turn to the environmental impacts caused by slash-and-burn agriculture, especially those studies that apportion it some of the blame for deforestation and biodiversity loss in tropical forests, for global warming and for soil erosion and impoverishment. In a world of constant environmental and socio-economic change, special attention is paid to studies that analyze changes in forms of land use and their effects on the subsistence of smallholders that practise fire-fallowing. Lastly, the article will run through the proposals being made to cither substitute slash-and-burn agriculture or promote its sustainability through low-impact technologies and alternative income sources for the traditional farmer.

Slash-and-Burn Agriculture

Definition and Range

Slash-and-burn agriculture can be broadly defined as any continuous agricultural system in which ground is cleared to make way for fields that will be farmed for shorter periods of time than they are left to fallow (Conklin 1961; Eden and Andrade 1987; Kleinman et al. 1995; Posey 1984). McGrath (1987) defines it as a resource management strategy in which plots are rotated so as to mine the energetic and nutrient capital of the natural forest soil/vegetation complex, which often provides the only source of nutrients for cultivation. Slash-and-burn agriculture is a highly efficient adaptation to conditions in which labour rather than land is the core limiting factor on agricultural production (Boserup 1965).

The process is given many names in the literature, including *swidden* (England), *rai* (Sweden), *coivara, milpa, conuco, roza, chacra, chaco* (Latin America), *shamba, chitemene* (Africa), *jhum* (India), *kaingin* (Philippines), and *ladang* (Indonesia and Malaysia), among others. Nevertheless, the most common designations for this process are *slash-and-burn*, *shifting cultivation* and, less frequently, *swidden*. Eden (1987) suggests the use of the term "swidden" to designate shifting cultivation in the strict sense, where the scorched clearings are farmed for a shorter period than their fallow length. Though the use of the term has been encouraged by anthropologists, it has won little favour with researchers from other areas, who tend to prefer "shifting agriculture" and "slash-and-burn agriculture". However, designations vary and confusion can occur. In an attempt to standardize usage and differentiate between the two systems, Sanchez and collaborators (2005) suggested the use of shifting cultivation to refer to the traditional rotation system with long periods of fallow, which would be equivalent to Eden's "swidden" (1987), and slash-and-burn agriculture to designate other scorched earth systems with short or even non-existent fallow.

As this article is a review, we have opted for the broadest possible term, slash-and-burn agriculture, which, by extension, also includes shifting cultivation in the strict sense. In the interests of coherency with the terminology used by the authors under discussion, the distinction will be made as necessary. However, the degree of shift or proportion between farming and fallow lengths can vary greatly, making such a distinction hard to draw, as suggested by the above mentioned authors.

In terms of range, slash-and-burn agriculture is currently practised throughout the tropics and even into some sub-tropical regions. According to a study published by Lanly (1982), slash-and-burn agriculture was responsible for the formation of two-thirds of the earth's secondary forests. Of all the area under secondary forest cover originating from abandoned plots, 47% is located in Latin America, with the rest distributed between Africa and Asia (Lanly 1982). Some authors argue that, depending on the production activities established on deforested areas, secondary forest may eventually become the predominant ecosystem in the Amazonian landscape (Pereira and Vieira 2001). Although these figures need to be updated, it is estimated that slash-and-burn agriculture ensures the livelihoods of somewhere between 300 and 500 million people worldwide, the majority in the Tropics (Attiwill 1994; Brady 1996; FAO 1985; Lanly 1982), together using 240 million hectares of dense forest and 140 million hectares of open forest, or approximately 21% of the world's total tropical forest cover

(Lanly 1982). In the Amazon, traditional slash-and-burn practices feed some 600 thousand smallholder families (Homma et al. 1998).

History and Concept of Cultural Forests

Slash-and-burn agriculture is probably the oldest system of cultivation (Nye and Greenland 1960), dating back as far as the Neolithic, when human populations gradually switched from hunter-gathering to the more sedentary habits of agropastoral activities (Dean 1996; Harris 1972; Iversen 1956; Mithen 2003). However, before the advent of agriculture, anthropogenic fires may well have contributed indirectly to changes in forest ecosystems and climate (Schüle 1990a, 1990b, 1992a, 1992b). Throughout Antiquity, slash-and-burn practices were widespread in temperate regions, but with population growth in Europe and Asia, especially from the 18th Century onwards, farming activities had to be intensified and the practice was abandoned (Boserup 1965, Worster 2003). According to Boserup (1965), the system's range was limited in these regions to the sparsely populated forest remnants of Eurasia and to the mountains of Japan and Korea. In Switzerland during the Middle Ages (9th to 16th Centuries), the practice was encouraged by the government as a means of ensuring the livelihoods of the poor and of converting forest into inhabitable land, generating more taxes for the kingdom as a result (Hamilton 1997).

In tropical forests like the Amazon, where many of the wild vegetal species are inedible or hard to harvest, slash-and-burn agriculture was an extremely important adaptive strategy to support the subsistence economy in place there (Sponsel 1986). However, some authors have contested the antiquity of slash-and-burn in the Amazon. Denevan (1991) suggests that slash-and-burn agriculture with lengthy fallow periods in the Amazon was a practice introduced shortly after the arrival of the Spanish and Portuguese colonists and adopted only once metal tools had become available. He argues that it would have been very difficult to open clearings in dense Amazonian forest using stone tools, which was why the cultivations were more perennial and were disintensified as slash-and-burn practices spread after colonization. The same would seem to hold for North-America (Doolittle 1992). The importance of slash-and-burn agriculture to the subsistence of pre-Columbian Amazon populations has also been questioned from the perspective of the processes of producing Indian black earth (German 2003; Lima et al. 2002).

One way or another, the fact is that many authors are today questioning the way western science has viewed the tropical forests, i.e., as pristine formations, when in reality they have been, or may have been, highly managed by people (Adams 1994; Balée and Campbell 1990; Brown and Lugo 1990; Denevan 1992; Lindbladh and Bradshaw 1998; Sanford et al. 1985; Uotila et al. 2002; Willis et al. 2004). Hence the development of new concepts, including those of cultural forest (Adams 1994; Balée 1989), anthropogenic forest (Peluso 1996) and secondary forest (Brown and Lugo 1990; Noble and Dirzo 1997). Gómez-Pompa and collaborators (1987), in Mexico, and Heckenberger and collaborators (2003), in Brazil, have demonstrated that large swathes of forest hitherto considered primary are actually secondary forests previously managed by indigenous societies. In the Atlantic Forest, the stewardship of Pre-Columbian populations can be inferred from evidence of the activities of hunter-gatherers in the region dating back as far as 11 thousand years (Dean 1996). This evidence also indicated a gradual shift from the gathering of vegetal foodstuffs to planting and harvesting through the development of the slash-and-burn technique (Dean 1996).

The Dynamics of the System

Soils and the Nutrient Cycle

When practised in traditional fashion in large forest areas, with low-impact technologies and long fallow periods, slash-and-burn agriculture can be managed in an ecologically sustainable manner, without drastically affecting soil fertility (Johnson et al. 2001; Kleinman et al. 1995; Mendoza-Vega et al. 2003). It is a practice particularly suited to many tropical soils, which are normally not very rich or lack certain nutrients (Adams 2000a). In the Amazon, for example, most of the different soil types are nutrient-poor, except the purple soils and the anthropic variants, such as Indian black earth (Denevan 1996). As such, the system depends on the burning of the biomass accumulated during the clearing in order to boost the nutritional status of the soil and prepare the area for plantation using the ashes, which can vastly increase the potassium, calcium and magnesium content (Andriesse and Schelhaas 1987b; Brinkmann and de Nascimento 1973; Oliveira 2008; Stromgaard, 1984).

The role of organic material and the nutrient dynamic under the slash-and-burn agricultural system have been widely studied in tropical regions of Africa, South America and Asia (Brubacher et al. 1989; Davidson et al. 2007; Frizano et al. 2003; Johnson et al. 2001; Markewitz et al. 2006;

Nakano 1978; Nye and Greenland 1960; Oliveira et al. 1994; Tulaphitak et al. 1985; Van Reuler and Janssen 1993; Zarin et al. 2005). Many of these studies have focused on changes in nutritional status after slash-and-burn (Palm et al. 1996), but few have compared the total nutrient stocks in primary forest soils with those of farmed lands and the successive stages of subsequent brushwood (Frizano et al. 2003; Johnson et al. 2001; Juo and Manu 1996; McDonald et al. 2000; Oliveira 2008; Zarin et al. 2005). Among those who have, Frizano et al. (2003) and Johnson et al. (2001) conclude that the effects of slash-and-burn on C, N, P, K, Ca and Mg stocks are not sufficient to compromise the growth of secondary Amazonian forest, despite the study area having suffered various fire-fallow cycles for agricultural purposes. In the Brazilian Atlantic Forest, Oliveira (2008) has verified that nutrient-capturing mechanisms (litter and end roots) recover relatively quickly (5 years) after the abandonment of a plot. On the other hand, Zarin et al. (2005) have shown that a history of successive burnings reduces the secondary forest growth rate in the Amazon Basin, mainly due to reduced nutrient stocks in cycle. Moreover, forest areas previously burned become more susceptible to fire (Malhi et al. 2008; Zarin et al. 2005).

Some more detailed studies have attempted to calculate approximately how long it takes soils to recover after cultivation. Based on a review of the literature, Brown and Lugo (1990) wagered a period of 40 to 50 years for the soil's pool of organic material to recover sufficiently to resemble that found in adjacent mature forest. This relatively long period of recuperation is due to the high productivity of the growing forest during the first 20 years after the abandonment of the farmed land, when the nutrient cycle is restricted to the top layers of live biomass and litter, without reaching the soil itself. The soil, in turn, will only recover and accumulate organic matter after the twenty-year mark, when the brush growth-rate diminishes and soil nutrient stocks are more efficiently replenished (Juo and Manu 1996). Nevertheless, the delicate balance of the nutrient cycle – biomass topsoil and litter – is compromised by the precocious burning of the cleared vegetation, as the nutrients that were not rapidly absorbed by the vegetation re-colonizing the area will be run off and irretrievably lost (Sanchez et al. 1982). The environmental damage caused to soil's nutritional status and nutrient cycle under the fire-fallow system will be discussed in the following section.

Brushwood and Vegetal Succession

As with the nutritional status of the soil, studies have also been conducted into the dynamic of secondary forests and the richness and similarity of their species load in comparison with the primary forests (Guariguata and Ostertag 2001; Swaine and Hall 1983; Uhl 1987). Brearley and collaborators (2004), for example, conclude that a period of 55 years after the abandonment of a plantation is long enough for the original forest structure to recover, but not for the majority of species found in primary forests to return. Other authors have investigated the issue, but, given the differences in environment and intensity and scale of cultivation, the recuperation periods vary considerably: 60 to 80 years (Brown and Lugo 1990), 150 to 200 years (Knight 1975; Saldarriaga and Uhl 1991), 250 to 500 years (Kartawinata 1994) and centuries (Whitmore 1991). According to Brown and Lugo's review (1990), secondary forests present wood species in similar number to those found in mature forests relatively quickly, within a timeframe of roughly 80 years. In some cases, species recovery took much less than 80s years, and in others the secondary forest presented a higher number of species than the primary forest it replaced. Post-abandonment succession has also been the subject of much study, with many papers focusing on vegetal succession in the initial stages (Aweto 1981; Uhl 1987), while others go on to more advanced stages (Saldarriaga et al. 1988). Some focus on floristic diversity (Saldarriaga et al. 1988; Smith et al. 1999; Stromgaard 1986; Tabarelli and Mantovani 1999), while others emphasize the accumulation of biomass and its relationship with soil recovery (Stromgaard 1985). Ethnoecological studies have sought to investigate the knowledge and use of vegetal species as indicators of different levels of disturbance to the brushwood after the plots have been abandoned (Slik et al. 2003), or even of soil degradation (Styger et al. 2007). Lastly, over the last two decades, the study of secondary forest succession has received a great deal of attention because of the debate on global warming and carbon uptake (Fearnside and Guimarães 1996; Guariguata and Ostertag 2001; Tschakert et al. 2007). This theme will be looked at more closely in the following section, on the damaging effects of slash-and-burn.

Soil use directly influences species composition in secondary forest for many decades to come, making predictions on the succession process extremely difficult to make (Guariguata and Ostertag 2001; Pereira and Vieira 2001). In brushwood, thirty years after the abandonment of a plot, the diversity is greater than that found on former coffee plantations, where the few species used to provide shade there end up predominating (Brown and Lugo 1990). In the Amazon, the brushwood on abandoned pasture is much richer than that on lands subjected to mechanized agriculture (Pereira and Vieira 2001). In terms of forest succession, Ferguson and collaborators (2003) affirm that succession occurs more quickly on land subjected to slash-and-burn, which renders counterproductive those conservationist strategies that want to take some of the pressure off mature forests by encouraging agricultural intensification.

With regard to the functional role played by the practice of fire-fallowing, many studies have proposed that human interference in the forest succession process through agricultural activity has actually served as a source of variability that not only maintained, but perhaps even enhanced the regional biodiversity (Altieri 1999; Andrade and Rubio-Torgler 1994; Gupta 2000; Neves 1995; Raman 2001).

Agrobiodiversity and Sustainability

Despite the many variants, the essential characteristics of the slash-and-burn system are similar throughout the humid tropics (Carneiro 1988). One such characteristic is the ample diversity of species and variety of cultigens brought about by the management of evolutionary processes, including past and present interactions between farmers and crops, the conservation of germplasm and environmental conservation (Altieri 1999; Brush 1995; Dove and Kammen 1997; Harris 1971; Martins 2005; Oldfield and Alcorn 1987; Peroni 1998; Salick et al. 1997). Farmers use local folk crop varieties as a key component of their agricultural systems, functioning as raw material for the development of modern varieties (Cleveland et al. 1994) and therefore aggregating great strategic value for those who keep them (Martins 1994). As such, inter and intraspecific diversity among cultigens offers the means of promoting a more varied diet, increasing production stability, minimizing risk and the occurrence of plagues and blights. In addition, such advantages ensures the efficient use of manpower, the intensification of production under limited resources, and the maximization of returns despite low levels of technology (Altieri 1999).

In the search for general models, Geertz (1963) proposes one that treats the slash-and-burn system as a "miniaturized tropical forest", according to which some of the methods employed in the practice mimic natural ecological processes, such as the protective structure and extraordinary diversity of species characteristic of tropical forests. The sustainability of the system associated with this analogy has been acknowledged by a series of authors (Altieri 1999; Harris 1971; Hiraoka and Yamamoto 1980; Meggers 1971; Moran 2000, Rappaport 1971; among others). On the other hand, many other studies have since criticized the presuppositions on which the Geertz model (1963) is based, such as Beckerman (1983a, 1983b), Hames (1983), Vickers (1983), Boster (1983) and Stocks (1983). These authors claim that these plots are normally irregular or zoned, geared towards one or two key species, such as cassava (*Manihot esculenta* Crantz) in the Amazon (Boster 1983; Eden 1987), yam (*Dioscorea* spp) and tanier (*Colocasia* spp and *Xanthosoma* spp), or banana (*Musa* spp), in New Guinea (Ohtsuka 1983), with more diversity of variants than species. For instance, in western Africa, irregular or zoned patterns of cultivation are also frequent, as shown by Igbozurike (1971). Under these circumstances, the idea of the plot cleared by slash-and-burn functioning as a "miniaturized tropical forest", with its own adaptive advantages, does not necessarily apply (Eden 1987).

It is also important to stress the relevance of the traditional ecological knowledge in terms of understanding, managing and handling this diversity of natural resources and cultigens, and of the agricultural practices and forms of household organization that go with it. As mentioned above, slash-and-burn agriculture is an important adaptation to the obstacles and limitations tropical forests impose, and a wealth of inter and intraspecific diversity is one of the intrinsic characteristics of this agricultural system (Peroni and Hanazaki 2002). Its complexity is reflected in the depth of knowledge its management requires and the social relations established around it. In addition to the sheer number of species cultivated in consortium, most of these, and cassava in particular, boast high intraspecific diversity, each variety with its own calendar and set of uses (Martins 1994). One of the central elements in maintaining the complexity of these agricultural systems is the social capital established by the local populations. Social capital is based on systems of trust; networks of exchange and reciprocity; shared rules, norms and sanctions; and organization into groups and associations (Bourdieu 1985; Coleman 1988; Portes 1998; Pretty and Ward 2001). In agricultural work, social capital can be discerned through social relations within and among domestic units, where it ensures the productivity of the traditional agricultural systems and plays an important role in the *in situ* management and conservation of local varieties (Martins 1994).

The knowledge that traditional and indigenous horticulturists have of the properties and qualities of the soil they manage presents some striking similarities and congruencies with the scientific knowledge (Barrera-Bassols and Zinck 2003; Paniagua et al. 1999; Winklerprins 1999). Studies have been conducted in the attempt to understand these similarities and associate them with soil use and conservation strategies (Birmingham 2003; Gray and Morant 2003; Saito et al. 2006).

Unsustainability

Impact Assessment

Soil Impoverishment and Erosion Processes

Ever since Nye and Greenland's article (1960) many researchers have been conducting detailed studies into the impact slash-and-burn activity has on soil fertility and erosion (Andriesse and Schelhaas 1987a, 1987b; Brand and Pfund 1998; Ewel et al. 1981; Kyuma et al. 1985; Nagy and Proctor 1999; McDonald et al. 2000; Ramakrishnan 1992; Sanchez and Hailu 1996). The clearing and burning of vegetation followed by tilling and planting ends up destabilizing the nutrient cycle, causing loss of nutrients in the soil/vegetation system (Davidson et al. 2007; Ewel et al. 1981; Gafur et al. 2003; García-Oliva et al. 1999; Hölscher et al. 1997; Lugo and Sanchez 1986; Sá et al. 2006-7; Uhl and Jordan 1984) and, depending on the size of the clearing, even soil erosion and degradation (McDonald et al. 2000; Singh et al. 1992; Soto et al. 1995;). Reduced soil fertility largely brought about by slash-and-burn activities is one of the main causes of food shortages in Africa (Sanchez and Leakey 1997). While slash-and-burn practices vastly increase potassium, calcium and magnesium content in the soil through the burning of vegetal biomass, most of the nitrogen and phosphorous input depends on the mineralization of the burnt organic matter (Frizano et al. 2003; Roder et al. 1997). However, large volumes of this organic matter can be run off after burning (Chidumayo and Kwibisa 2003), as much as 10 t ha^{-1}, according to a study conducted in northern Laos (Roder et al. 1994), or 13 t ha^{-1}, as measured in the north of Thailand (Tulaphitak et al. 1985).

Other studies have focused on the role slash-and-burn processes play in the deterioration of the physical properties of the soil, such as porosity, which can lead to compaction, increased leaching and erosion (Alegre and Cassel 1996; Hernani et al. 1987; Pereira and Vieira 2001; Weeratna 1984), as well as the pollution of soils and watercourses through mercury run-off (Farella et al. 2006). In many places, the proliferation of weeds is a more serious limiting factor for agriculture than soil impoverishment itself

(Fujisaka et al. 2000; Nye and Greenland 1960; Warner 1991). In Laos, for example, public policies were implemented to discourage the practice of slash-and-burn agriculture (Roder et al. 1994). These interventions, along with population growth, resulted in agricultural intensification and consequently shorter fallow length, thus exacerbating problems with weeds and soil degradation (Fujisaka 1991; Roder et al. 1994).

Nevertheless, other studies have shown that hydrological impacts such as soil erosion and run-off under commercial agriculture could be considerably more severe than those experienced under secondary vegetation of the traditional slash-and-burn system (Zinke et al. 1978; Alford 1992; Forsyth 1994). Thus, as shown by Gafur et al. (2003) in Bangladesh uplands, in a small scale slash-and-burn cultivation system associated with a high degree of hydrological resilience, the flow regime and the rate of soil loss return to near "normal" levels already in the first years of fallow following one year of cultivation, because of a quick establishment of secondary vegetation cover. The negative impacts such as depletion of soil fertility in the uplands and increased downstream flooding, erosion and sedimentation only appear when the system is intensified.

Deforestation and Biodiversity Loss

As seen before, slash-and-burn cultivation is compatible with tropical forests as long as it remains small scale and there is enough forested area available for swidden. On the other hand, along with political, ecological and socio-economic factors, such as population growth, forest area reduction and agricultural intensification, slash-and-burn agriculture has also been considered by many the main proximal cause of deforestation in tropical regions (Geist and Lambin 2002; Myers 1993; Ranjan and Upadhyay 1999), as in Asia (Angelsen 1995; Sam 1994; Fox et al. 2000; Lawrence et al. 1998; Le Trong Cuc 1996; Lianzela 1997; Rambo 1996), Africa (Chidumayo 1987; Chidumayo and Kwibisa 2003; Zhang et al. 2002) and South America (Fearnside 1996; Homma et al. 1993; Houghton et al. 1991; Serrão et al. 1996). Nevertheless, Angelsen (1995) points that environmental consequences of deforestation caused by traditional slash-and-burn agriculture, such as global climate change, biodiversity, and soil erosion, are smaller compared to many other possible uses of forestland.

Deforestation leads to habitat fragmentation and loss, while soil degradation hampers the recovery of secondary forests, which in turn become species-poor and less diverse. These two factors are currently the main culprits behind the loss of biodiversity in tropical forests (De Jong

1997; Gupta 2000; Gupta and Kumar 1994; Miller and Kauffman 1998; Nakagawa et al. 2006; Raman 2001; Wilkie and Finn, 1990). But Angelsen (1995) shows that in the lowland rainforest of the Seberida district, Sumatra, fauna diversity of long-fallow forest (with or without rubber) is only slightly lower than for logged or unlogged primary forest, and well above that of plantations. Padoch et al. (2007), based on several studies, argue that in contrast to commercial agriculture, secondary vegetation following slash-and-burn agriculture often has a diversity of species that is comparable with more mature forests, and its disappearance may also be detrimental for the gathering of foods, medicines, firewood and other forest products that poor people depend on.

The deforestation rates attributed to slash-and-burn vary. Houghton and collaborators (1991) estimated that, between 1850 and 1985, approximately 10% of Latin-American forests were deforested for these purposes, while De Jong (1997) reckons on 50% for Indonesia. In the Amazon, Serrão and collaborators (1996) reached a figure of somewhere between 30 and 35%. Much confusion about the role of slash-and-burn agricultural systems arises because no distinction is made between permanent and temporary conversions, between conversion and alterations, or between deforestation and forest degradation (Angelsen 1995). The author also points that distinction between different types of forest conversions or alterations under the deforestation umbrella is important because the environmental effects and social costs may be very different. Brown and Schreckenberg (1998) warn us that, although ICRAF (1996) states that slash-and-burn agriculture results in the cutting and burning of over 10 million ha of tropical moist forests every year, Fujisaka et al. (1996) found that primary forest is only cleared in a relatively small proportion of reviewed cases (17%) and that even fewer entail a permanent conversion of forest to a completely treeless landscape. The authors conclude that many international agencies such as the World Bank now acknowledge that the causes of deforestation are much more complex than had previously been thought, and that it is a mistake to blame the small cultivators and the expansion of traditional slash-and-burn agriculture practices. Deforestation under modern conditions is much more likely to be the result of market and policy pressures arising outside the traditional farm economy (Brown and Schreckenberg (1998).

In the Amazon region, for example, deforestation has been accelerated by public policies, fiscal incentives and the agricultural credit system (Binswanger 1991; Fearnside 2005). Other causes are related to institutional adjustments related to agrarian system and macro-economic change (Brondízio 2006; Brondízio et al. 2002; Fearnside 2001; Malhi et al. 2008; Nepstad et al. 2008), land and resource ownership, the experience and technologies available to the horticulturist (Brondízio 2005, 2006), microeconomic processes at the household level and opportunities created by the market (Brondízio 2005, 2006; Homma et al. 1993).

Greenhouse Gases and Global Warming

One of the biggest political and scientific issues of our day is global warming. Many authors (Andreae 1991; Brady 1996; Malhi et al. 2008; Moran et al. 2000; O'Brien 2002; Palm et al. 1986; Tinker et al. 1996) believe that the burning of forests, especially in the tropics, is one of the main causes of greenhouse gas emissions into the atmosphere. Carbon dioxide, for example, is not only emitted during burning, but also through the oxidation of the vegetation cleared and left to decompose on the deforested ground, a process that may be responsible for up to a quarter of the total carbon released into the atmosphere (Detwiler and Hall 1988). However, some of the carbon released is taken up by the secondary forest growing on abandoned plots, which is an important contributor to balancing out the carbon in deforested tropical areas (Fearnside 1996).

As slash-and-burn agriculture is one of the causes of forest convertion, many studies have sought to understand its role in the emission of carbon and other greenhouse gases and, by extension, in global warming. However, despite the interest in the theme (Gash et al. 1996; Gupta et al. 2001; Kotto-Same et al. 1997; Palm et al. 1996; Prasad et al. 2001; Tinker et al. 1996;), little is actually known about the relationship between slash-and-burn agriculture and rising global temperatures. Davidson et al. (2008) compared greenhouse gas emissions from Amazonian soils subjected to traditional slash-and-burn and other plots in which mulching[2] replaced burning. Total greenhouse gas emissions on plots where no burning took place were no less than 5 times lower than on scorched-earth plots, albeit with slightly higher partial soil emissions for CH_4. But the net carbon flux from "deforestation" will be small if most of the cleared forest is secondary/fallow forest which also regenerates into secondary forest (Angelsen 1995). Conversions of old growth forest to slash-and-burn agriculture or plantations will typically loose 30-60% of the initial carbon stock in the vegetation, whereas conversion to permanently cultivated land or pasture looses more than 90% (Houghton 1993). Indeed, carbon sequestration under swiddening – with the landscape mosaic of fields and fallows in various stages of regrowth – may be approximately the same as under intensive tree-crop systems while it is negligible under permanent agriculture and horticulture (Tomich et al. 1998).

On the other hand, global warming may catalyze change in agricultural systems currently or previously based on slash-and-burn. In the Congo, for example, global warming has meddled with the seasonality of precipitation, increasing the frequency of storms once rare during the dry season. These unseasonal storms can reduce arable land area by cutting the drying time of the cleared biomass and thus further undermining the food security of the region's rural poor (Wilkie et al. 1999). In the Amazon, forest burning lowers annual rainfall, making the forests drier and more susceptible to wildfire, a vicious circle that has been aggravated in years that have felt the effects of other large-scale phenomena, such as El-Niño (Nepstad et al. 2008; Sá et al. 2006-7). Furthermore, the smoke created by the burning can also hamper productive potential, especially for C4 plants, by preventing sufficient sunlight from reaching the soil (Sá et al. 2006-7).

Land Use Changes and Alternatives

Land Use Changes

Slash-and-burn agriculture is a form of land use in which annual cultivars are produced on plots, resulting in vegetal cover and secondary forest in various stages of regrowth. Such diverse factors as population growth, market insertion and restrictions imposed by environmental legislation have led to the intensification of traditional agriculture and, consequently, changes in soil use and cover (Brady 1996; Byron and Arnold 1999; Cairns and Garrity 1999; Mertz et al. 2005; Metzger 2002; Netting 1993; Padoch et al. 1998; Pedroso-Jr. et al. 2008; Walker and Homma 1996). This process can lead to permanent intensive agricultural or agroforestry systems. In most parts of Southeast Asia, for instance, it appears that swidden is gradually disappearing, only remaining stable in few areas (Padoch et al. 2007). In the community of Tae, in West Kalimantan, Indonesia, for example, Padoch et al. (1998) documented the gradual, fitful demise of shifting or swidden cultivation over the last decade of the 20th century. Similar event was described by Pedroso-Jr. et al. (2008) for traditional communities of Ribeira Valley, Brazil, where restrictions imposed by environmental laws, conflict over land, construction of a major road in the region, growing insertion into market economy, and intervention programs of governmental and non-governmental development agencies were the main factors behind the changes observed in their subsistence system and, consequently, in the demise of shifting cultivation.

One of the most significant consequences of the process of agricultural intensification is the reduction of fallow length, which generates the need for alternative management practices if the horticulturists are to adjust to the change and ensure the sustainability of the system (Styger et al. 2007). Roder and collaborators (1997) managed to identify a drastic reduction in fallow time in Laos, from an average of 38 years in the 1950s to a mere 5 years in 1992. In 2002, Troesch (2003) noted that the fallow length among this same group had reduced still further, to 2 to 3 years, as was to be expected in the wake of the public land allocation policies to encourage agricultural intensification. The same is occurring in the Malaysian states of Sabah and Sarawak, where farmers are under increasing pressure from government and private companies to convert swidden land to large scale plantations (Ngidang 2002). Adams (2000b) and Queiroz (2006) have shown that conservation, infrastructure and development policies were responsible for a considerable reduction in the size of the traditional Brazilian Atlantic Forest plot during the latter half of the 20th Century. Although some studies, as discussed by Mertz (2002) have foreseen a social and environmental colapse caused by population growth and reduction of fallow period, or even the demise of traditional slash-and-burn agriculture, a considerable amount of research has recently shown that, rather than collapse, shifting cultivators are gradually modifying their practices to guarantee their subsistence (Padoch et al. 2007).

Land tenure and use decisions (especially concerning fallow length) are associated with both specific factors related to the internal dynamic of the household, such as differences in access to land and manpower, the age of the family head and ownership (or otherwise) of plots, and external factors, such as the impositions of environmental legislation, fiscal incentives and access to credit (Abizaid and Coomes 2004; Brondízio 2005; Brondízio et al. 2002). Fujisaka and collaborators (1996) analyzed to what extent external factors have forced changes in soil use among traditional Amazonian communities. According to the authors, rather than allowing brushwood to grow on abandoned plots, the current trend is to convert the cleared land into cattle pasture or simply leave it use-free so as to allow the "improvement" to drive up the market value of the property. However, other studies have shown that plot management associated with agroforest practices (Loker 1994) or even horticulture (Hohnwald et al. 2006) can be alternatives to slash-and-burn agriculture and plot clearing without causing soil degradation.

Another result of intensification is a diminishing of the agricultural diversity under the slash-and-burn system (Almeida and Uhl 1995). This process has caused environmental disturbances and eroded the genetic diversity of the cultivars, impacting negatively on the productivity and sustainability of the system. Another factor behind this loss of diversity is the substitution of species for domestic consumption with others of market value (Altieri et al. 1987). These processes could have serious implications on the health, nutrition and economic survival of the contingent of the rural population that practises slash-and-burn agriculture (Altieri et al. 1987). One extreme example occurred in Lampung, Sumatra, where slash-and-burn agriculture was predominant up to 1930, but was gradually phased out in favour of intensive agriculture, such that, by 1970, there were no remaining swathes of forest and the fire-fallow practice had fallen into complete disuse (Imbernon 1999).

Myers (1993) differentiates between "*shifting cultivators*" and "*shifted cultivators*", the latter being migrants who have only recently taken to practicing slash-and-burn agriculture, but without taking the region's environmental potential and limits into consideration. As the number of shifted cultivators has grown considerably over the last four decades, public attention tends to focus on this contingent rather than the traditional shifting cultivators. In a comparative study between the agricultural dynamics of the Tawahka Indians in Honduras and their immigrant peasant neighbours, House (1997) attested to the sustainability of the former and the ecologically harmful nature of the latter. His comparison showed that it is not the agriculture itself, but the lack of technical and ecological knowledge of the practitioner, in addition to a number of other external factors, that can lead to the permanent conversion of forest into other land uses, such as pasture or intensive plantations. Similar factors determine land-use developments in the Amazon (Brondízio 2005, 2006) and other regions (Brady 1996). Much of the forest destruction in Amazonia, for instance, may be accounted for by frontier-type land markets which promote forest clearance by migrants (shifted cultivators) who remain on one plot of land until it is exhausted, and then sell it to accumulate the necessary capital to buy higher value land with better access to social infrastructure (Richards, 1996)

Alternatives to Slash-and-Burn Agriculture

Many alternatives have been put forward as ways of reducing the damage caused by the practice of slash-and-burn agriculture (Cairns and Garrity 1999; Palm et al. 2005; Pereira and Vieira 2001; Sá et al. 2006-7), such as incentives to agricultural intensification conditioned upon the abandonment of forest plots (Brand and Pfund 1998), the implementation of agroforest systems (Dash and Misra 2001; Fagerström et al. 2002), the shift away from fire in favour of mulching (Denich et al. 2005; Norgrove and Hauser 2002), the improvement of present techniques (Barrow and Hichmann 2000) and the management of secondary forests (Fox et al. 2000), among others.

Agroforest systems encompass an array of different techniques whose successful implementation depends on their economic viability (Hardter et al. 1997), understood not merely in terms of production stability, but also of agroecosystemic efficiency in recycling mineral nutrients (Juo and Manu 1996). Under a rotational system, where various areas are at different stages of succession at any one time, agroforest systems can generate a series of products at the same time of year whilst maintaining some of the land under secondary forest cover of various different ages (Perz and Walker 2002). In a region of Camaroon, for example, Kotto-Same and collaborators (1997) suggested three agroforest alternatives to slash-and-burn: commercial cassava production (more intensive and intercalated with pockets of tree cover at a ratio of 3:1), better land preparation (the protection and selective harvesting of market-grade woods and the management of soil fertility) and the formation of stratified agroforest. Sanchez (2000) presents two examples of studies that identified agroforest alternatives to slash-and-burn practices that centred on policies to provide smallholders with lucrative soil use alternatives in return for carbon sequestration services. These alternatives involved the use of nutrient inputs to guarantee greater agroforest sustainability, such as: 1) the planting of leguminous tree species (replenishing approximately 35% of the original carbon stock of the cleared forest), 2) the application of phosphorous from indigenous phosphoric rock and 3) the transfer of biomass from nearby nutrient-rich ruderal species, which mineralize rapidly, so as to fertilize the soil under the agroforest regime. The author concludes by estimating that low-productivity plots can be transformed into sequenced agroforest, which could triple sub-Saharan tropical carbon stocks in twenty years.

On the other hand, the main limitations upon the implantation of agroforest systems are: low income generation; difficulties of finding institutional mechanisms to promote agroforestry; and insufficient infrastructure to increment the agroforest. In relation to the last-mentioned factor, of particular note are the lack of agricultural credit, good quality seeds or seedlings, facilities for stockpiling and preservation, means of transport, stable markets and opportunities to aggregate value (Perz and Walker 2002). Rasul and Thapa (2006) argue that the economic returns from agroforestry are actually higher, but that its low level of adoption is due to political and institutional limitations, such as unstable land tenure, double taxation on agricultural produce and the precarious socio-economic conditions that surround the low-income rural smallholders.

As mentioned earlier, forest burning is one of the chief culprits in terms of deforestation, global warming and related economic losses (Mendonça et al. 2004). For this reason, some alternatives to fire have been suggested, such as mulching and the plantation of leguminous crops using organic fertilizer (Davidson et al. 2008; Denich et al. 2005; Eastmond and Faust 2006). Davidson et al. (2008) demonstrated that mulching can match slash-and-burn for agricultural productivity, but without affecting soil fertility or causing greenhouse gas emissions. However, the shredded biomass used in this system is slow to decompose and cannot provide the necessary nutrients as quickly as ash, which means fertilizers must be used to give the initial nutrient input for the first crop after the vegetation has been cleared (Davidson et al. 2008; Kato et al. 1999). Furthermore, some authors argue that mulching may not be economically viable for the smallholder. Ketterings and collaborators (1999), for example, allege that mulching cannot equal the range of benefits provided by the use of fire, such as opening up space for cultivation, rapid fertilization through ash cover, and reduced competition from weeds and trees, among others. Another alternative would be slash-and-removal, but while this method also serves to clear space, it is an enormous drain on the available manpower. In other words, by accepting the alternatives, the farmer can expect lower productivity and smaller plot size in the medium-term, not to mention higher labour costs. The high costs of clearing without burning (CWB) indicate that the future of such techniques in the region will depend on new and accessible lines of credit for the small producer (Davidson et al. 2008). The incentives associated with the CWB proposal are obviously important to cushion the high initial costs and problematic yields over the short-term, but only long-term results will really be able to determine the viability of the CWB scheme (Blum 2007).

Incentives for agricultural intensification could be made through the creation of public policies and laws that grant the smallholder greater security and better opportunities to negotiate agricultural produce, as well as access to the minimum inputs required to maintain and boost production (Brand and Pfund 1998). Pascual's work in the Yucatan region of Mexico (2005) suggests that improved technical efficiency could be key to land use intensification, as it ensures greater productivity associated with shorter fallow lengths and, consequently, roughly a 24% decrease in deforestation and lower manpower investment. If the processes of agricultural innovation and diffusion and the extension of agroforest practices are to be successful, it will require effective modes of communication that mainly depend on the intervention of extensionists and agricultural associations, as well as self-management after the initial cooperation period (Blum 2007; Glendinning et al. 2001). Intensification programs in a subsistence model have the potential for reducing forest clearing, as people can get their subsistence income from a smaller land area. The increased profitability of farming, however, will attract more people, both through a shift from alternative income generating activities among those already living in the area, and through immigration (Angelsen 1995). Even so, many of these alternatives and interventions may not obtain success. Brondízio and collaborators (1994) showed that the intensive, mechanized agricultural model promoted by a Catholic Church-run cooperative in an Amazonian community not only had a severe impact on the landscape, but also proved more costly in terms of labour and capital whilst failing to achieve any form of sustainability. This is made clear by the vast areas that have been abandoned to succession, despite access to tractors with which to control weed infestations (Brondízio et al. 1994). In response to such difficulties, Cairns and Garriy (1999) advocate the adoption of a local participative approach in looking for alternatives to slash-and-burn agriculture.

Upon accepting that the use of fire is one of the main causes of environmental degradation and biodiversity loss in Mexico, the local government made a concerted effort to identify alternatives to the slash-and-burn system. However, it must be noted that outlawing contained fires could actually partly compromise the biodiversity and increase the likelihood of catastrophic wildfires (Eastmond and Faust 2006). Ferguson et al. (2003) and Pereira and Vieira (2001) suggest that conservationist strategies, proposed by governmental and non-governmental agencies, designed to discourage the practice of slash-and-burn in favour of a more sedentary, intensive and fertilizer-dependent agriculture in the hope of relieving some of the pressure on mature tropical forests could indeed prove counterproductive in the long run. Intensified agriculture (whether based on agroforestry or annual monocultures) can reduce or curtail agricultural expansion in the short-term, but may destabilize the ecological dynamic of the landscape over the long term. In a village in northern Vietnam, the forest rangers limited the area on which slash-and-burn could be practised in order to prevent further deforestation and soil degradation. However, this reduction has caused a severe fall in upland rice production and the system has intensified through the cultivation of rice varieties grown on paddy terraces, increased cattle-raising and the commercialization of non-timber forest products. Nevertheless, as the latter is a source of extra income for the smallholders, rice production has widely decreased because of higher capacity to purchase the product at markets (Jakobsen 2006).

Conclusion

Slash-and-burn agriculture, as discussed above, encompasses a complex system of practices and is responsible for the subsistence of millions of people worldwide, but especially in tropical forests. From this review of the literature on the theme we can observe that the scientific community no longer holds a synoptic view on the matter. For a long time, slash-and-burn was treated as a traditional and ecologically sustainable agricultural system whose practices and techniques seemed to be adapted to the surrounding soils and biodiversity. As such, many of the studies reviewed here converged in an attempt to understand the structure and functioning of the various elements that comprise the system.

However, since the 1970s there has been a growing number of studies on the damage slash-and-burn agriculture has been causing to the forests and, more recently, on the extent to which it is an agent of deforestation and global warming. In the face of accelerated population growth and agricultural intensification, considerable effort has been made to comprehend the nature of the changes in soil use and cover and their impact on the subsistence systems and nutritional security of traditional populations in general and those in the tropics in particular. Within the framework presented in the literature, about the role of slash-and-burn agriculture, is possible to assess this system both as a factor in the degradation and loss of tropical rainforests, and as a system of management for their conservation or rehabilitation (Lawrence et al. 1998).

Of particular note is the increase in academic interest in the process of secondary succession that takes hold once a plot has been abandoned (Guariguata and Ostertag 2001; Lu et al. 2003; Pereira and Vieira 2001) and its significance in carbon uptake, biodiversity maintenance, soil recovery and the creation of ecological corridors (Davidson et al. 2007; Moran et al. 2000; Oliveira et al. 1994; Pereira and Vieira 2001; Zarin et al. 2005).

Another theme that has been drawing considerable attention in the research on slash-and-burn agriculture concerns the negative externalities caused by the use of fire (Mendonça et al. 2004), though not only those linked to traditional agriculture. In addition to damage caused to soil and biodiversity, which is widely known, other associated economic losses have also been discussed, such as wildfires, the effects of smoke on public health and disruption to airlines and electrical companies, as well as its impact on macro-climatic processes, attracting the interest of the public to the discussion (Malhi et al. 2008; Mendonça et al. 2004; Sá et al. 2006-7). A conservative estimate of the losses incurred through these factors in the Amazon alone is somewhere in the region of US$ 90 billions to US$ 5 trillions (Mendonça et al. 2004). While the technological alternatives to fire available in the Amazon region today are promising, their results remain incipient, especially as they have not been incorporated into land use policy and crop financing mechanisms (Pereira and Vieira 2001; Sá et al. 2006-7).

More recently, the expansion of cropland to cater to the production of biofuels and other agro-industrial commodities in hot demand has been raising concerns about a possible hike in deforestation rates (Nepstad et al. 2008; Sachs 2008). An analysis of the erosion in commodity prices paid to developing nations since the 1970s also suggests, in the long run, that a commitment to commodity production poses real risks for swidden farmers who convert their swidden fields to commodity crops (Xu et al. 2005).

In short, despite the advances made in understanding slash-and-burn agricultural systems in recent decades, many of the alternatives that look to mitigate their negative impact end up steering agricultural activity toward intensive cultivation systems and market-bound production, contributing to a proliferation of accounts of failure in the literature (Brondízio et al. 1994; Jakobsen 2006; Ketterings et al. 1999). However, as Padoch and collaborators (2007) conclude, integrating market crops into the slash-and-burn agriculture system, thus allowing for more flexibility of land and labor, is often a more appropriate and economically safer strategy, as smallholders throughout Southeast Asia have shown over the last century (Dove 1993; Ducourtieux et al. 2006). In this sense, slash-and-burn becomes a part of a wider agricultural system alongside market-bound perennial cultivars associated with other forms of soil use. Together, they form a system that is sustainable on two fronts: resilient to exogenous disturbances, and better geared toward stable agricultural productivity (Berkes and Folke 2000; Byron and Arnolds 1999; Cramb 1993; Diaw 1997; Muller and Zeller 2002). These changes could be stimulated by public policies, but their success will largely depend on the innovativeness and adaptability of the populations involved, and will naturally run more smoothly the greater the social capital (Barrow and Hicham 2000). This occurs when articulations within the groups and between the groups and external agents make them better able to successfully respond to changes in their ways of life (Berkes and Folke 2000).

Thus, slash-and-burn agriculture is not as a single agricultural practice, but a variable element within a wide variety of farming systems encompassing traditional rotational systems, extensive forest fallow cultivation, more intensive market crop plots, agroforests and other ones. For this reason, the associated environmental impacts of slash-and-burn vary, and public policies should take into consideration the complexity of this system. A more efficient, as well as humane, policy would be to invest in research on methods of maintaining the biodiversity associated with swidden fallows while increasing their productivity and soil-sustaining properties (Fox et al. 2000). Hence, Padoch et al. (2007) suggest that slash-and-burn agriculture can be considered an important provider of environmental services compared to alternative land use systems, but quantification of these services is not possible unless we know the extent and dynamism of swidden. To put it in other words,

> until ways are found to address the institutional and legal constraints in a manner acceptable to the shifting cultivators, those responsible for development interventions may be better advised to support innovative capabilities within the constraints of their existing land use systems rather than attempt to introduce alternative systems of permanent cultivation with uncertain environmental and social effects (Brown and Schreckenberg 1998:13).

It is clear that some of the traditional features of slash-and-burn agriculture must be adapted and adjusted to modern socio-demographic and political-economic conditions encountered in modern times. However, such necessary changes do not imply into a complete eradication of the system. Instead, intervention policies and research should focus on mixed models that combine characteristics of intensive agro-forestry with slash-and-burn practices. In order to achieve some degree of success, these attempts should include action research processes and detailed studies of local ecological features, and socio-cultural, political and economic constraints. With these concerns in mind, we might be able to develop solutions that give some chance of socio-economic and ecological sustainability for slash-and-burn agriculture in tropical forests around the world as well as to those who rely upon it.

Acknowledgements

We thank Dr. Walter Neves, Dra. Regina Mingroni Netto, and Dra. Sabine Eggers, of the Institute of Biosciences, University of São Paulo, for useful criticism and suggestions. We also thank Maissa Bakri, Lucia Munari, Carolina Taqueda and Juliana Bortoletto for editing and formatting previous versions of this chapter.

Endnotes:

[1] A previous version in Portuguese of this paper has been published in the Boletim do Museu Paraense Emilio Goeldi, 2008, 3(2): 153-174.
[2]A covering of rotting vegetal matter, such as straw, rice husk, dry leaves, shavings and other substrate.

CHAPTER TWO

Climate and Culture[1]

SUSAN A. CRATE

Introduction

The central argument of this chapter is that 21st century global climate change[2] has cultural implications for the earth's many human inhabitants and, accordingly, anthropologists are strategically well-suited to interpret, facilitate, translate, communicate, advocate and act in response. In fact, anthropologists and other social scientists are increasingly engaging the issue of culture[3] and climate change, and doing so in many contexts be they encountering and facilitating research on the local effects and broader social, cultural, economic and political issues of climate change with their field partners or working as practitioners and educators, reframing their work in the discourse and application of concepts such as sustainability and carbon-neutrality. This chapter explores some of these diverse roles for 21st century social science and specifically anthropology. My intent is not to speak just about and to anthropologists but, perhaps more importantly, to provide an overview of recent research and practice by anthropologists and to increase the multi-disciplinary understanding of the important place of culture in the climate change discourse.

To these ends, I first substantiate the chapter's overarching climate-culture argument. With that as a frame, I provide an inventory of recent research in anthropology and other fields, beginning with a brief overview of some of the more recent edited volumes and monographs of the new millennium, some of an anthropological and some of an interdisciplinary focus, that engage climate and culture. The chapter then explores research efforts in a diversity of world regions to illustrate, through "anthropological encounters", how place-based peoples are perceiving and responding to climate change and the research interventions of anthropologists collaborating with these groups. I next discuss anthropologists' roles as field researchers, academics and practitioners. I conclude with a discussion of the importance of engaging in interdisciplinary collaborative projects to effectively engage in issues of climate-culture.

First several disclaimers: The anthropological engagement with the cultural implications of 21st century climate change is a fast-moving target.[4] By the time this chapter is in print, much of its content will be old news. Albeit this is true with many essays once they reach publication, I would argue that it is even more the case for a rapidly unfolding issue such as climate change. Secondly, this chapter is not an exhaustive review of the ways anthropologists and other social scientists are addressing and called to address the cultural implications of climate change. My aim is to provide an overview of some of the existing efforts, open the dialogue to expand our role(s) and effectiveness, and emphasize anthropology's, and other social sciences', call to action.

The Climate-Culture Argument

International forums use a specific terminology to talk about how both humans and biological systems will respond to the local effects of climate change. In brief,[5] ***adaptive capacity*** is the potential or ability of a system, region or community to adapt by adjusting to, moderating potential damages of, taking advantage of the opportunities created by, or coping with, the effects and impacts of global climate change (Smit and Pilifosova 2003). ***Vulnerability*** is defined as the extent to which a natural or social system is susceptible to, or unable to cope with, global climate change's adverse effects (IPCC 2007: 20). ***Resilience*** is the flip side of vulnerability—a resilient system or population is not sensitive to climate variability and change and has the capacity to adapt (McCarthy et al. 2001: 89).

Albeit these concepts are useful, they do not fully address the human predicament. The effects of global climate change are not just about a person's or populations' capacity to adapt and exercise resilience in the face of unprecedented change. In contrast to other biological systems, human-environment interactions are more highly complex, involving multiple feedbacks that generate new effects and outcomes (Gunderson and Holling 2002). Humans, most evidenced in the cases of place-based and indigenous peoples, often negotiate the more complex human-environment relationships via intricate systems of belief (Berkes 1999; Rappaport 1984). Furthermore, "Culture frames the way people perceive, understand, experience and respond to key elements of the worlds which they live in" (Roncoli et al. 2009).

Place-based and indigenous peoples possess a relatively high capacity for adaptation to uncertainty and change due to both a generalist and time-tested knowledge of subsistence survival, and a propensity for innovation in the context of environmental, socio-cultural, political, and economic change (Berkes 1999). However, these same populations are also highly vulnerable when overwhelmed and exploited by more technologically advanced groups.[6] The 4th assessment of the Intergovernmental Panel on Climate Change (IPCC) states that the most vulnerable societies are those in high risk areas (coastal and river flood plains, extreme environments, and the like) and whose economies are closely linked to climate-sensitive resources (IPCC 2007: 7). In this context, indigenous subsistence-based communities are among the highly vulnerable.

Take the example of my research partners, rural Viliui Sakha of northeast Siberia, Russia, who, like the Nuer, construct their daily activities and spiritual orientations around the needs of their bovine companions (Evans-Pritchard 1940; Crate 2006, 2003).[7] Global climate change is transforming the natural grassland and taiga ecosystem of Viliui Sakhas' homeland to the extent that cows may not be able to live there. Considering the centrality of cows to rural Viliui Sakha subsistence and cosmology, how can my research partners adapt to the loss of an animal that is the foundation of their culture? Similarly, northern reindeer-herding peoples are facing the same questions as they experience increasingly difficult conditions for their herds due to ice layers beneath the snow that prevents animals from reaching their fodder.

These peoples *can* adopt other subsistence modes—perhaps turn to fishing as more and more of their landscapes give way to increased saturation and lakes expand to encompass more and more of the land. But I argue less is known about how that such changes in subsistence mode could play out in terms of a people's cultural predilections, the restacking and appropriating of belief systems and cognitive orientations—perceptions and understandings of "home" and the cyclical annual changes that support the earth's variety of plants, animals and ecosystems. The cultural implications of unprecedent changes in ecosystems, plants and animals could trigger disorientation analogous to the alienation and loss of meaning in life that happens when any people are removed from their environment of origin, like Native Americans moved onto reservations (Castile and Bee 1992; Prucha 1985; White 1983). One difference is that with the local effects of climate change, affected communities are not the ones moving—their environment is.[8] With these changes it is vital to understand the cognitive reverberations and cultural implications to a people's sense of homeland and place.

If researchers do not fully consider the cultural implications of global climate change, they will have a distorted idea of their research partners' capacity to adapt. "Eveny are highly adaptive. Sometimes they joke, 'This is our home. If the climate gets too hot, we'll just stay and herd camels'" (Vitebsky 2006:10). Although completely plausible that such highly adaptive cultures as the reindeer-herding Eveny of northeastern Siberia will find ways to feed themselves even if their reindeer cannot survive the projected climactic shifts, as anthropologists we need to grapple with the implications of reindeer-centered cultures (or other groups dependent on species that may not adapt) losing the animals and plants that are central to their subsistence, daily practices, annual events and sacred cosmologies. Global climate change is about the relocations of human, animal and plant populations to adjust to change, as witnessed by the resent resettlements of indigenous refugees to safer ground (Tuvalu, Shishmaref, etc.). Lost with those relocations are the intimate human-environment relationships that not only ground and substantiate indigenous worldviews, but also work to maintain and steward local landscapes. In some cases, moves also result in the loss of mythological symbols, meteorological orientations and the very totem and mainstay plants and animals that ground a culture.

Keith Basso's argument, that "human existence is irrevocably situated in time and space, that social life is everywhere accomplished through an exchange of symbolic forms, and that wisdom 'sits in places'" (1996:53), reinforces this need to consider cultural implications. Since global climate change is and will increasingly transform spaces, symbolic forms and places, then it follows that the result will be a great loss, of wisdom, of the physical make-ups of cosmologies and worldviews, and of the very human-environment interactions that are a culture's core (Steward 1955; Netting 1968, 1993). As anthropologists, we need to look closely at the cultural implications of the changes global warming has and continues to bring.

As a discipline grounded in the study of culture, anthropology has much to offer the investigation of climate change. With their seminal publication, *Weather, Climate, Culture*, Strauss and Orlove underline both anthropology's legacy and its continuing work in deciphering how peoples, as cultural beings, relate to, talk about, and frame their perceptions of weather and climate.

> Unlike other animals, humans have unusually varied and elaborate forms of social life and communication that are made possible by language and by culture. Our complex forms of collective life influence the way we are affected by weather and climate, creating both forms of vulnerability and capacities to reduce impacts. Our highly developed cognitive capacities allow us to recall the past and anticipate the future (Strauss and Orlove 2003:3).

Anthropology, as a discipline founded on the interpretation and comprehension of culture, has many tools to use in the climate-culture endeavour. In the field, anthropologists can assess how people(s) are framing their concerns of global climate change—what they are seeing the climate perturbations *with* – how they are framing the local effects of global climate change in order to tease out the cultural implications (Quinn and Holland 1987). For example, research on cultural models and cognitive framing provides the tools to, as closely as possible, enter the world of "the other" and access such cultural frames of reference. The theoretical frame of cognitive anthropology, enables us to "study the relationship between human society and human thought, and how people in social groups conceive and think about objects and events, both abstract and concrete, which make up their world" (Kempton 2001; D'Andrade 1995).

It is also critical that field researchers have an in-depth understanding of locales and cultures in order to fully gauge their research partners' views and insights and to locate potential distortions where they exist (Henshaw 2003:219). Additionally, oral history is a key methodology for understanding past and present adaptive strategies and to gain a temporal stance on a people's collective understanding of global climate change (Cruikshank 2001; 2005). Lastly, there is also an urgency to work collaboratively with communities and to revive the applied, public and activist roots of the discipline (Lassiter 2005:84).

With these basic understandings in mind of the climate-culture relationship and anthropology's unique capacity to contribute, I next ground these understandings in some of the recent literature.

The New Millennium

In the wake of the fast moving target of climate change, the new millennium has ushered in a number of notable edited volumes and monographs addressing the socio-cultural, human-ecological and ethical issues of climate change[9]. Many of these publications interweave a variety of interdisciplinary fields, including anthropology, geography, environmental science, public policy, business, economics, communication, psychology, and the like to lay the groundwork for the vital multi-disciplinary dialogue we must initiate and act upon to comprehensively address climate change.

Of works specifically engaging anthropology, there are several edited volumes of note. In the aforementioned *Weather, Climate and Culture*, Strauss and Orlove structure their volume on two key issues of weather and climate to emphasize how both human perceptions and reactions to these phenomena are ultimately shaped by culture: 1) how humans experience weather and climate in differing time frames; and 2) how people use language in response to weather and climate. In their analysis the authors integrate understandings of two camps within anthropology, the materially-grounded ecologists and the meaning-centered symbolic anthropologists, to examine how humans think about and respond to meteorological phenomena (2003). The volume shows how time and talk are central to both how the overall public understanding of unprecedented contemporary climate change is increasing and to the cultural patterns centred on language that underpin international climate policy negotiations. Contributors to the volume forefront the value of using the "anthropological eye and ear" to understand how societies perceive and act in response to climate perturbations.

A second edited anthropological volume is *Anthropology and Climate Change: From Encounters to Actions* (Crate and Nuttall 2009). The volume's four introductory chapters provide readers with foundational materials about humans' historical relationship with climate and culture, anthropology's unique skill sets for investigation, how climate change is affecting different world populations, and the evolving anthropological discourse(s) of climate change causes and effects. Part two includes eleven case studies by anthropologists conducting research in a diversity of global ecosystems. Here the important message is how place-based peoples are perceiving, understanding and adapting to unprecedented change and how anthropologists find themselves and their work in this process. Part three includes chapters by anthropologists working in a variety of contexts on the issue of climate change. The intent of the editors is to open dialogue among anthropologists to questions concerning the extent of our roles as advocates, communicators, educators, practitioners and activists.

At least two recent anthropological monographs address the climate-culture relationship. Archaeologist Arlene Miller Rosen gives an in-depth understanding of how different Terminal Pleistocene through Late Holocene communities were impacted by and responded to historical climate change (2007). Her analysis of medium to small temporal scales of environmental change in ancient societies of the Near East highlights the importance of working at the local scale to appreciate how, for example, small shifts in rainfall can affect people living without the benefit of world market systems and international aid programs. She argues that societies do not interact directly with their environment but with their perceptions of that environment. The environment is only one of many actors in determining social change and plays a less important role than perceptions of nature.

Accordingly, societies overcome environmental shifts in a diversity of ways, and failure to do so signals a breakdown in one or more social and political subsystems. Success or failure is most often related to internal factors: social organization, technology and the perception of environmental change. A second monograph addressing the climate-culture relationship is Julie Cruikshank's monograph, *Do Glaciers Listen: Local Knowledge, Colonial Encounters, And Social Imagination*, which I will discuss in a later section on high latitude research.

Several volumes engage an interdisciplinary social science focus. In *The way the wind blows: Climate, history, and human action*, McIntosh et al. assemble papers representing the broad geographical, cultural and temporal coverage of global change in history and prehistory. The volume focuses on the need to bridge the social and biophysical sciences—to build "a common language to appreciate the full history and prehistory embedded in human responses to climate" (2000). The authors emphasize understanding our species' symbolic past, retained in social memory, or "the long term communal understanding of landscape and biocultural dynamics that preserve pertinent experience and intergenerational transmission; the source of metaphors, symbols, legends and attitudes that crystallize social action". The deliberate analysis of social memory clarifies how communities curated and transmitted past environmental states and responded to them and this can inform the present. The editors argue that economists and policy-makers need to understand and make allowances for how social memory works because it 1) is relevant to small-scale subsistence producers who are most vulnerable to climate change but most often left out of macro-policy decisions and their effects, and 2) implicates cultural conservation to the extent that indigenous social memory is a great repository of human experience and a vital resource for resilience and adaptation in our rapidly changing contemporary global context. The editors emphasize the need to tap this great reservoir of human experience for legitimate, appropriate, and economically and culturally sustainable responses to climate change.

In a second interdisciplinary volume, *Creating a Climate for Change: Communicating Climate Change and Facilitating Social Change,* editors Moser and Dilling forefront stories of success in communication and social action on climate change (2007). They justify their volume on the, to date, ineffectiveness of communication that has in turn prevented the general public from understanding and taking action against climate change. They argue that effective communication that mobilizes action must engage relevant cultural, social and cognitive characteristics. The volume distills the scholarship of both practitioners and an interdisciplinary research group to offer improvement on current communication strategies that empower individuals and communities to act in response to climate change.

In addition to these volumes and monographs of the new millennium, I make reference to several others in the context of discussing specific world areas in the sections that follow.

Anthropological Encounters

Anthropologists conducting research in the diversity of world ecosystems encounter how environmental change interacts and affects natural and socio-cultural systems of their field site. This section highlights some of this recent work, organized into the broader ecosystem categories of high latitudes, high altitudes, and low latitudes.

High Latitudes

High latitude contexts are world areas where the effects of climate change have proceeded more rapidly than other world areas due to the phenomenon of polar amplification[10]. Northern inhabitants are increasingly challenged by a changing environment. Some changes require adjustments to centuries-old adaptations including revising hunting calendars and migratory moves as seasons become less predictable. More extreme changes require relocations of entire communities as land diminishes. The heightened environmental change has brought global attention to the Arctic, often referred to as the "poster child of climate change." Case studies reveal that faster environmental change has also galvanized Arctic indigenous communities to bring attention of their plight to the world arena. Inuit have been outspoken about how the rapid loss of sea ice and other unprecedented changes represent threats to their basic human rights. Inuit Circumpolar Council (ICC) past chair, Sheila Watt-Coultier's testimony explicitly posits climate change as a human rights issue,

> Inuit are taking the bold step of seeking accountability for a problem in which it is difficult to pin responsibility on any one actor. However, Inuit believe there is sufficient evidence to demonstrate that the failure to take remedial action by those nations most responsible for the problem does constitute a violation of their human rights—specifically the rights to life, health, culture, means of subsistence, and property (Watt-Coultier 2004).

In areas not yet facing evacuation, anthropologists are documenting how local communities are coping and finding various ways to adapt to the rapid changes of climate change. As these highly resilient cultures continue to face new challenges, researchers explore issues of vulnerability, or the point at which the communities they are working with can no longer adapt to a rapidly changing environment and continue their livelihood (Marino and Schweitzer 2007). Alaska, if compared to other high latitude Arctic areas, has had a groundswell of media attention about climate change which itself challenges anthropological investigations. Some anthropologists conducting climate change research in Alaska are finding that they can either make or break their investigation depending on how they use the term "climate change" in their field inquiry. In this case, the climate-culture dynamic is influenced by outside assumptions. Working in five Inupiaq villages of northwestern Alaska, Beth Marino and Peter Schweitzer found that when they used the phrase "climate change", local level patterns of speech were altered (2009). They further document how the Alaskan and Canadian Arctic is inundated with "the photo snatchers", including members of the media, tour groups, and the like who, in addition to snatching the world's images of climate change, also leave in their wake specific ideas and concepts about the global process with local inhabitants. Based upon their research in native communities and experimenting with various terminologies, the team found that when they used the term "climate change", consultants responded by giving summaries of information they had gleaned from scientific and other media outlets. However, when they asked about change in the context of the local environment, consultants shared their personal experiences based on daily and seasonal activities. Marino and Schweitzer's research argues that the anthropological investigation of climate change will proceed much farther if "we stop talking about it".

Other research in Alaska reveals the importance for researchers investigating climate change in other cultural contexts to, before entering the field, take into consideration the culturally-based assumptions informing their research methodologies. David Natcher's research shows the extent to which notions of time and sentience can differ greatly (2007). Working with the Koyukon community of Huslia, Alaska, Natcher's research group found that they needed to frame the issue in *emic* time scales, in this case, around immediate needs and seasonal cycles of change, in order for their research partners to grasp their project's intent. Additionally, they concluded that researchers need to be sensitive to the extent their methods dismiss the agency a community ascribes as inherent within the physical environment. A main recommendation from this study is to encourage greater collaboration with local communities and to use more equitable approaches when designing research.

Anne Henshaw's research in the Canadian Arctic highlights how local communities meet the climate change issue as both a blessing and a curse. Inuit communities on southwest Baffin Island, Nunavut, Canada, depend on sea ice, which dominates the Inuit seascape for most of the year and is a critical element of Inuit travel and hunting. Anne Henshaw's long-term research with these communities shows how sea ice has become a critical barometer of environmental change (Henshaw 2009). She explores this dynamic through the careful analysis of interviews with community members and historical records. On a local level the loss of sea ice both poses threats to travel and hunting and brings new subsistence resources. Henshawa argues for anthropologists to facilitate collaborative, community-based projects that can work to mediate the complex and rapidly changing social and political environment in which climate change is taking place.

Based on his long-term research in Greenland, Mark Nuttall illustrates just how climate change means different things for different people. Nuttall emphasizes that communities' ability to adapt to climate change has more to do with issues of autonomy—their capacity to make decisions on their own and to continue to exercise their ancestral way of *becoming* with the environment around them—in order to adapt and be flexible in coping with climate variability and change (Nuttall 2009). In the context of Greenland's Home Rule government, politicians see climate change as an opportunity for mining and hydrocarbon development that for them translates to greater political and economic independence from Denmark. Conversely, Greenlandic peoples interpret climate change not as a change to some environment outside themselves but as a change to their personhood. Although these same communities have an historical precedence for adaptation to environmental change, Greenlanders' capacity to adapt to change is highly dependent on the strength of their sense of community, kinship and close social associations.

In a similar vein, the Saami people of northern Sweden are boxed in by national narratives and welfare state policies that have redefined their rights and identities. Anthropologists Broadbent and Lantto argue that at the national level, historical narratives and social policies can either enhance or limit societal responses to climate change, particularly when indigenous minorities are involved (2009). Central to the problems Saami face is the fact that the Swedish government has not recognized them as indigenous people in accordance with United Nation policies. In recent court cases Saami land use rights have been challenged and lost. The Swedish nation-building process incorporated myths about the origin and the nature of both

Swedish and Saami identities, and these cultural and political myths still influence environmental and cultural policies. Although the authors' archaeological evidence provides new evidence of Saami territories and identities, government policy reports do not allow archaeological evidence to be used in court cases. Broadbent and Lantto discuss the role of anthropology in helping to contextualize contrasting narratives and unravelling the complex webs of discourse emerging from them. Based on this, the authors conclude that Saami may well survive as fully acculturated Swedish citizens but face major political obstacles to sustaining land use practices and traditions that make them part of the indigenous world.

In yet another part of the Arctic, in northeastern Siberia, Russia, my own research shows how Viliui Sakha, Turkic-speaking horse and cattle breeders, have adapted to a sub-arctic climate, to Russian colonization and to the challenges of both the Soviet and the post-Soviet periods (Crate 2006). Their newest adaptive challenge is climate change. Ninety percent of 2004 survey participants confirmed that climate change is causing unprecedented change in their local areas and threatens to undermine subsistence. The question central to my research is how cultural perceptions and frames are transforming along with a changing environment. Communities report unprecedented seasonal changes and that they can no longer read the weather. Sakha relate their perceptions of natural variability and change with their perceptions of anthropogenic influences, including technological change in Soviet and post-Soviet Siberia. Most disturbing is the extent to which environmental change caused by climate perturbations is affecting cultural symbolic forms, such as the Viliui Sakha's Bull of Winter (Crate 2008).

In *The Earth is Faster Now*, Igor Krupnik and Dyanna Jolly have assembled reports from across northern Canada and Alaska that show how indigenous people are seeing changes and what they are saying about those changes (2002). The editors intended the volume to both inform public policy, through its translation of local, place-based perceptions of change, and to balance scientific research focused on futuristic models of what *might* happen with local experience that shows what *is* happening. Contributions describe projects that are in many ways exploratory by incorporating both new methods and tools for learning and sharing about data based in indigenous knowledge and collaborative forms of research.

Julie Cruikshank, in her monograph, *Do Glaciers Listen*: *Local Knowledge, Colonial Encounters, And Social Imagination*, draws some important lessons for our pursuit of defining anthropology's role in climate change (2005). Her central point is that there is a diversity of ways of understanding climate change and they each deserve attention and consideration. She illustrates that point with a focus on how glaciers, previously considered eternally frozen, largely inert and safely distant, gain new meaning for her research partners in the context of contemporary climate change. Glaciers now are understood as a new endangered species that melt and fail to reproduce themselves. They now serve as "cryospheric weather vanes for potential natural and social upheaval". Cruikshank also contextualizes her local analysis in the world system of glaciers by discussing the plight of other world peoples and thereby illustrating how glaciers undergoing rapid environmental change are interpreted differently. In the Andes, an ancient ritual practice involving pilgrims removing pieces of their glacier away from the source has recently stopped to help slow the glacier's unprecedented diminishing. In Asia, local men "plant" ice in opposite gender ice fields in order to grow the glacier and prevent drought. And in Peru's Cordillera Blanca, campesinos take scientists' measuring devices because they believe those instruments are what are causing drought.

High Altitudes

Similar to high latitude world regions, high altitude areas are also undergoing more rapid environmental change compared to other parts of the world. Research with the Quechua people of the high Peruvian Andes shows how a highly adaptive and resourceful people is challenged by the impacts of climate change due to the unprecedented retreat of glaciers that supply all drinking and agricultural water (Bolin 2009). Water scarcity issues in the Andean Cordilleras have challenged their human inhabitants for millennia and ancient ways of adapting are documented in myths, legends and spiritual practices. In the face of contemporary climate change, Quechua reinstate ancient water conservation practices including 1) the Inca practice of terraced gardens, both pre-existing and newly built, to prevent water run-off and erosion; 2) the pre-Columbian practice of conservation tillage, which also lessens water loss; and 3) the revival of ancient Andean subterranean water channels to transport water to needed dry areas. Bolin argues that much more must be done to ensure that the Quechua and others who make up the one-sixth of the world's population that relies on glaciers and seasonal snow pack for their water supply, will be able to continue to inhabit their ancestral lands.

Sarah Strauss's research in the Swiss Alps considers climate change impacts in a different high altitude context (2009). Central to Strauss's work are the stories of glaciers, which have historical, social, cultural and economic importance for Swiss Alps inhabitants. Throughout history and today glaciers continue to provide water for drinking and power generation, and ice for refrigeration. In the recent times glaciers also are tourist attractions, markers of environmental change and are considered repositories of lost souls[11]. It is in this cultural context that Strauss discusses how concern over glacial retreat and demise is forcing people to think about the implications of an ice-free future. While some are not too perturbed about a glacier-free scenario, others are anxious about decreased water supplies and electricity generation, a decline in tourism with its consequences for the local economy, and more hazards from flooding and avalanches. While locally-specific, these concerns, teased out through cultural analysis and interpretation, highlight issues that also have implications for people who live far beyond this narrow Alpine valley.

The edited volume, *Darkening Peaks: Glacier Retreat, Science and Society*, compiles essays addressing the nature, history and consequences of the recent unprecedented retreat of the world's mountain glaciers (Orlove et al. 2008). By weaving the concepts of perception, observation, trends, impacts and responses, volume authors emphasize how human responses to glacier retreat are shaped as much by cultural attachment as by economic issues. In conclusion the editors argue that it is the cultural and iconic power of glaciers for humans who both inhabit their presence and who live afar that will be the force to motivate action on climate change.

Low-Latitudes

Low-latitude and the humans, plants and animals that inhabit them are also important foci of anthropological research and intervention. Drawing on her rich ethnographic material in the remote islands of Tuvalu, Heather Lazrus argues that Tuvaluans not only draw upon their locally-grounded traditional knowledge to understand atmospheric and climatic disturbances but also upon different forms of knowledge derived from participation in broader transnational networks (2009). Local knowledge about ocean currents, wind and precipitation intersects with scientific and universal ways of knowing about anthropogenic influences on the atmosphere. The scientific consensus is that climate change will dramatically transform this small South Pacific country, a mix of atolls and table reef islands. Locals are concerned about sea level rise, increases in sea surface and sub-surface temperatures, ocean acidification, coral bleaching, coastal erosion, increased intensity (but decreased frequency) of rainfall, and increased frequency of extreme weather events including drought. Lazrus frames the

Tuvaluan discussion of climate change within the global debate and emphasizes three ways anthropology can contribute by 1) highlighting how impacts of climate change are culturally perceived; 2) how cultural agency is retained; and 3) how governance can promote autonomy and sovereignty to respond more effectively to climate change and its impacts.

Similarly, in the Torres Straits between mainland Australia and Papua New Guinea, sea-level rise is inundating islanders' lands. Donna Green's research with Torre Strait Islanders shows how the interplay of an accelerated rate of environmental change and profound government mismanagement and neglect render some historically highly adaptive and resilient cultures vulnerable in the face of climate change (Green 2009). Torres Straits Islanders are adapting by building houses on stilts and away from lower areas—but it seems that eventually relocation will be inevitable. Green explores the extent to which traditional environmental knowledge (TEK) can both guide appropriate local-level adaptation strategies and provide needed historical observations for climate scientists. Although the government funded the initial studies of Green and her colleagues, they are not forthcoming with more funds to continue the work. Green argues the need to fund more projects focusing on TEK and climate change because, as the author emphasizes, when people have information and see change occurring around them, they act.

In yet another low-latitude area, anthropological research illustrates how anthropology can assemble the necessary toolkit to understand how changes in the natural system will revise current terms of engagement at the level of communities and households (Finan 2009). For coastal Bangladesh shrimp aquaculture moderate flooding is, from a livelihood perspective, important for fertility and the replenishment of freshwater fish stocks. However, excessive flooding due to, for example, sea-level rise caused by climate change, can be catastrophic, causing death and disruption to livelihoods. Finan argues that the human and natural systems have negotiated an uneasy balance. While, on the one hand, people's livelihoods are dependent upon the annual renewal of the resource base, extreme events are devastating.

Jerry Jacka works with Porgeran communities, several thousand Ipili and Enga speakers in the western central highlands of Papua New Guinea (PNG). Through a careful anthropological analysis of development, migration and land cover change, Jacka illuminates the importance of considering how local peoples perceive medium-term climatic trends, in this case more frequent El Nino events, and how these trends affect the moral, agricultural, environmental and cosmological dimensions of their livelihoods (2009). Jacka's research shows how the last twenty years of gold mining development, which has brought increases in population and deforestation to Porgerans, has affected indigenous subsistence practices and, perhaps more profoundly, his consultants' cosmological orientations. In the context of their enduring cosmological belief system, Porgera people interpret the physical manifestations of environmental change as signs of the world's end. Jacka clarifies how anthropological engagement plays an important role by taking into account emic perceptions and understandings of what the local effects of climate change mean, in this case, the "societal breakdown between native Porgerans and the rituals oriented toward more powerful spirits that control the cosmos".

Another low-latitude study that highlights local peoples' beliefs and understandings of environment and climate change is Daniel Peyrusaubes research with Merina farmers of Madagascar (2007). Here the population has adapted to an already climatically-extreme area, characterized by a wet tropical climate of cyclones, hail, and drastically alternating annual precipitation cycles, by developing what the author calls, a "meteoroclimatic" culture. Peyrusaubes argues that analysis of the Merina farmers' vernacular knowledge linked with climate features, including perceptions, faiths and prohibitions, reveals similarities and complementarities with western scientific data. To these ends, vernacular knowledge can inform and make more effective the public policies needed to facilitate effective local responses to climate change.

Two studies in northeastern Brazil similarly illustrate the rift between western scientific and local knowledge. Karen Pennesi's work explores how meteorologists and traditional "rain prophets" use different linguistic devices to construct their authority (Pennesi 2007). Meteorologists use a linear approach and report institutionally-derived data in a narrow linguistic style, while rain prophets provide needed information for agricultural production and advice on attitude, morality, and behaviour. Pennesi argues that these two delivery styles contrast sharply when it comes to affecting and informing local meanings and understandings. Renzo Taddei's work examines the contrasts between academic climate articles published in the United States and ethnographic data (Taddei 2007). Again we are reminded of the limitations of western science in effectively investigating the local contexts of climate change. Taddei's study suggests that an economic approach, where "decision making" and material production are paramount, predominates in academic analyses and thereby overshadows the cultural and symbolic interactions of local communities. This is critical because, as Taddei argues, it is exactly via the symbolic forms that communities appropriate and use scientific data.

The edited volume, *A Change in the Weather: Climate and Culture in Australia* explores the cultural space between weather and climate on the low-latitude continent of Australia (Sherratt et al. 2005). The editors predicate the volume on how the very climatic feature that first attracted whites to Australia, that "the sun offered a new source of light and energy to escape the gloom of Britain", is reframed in the context of climate change to serve as a reminder to its people of global responsibilities. With increasing effects of climate change, the entire world is sharing what was once distinctive to the Aussie experience and identity—irratic flood and drought cycles. The editors argue that Australians need to rise to the occasion with this bell weather for local to global action albeit a loss of national identity and, in many ways, of innocence for the weather that once surrounded inhabitants in familiar and reassuring ways.

US Encounters

Anthropologists in the United States are increasingly working on issues of climate change on a domestic level, often as a natural trajectory in disaster anthropology studies. Hurricane Katrina brought the issue of climate change into a domestic discourse and provided fertile ground for anthropologists interested in practical and applied work. Anthropologists Gregory Button and Kristina Peterson, part of a team of social and physical scientists collaborating with the community of Grand Bayou, Louisiana, show the importance of participatory research to validate local knowledge (2009). Their participatory approach works to enhance community capacity for adaptation and to increase the community's understanding of environmental disasters and climate change. Button and Peterson also emphasize the sociopolitical and economic relations that play a part in making a community vulnerable to risk Grand Bayou residents have been witnesses to the dramatic topographical transformation of their locality over the past few decades, a transformation attributable in part to the building of transportation canals for the petrochemical industry, increased coastal erosion resulting from clear-cutting of cypress, the digging of the Mississippi Gulf outlet, and oil and gas development. The environmental change and socioeconomic and geographical marginalization of the Grand Bayou community overexposed them to the impact of Hurricane Katrina and hid them from the sight of federal officials offering aid for debris clean up afterwards. Button and Peterson's research shows the critical role anthropologists can play to reduce vulnerability and empower communities by engaging them in the planning process for disaster preparedness, response and mitigation.

Anthropological research with Nez Perce of the Columbia River basin shows how the present and anticipated effects of climate change are prompting local discussion about adaptive strategies and the building of sustainable indigenous economies (Colombi 2009). Fish and water are of central importance to Nez Perce social and cultural life, with salmon bringing the energy of the ocean inland to plants, animals and people. Local people say that without the salmon, the river would die. Colombi argues that salmon are not only a keystone species in the ecosystem but are also fundamental to the ideological and material culture of Nez Perce. Additionally, salmon are cast in the role as an indicator species of the health of the Columbia River and its ecosystems within the contemporary discourse of climate change. Colombi underscores the need to understand the contemporary phenomenon of climate change within the context of multiple stressors, in this case, two centuries of social and ecological change brought about by settlement, commercial growth, urban and agricultural development, and the impacts that timber cutting, grazing, fire suppression and hydroelectric development have had on the ecosystems of the northwest.

Understanding Anthropological Encounters

The vignettes above illustrate how, in climate change's wake, more and more of the intimate human-environment relations, integral to the world's cultural diversity, lose place. Climate change brings different kinds of risks and opportunities for indigenous peoples around the world. It is increasingly threatening cultural survival and undermining basic human rights. In *Fairness in Adaptation to Climate Change*, Adger et al. (2005) comprehensively discuss how climate change's unprecedented forces raise moral and ethical questions about vulnerability, fairness and equity. Contributors detail how justice issues are key considerations in international negotiations where power relations between the rich and powerful and marginalized desperately need to be brought into question and shifted. In the developing world, where the most vulnerable, the old, young, poor and those dependent on climate-sensitive resources, can either benefit or become more vulnerable as a result of international action, fairness is central to adaptation. This volume bolsters the need to recognize and act upon the justice and equity issues of the causes of and responses to climate change and also challenges us to think about the higher purpose to politics and law. It is also a valuable volume for social scientists and anthropologists especially, who have a strong track record for working on issues of fairness, justice and human rights.

Other Roles for Social Scientists in Climate Change

In addition to anthropologists' roles in their field contexts, they also have many roles working on the climate-culture relationship in their own culture(s). Because one of the main drivers of climate change is western consumer culture, western anthropologists have roles in helping to transform that consumer culture into a culture of environmental sustainability. The question becomes how *can* we translate field experiences into effective messages and assist in re-evaluating consumer lifestyles and moving towards a carbon-free, sustainable society? Willet Kempton has a long track record of working on American environmental values and more recently, on issues of climate change (Kempton, et al. 1996; http://www.ocean.udel.edu/people/profile.aspx?willett#Publications).

Anthropologist Richard Wilk is interested in consumer culture. Wilk questions the extent to which anthropology and anthropologists have been successful in interpreting and addressing the issue of consumer culture in the context of climate change. He critiques the commonly-accepted perspective that western consumers just need to "tighten up their belts" to effectively counteract climate change (2009). On the contrary, Wilk argues that individual consumption is not the source of a majority of greenhouse gas emissions. The real culprit is a series of pivotal political and historical decisions that has and continues to cultivate western consumer culture and are deeply rooted in the variety of "cultural ideals about justice, comfort, needs and the future". Wilk critiques anthropologists' inability to form collaborative unions with other anthropologists that could be key in developing a comprehensive and sophisticated theory of consumer culture. His point is that anthropologists could be more pro-active but only by first creating new models of comparison and collaboration so that the whole truly may be more than the sum of the parts.

In the last decade or so anthropologists have also been increasingly involved in transforming the culture at their home institutions to both contribute to campus greening and provide a hands-on learning environment for their students. Peggy Barlett and Benjamin Stewart look at how academic anthropologists can effectively contribute to educating about climate change (2009). They argue that it is not just about making students aware of climate change and its causes and consequences, but also about expanding awareness and galvanizing action within the life and culture of the institution itself. Drawing on their work in these efforts at Emory University Barlett and Stewart argue that institutions of higher education not only should play a major role in innovative climate change research, debate and policy, but also take responsibility for their own large carbon footprints. They espouse the need for personal and institutional-level audits which can reveal that even the most “environmentally-friendly” practices can still be carbon-intensive due to trans-national networks of production and consumption. Albeit there are abundant resources detailing strategies for teaching and educating students about climate change, materials are fewer that address issues of culture, human agency and engagement with the local. Similarly, they emphasize the need to teach about the moral, justice and ethical dimensions of climate change on both personal and societal scales. In many ways climate change is challenging higher education institutions and their faculty to rethink and restructure how to teach students for the future.

Anthropologist Pamela Puntenney works on climate change in a global policy context, focusing on the complexities inherent in implementing multilateral and other policies needed to tackle climate change. Puntenney argues that only through the creation of comprehensive agreements founded on principles of sustainability, equity and justice can we effectively address climate change (2009). We have only relatively recently witnessed a growing concern and sense of urgency to bring the human dimensions of climate change to the policy forefront. Many new international and national forums and institutions are tasked with investigating the cultural implications and social dimensions of climate change. The most vulnerable countries need action now. This new paradigm has called social scientists, anthropologists included, to the table to affect decision making. Again, we see how anthropological training in the area of culture brings with it an irrevocable responsibility to engage in discussions and directives on international policymaking, debates, policies, and decision-making.

Shirley Fiske’s work engages both U.S. national legislation and international implementation of strategies for global carbon markets and emissions trading (2009). She shows that anthropologists and their research with different levels of power, have clear roles in policy debates in the U.S. Congress, and in analyzing the link between local communities and policy instruments, such as carbon offsets and sequestration. Fiske herself bases her claims on her unique experiences in working as an observer and participant in climate change policy discussion and development in legislative and executive branches of the U.S. government. Based upon this history, she emphasizes the changing nature of policy dialogue about climate change and various attempts to make it a legislative priority. The shifting terrain of the negotiation of policy questions about mitigating and controlling carbon emissions is fertile ground for the anthropologist. Fiske argues that the domestic discourse in the U.S. revolves around reducing emissions, cap and trade, carbon taxes, credit allocation, carbon reduction technologies, and so on. Central to anthropology's role is to bring to light the consequences of forestalling effective policy measures not just as a problem of and for technology but for people, families and communities.

At least one group of social scientists are turning their gaze to the complex relationships between cultural models, climate change and biofuels (Halvorsen et al. 2007). This group's study involved conducting interviews with specialists and citizens representing a diversity of stakeholder groups in Michigan, Wisconsin and Minnesota, to assess the cultural models used to understand climate change and possible solutions. Researchers in this study analyzed how policy favours more unsustainable sources of alternative fuels (corn ethanol and the like) that are not carbon neutral instead of more sustainable sources like ethanol derived from cellulose. Findings indicate that although participants correctly perceived the seriousness of the climate change issue, they had faulty information about alternative fuels, making a case for better information and communication about biofuels.

By Way of Conclusion

Based on this brief overview, it is clear that social scientists, and especially anthropologists, have an ever-increasing number of roles in contemporary climate change issues specifically because of their training in and skill sets for working on issues of culture. However, we have yet to commit comparable energy, in the form of funding, research, employment and the like, to understanding and grappling with the human dimension of climate change as we have to the understanding of the physical dimensions.

One way to balance this equation is for anthropologists and other social scientists to become strong advocates for place-based and global ethnodiversity, to the same extent that natural scientists promote biodiversity. Like the diversity of plants and animals makes for a healthy ecosystem, the diversity of cultures and ethnicities make for a healthy ethno-system. Anthropologist Wade Davis has been most outspoken about the need to acknowledge and enhance global cultural diversity (Davis 2001). Davis points out that it is in fact the very "sacred cosmologies" and the being "grounded" in "mythological symbols" and "totem" animals and plants that are part and parcel to that ethnodiversity.

Another area that needs social scientists' attention is addressing the cultural implications in contexts of powerbrokers, policymakers, and others in places of power (Lahsen 2007). Corporations, institutions, governments and private investors all have specific cultural ways of operating in the world. How can powerful institutions both fund activities that cause climate change and, at the same time, efforts to ameliorate the issue. For example, in March of 2008, The World Bank held a workshop on the Social Dimensions of Climate Change, and yet the organization continues to fund oil, gas and coal projects, which contribute to the problem.

Lastly, social scientists need to develop skills in collaborating within and across disciplines and to create more interdisciplinary projects. Social scientists have unique understandings of the climate-culture relationship, time-tested skill sets, and an innate responsibility to be part of the solution. By engaging in this issue, it also becomes clear that climate change is not all gloom and doom. As Ben Wisner and their co-authors state,

> Ironically, climate change offers humanity an opportunity for a quantum leap in sustainable development and in peace making. If international cooperation (as opposed to competition) is strengthened (2007).

Similarly, I argue, it can galvanize social scientists to rise to the occasion—to use our unique skills and tool kits to do the work before us and stop engaging in petty skirmishes and sub- and even inter-disciplinary squabbles that take our energy and talent away from where they are really needed.

Acknowledgments

I first acknowledge all the people with whom I have worked, especially the people of the Viliui regions of western Sakha, Russia, since it was due to my encounters of the changes to their physical and cultural worlds that my research in climate and culture began. Thanks go to the National Science Foundation (NSF) and Anna Kertulla de Echave, Office of Polar Programs, Arctic Social Sciences Division for funding support. The chapter also would not have been possible without all the panelists who participated in panels and policy forums at the 2007 SHE conference in Rio and at the 2007 American Anthropological Association and 2007 Society for Applied Anthropology meetings.

Endnotes:

[1]This chapter draws on material from presentations in the session "Climate and Culture" at the SHE XV conference in Rio de Janeiro, October 2007, from the edited volume, *Anthropology and Climate Change: From Encounters to Actions* (Crate and Nuttall 2009), and from the article "Gone the Bull of Winter? Grappling with the Cultural Implications of and Anthropology's Role(s) in Global Climate Change" (Crate 2008).

[2]I am using "climate change" throughout this chapter to refer to the contemporary phenomenon of anthropogenic global climate change.

[3]"Culture" is not a term I intend to use loosely but also one that I feel does not necessitate too rigid of a definition. In this chapter I use the term "culture" to refer to both the series of prescribed human activities and the prescribed symbols that give those activities significance; both the specific way a given people classify, codify and communicate experience symbolically and the way that people live in accordance to beliefs, language, and history. Culture includes technology, art, science, and moral and ethical systems. All humans possess culture and the world is made up of a diversity of cultures. Accordingly, I use the term in both its singular and plural forms. Culture change is a given and necessary process for all human societies, as they adapt, in our era, to a variety of effects including globalization, environmental degradation, land issues, climate change, and others. But what are the implications if change occurs at the level of what is essential to a culture—the core aspects, for place-based peoples (and I would argue for all peoples), their significant human-environment interactions with specific lands, plants, animals and even ecosystem characteristics, that shape their ways of being, cosmologies, beliefs and daily/cyclical orientations (Steward 1955; Netting 1968, 1993).

[4]This is not to say that social scientists have no legacy of studying climate and historical climate changes. I am, as I defined earlier on, referring here to the 21st century phenomenon of unprecedented anthropogenically-driven climate change.

[5]I am using the framework for these terms laid out in the IPCC 4th report (IPCC 2007). These terms are also used extensively in recent special issues of journals focusing on the issue of climate change. See, for example, Global Environmental Change, 17(2).

[6]History clarifies how events over time interplay with adaptive capacity and vulnerability (Crumley 2001, 1994; Egan and Howell 2001).

[7]I fully recognize that this sentence portrays contemporary rural Viliui Sakha as unengaged in the modern world. This is not my intent. I am drawing parallels with the Nuer to emphasize Viliui Sakhas' continued dependence upon cows (for subsistence and in their cosmology) to emphasize how the loss of those animals needs to be taken into account.

[8]I take poetic license here by saying that "the environment moves." It works well within the analogy. I fully acknowledge that the environment cannot move but that it changes.

[9] I emphasize again that by the time the time you are readings this chapter, there will be many more such edited volumes, monographs and documentation.

[10]"Polar amplification" results because with any temperature increase of the entire earth system, there is a correspondingly greater temperature increase in the Arctic due to the collective effect of environmental feedbacks and processes.
[11]According to local siren tales, entrapped souls call out to the living, reminding people to stay close to home and not wander too far into the mountains. The cautionary nature of these tales reveals local perspectives on the dangerous nature of glaciers, as well as the importance of following local religious traditions.

CHAPTER THREE

Culture and the Natural Environment

THOMAS JEROME BURNS

As students of culture such as Max Weber and Clifford Geertz have suggested, we live "suspended in webs of meaning that [we ourselves have] spun" (Geertz 1973). These collective meanings serve as the basis of an ethical framework informing how people perceive, remember and communicate. The natural environment historically has influenced how cultures develop, and in turn has been affected by human action and belief systems. Research on culture and the environment offers a number of insights into ways in which these processes occur. Particularly, but not exclusively, with technological innovation and its cultural incorporation, human interactions with the environment have varied dramatically. Put another way, while humans have impacts on the natural environment, they often do so in ways that are influenced profoundly by the cultures of which they are part.

This article explores a number of aspects of culture, including the most deeply held ethical values. While various attitudes, such as those toward the environment, are rooted in ethical orientations, attitudes are potentially more quickly changing while culture changes very slowly, typically over the course of generations. This work examines what causes cultures to constrain and inform human action—particularly relative to the natural environment—while themselves changing slowly over time.

Issues of social change and cultural lag are crucial here. While it is not uncommon for cultural change to follow major technological innovation, that change may unfold over years or even generations. While there has always been technological innovation, its interaction with culture is now more problematic than in previous times because the rate of change is itself accelerating (McNeill 2000). This article explores linkages between culture and the environment, and suggests directions for research and informed public policy.

What is Culture?

If institutions are where individuals and society come together, culture is the milieu in which those interactions occur and have meaning (Wuthnow 1987; Handel 1982). Culture can be seen as a way of organizing and prioritizing thought, value, belief and action around an ethical framework (for extended treatments, see Weber 1978 [1921], 1948, 1985 [1904]).

An important aspect of culture is that it often serves as the backdrop against which judgments of appropriateness are made (Archer 1988). Much of the power of culture lies in its taken-for-granted quality (Brown 1987). As Aristotle recognized, what made human communication meaningful was a common ethos, or set of values, norms and beliefs. Institutions and actions may be judged relative to culture, but rarely is the converse true.

As various thinkers in the Platonic tradition, most notably Immanuel Kant (1958 [1781], 1950 [1783]) and Phenomenologists such as Heidegger (1966 [1932], 1999), Husserl (1965), Schutz (1972) and Berger and Luckmann (1967) have pointed out in various ways, there is a human tendency to perceive, to remember and to make sense of one's life-world in terms of underlying forms. These go by a number of different names, each emphasizing some particular aspect of how information is perceived, prioritized, remembered and communicated; what is seen as important and what is ignored or minimized; what is explicit and what is implied. Kant himself used the word "schema", which was adopted by Twentieth Century thinkers, most notably Frederick Bartlett (1932) and Jean Piaget (1951, 1954). Schema has been used primarily in terms of how *individuals* store information.

When applying these ideas to larger communities, particularly but not exclusively scholarly communities, the term "paradigm" has been used. The metaphor of "frame" is also used by social scientists, particularly since Erving Goffman's (1974) popularization of it. It is also used in computer and cognitive sciences in a somewhat analogous way. The "frame" problem in a computer's "understanding" of natural language, for example, refers to how much of a problem is actually unstated. A human being tends to perceive partially ambiguous information by "filling in the blanks" in a certain way, typically in concert with one's cultural expectations. This tendency is referred to by some students of human communication (particularly those identifying with the ethnomethodological school), as the "et cetera principle" (Mehan and Wood 1975; Handel 1982)[1].

By observing ways in which people perceive, prioritize, think about and communicate information, we gain valuable insight into culture itself (DiMaggio 1997). We can then look at how people act on the environment, and how they make sense of and justify acting within a cultural context. This in turn will give us the diverse arrays of insight into why some cultural ways of seeing and tendencies for acting have differential effects on the environment.

Cultures as Networks of Meaning

Taken together, much of the aforementioned literatures converge to tell a compelling story. Culture can be seen as an interrelated set of complex networks of information and values (Alexander 2003). These networks serve as a backdrop against which behavior, mores and folkways, and other information are judged. Perceptions and ideas are found meaningful in relation to these networks (Douglas 1970). When judgments do occur, either explicitly or implicitly, they typically are made such that if there is dissonance between the cultural networks and the object of perception, it is the *object* that is found wanting and judged as strange, foreign, and worthy of rejection (Cooper 2007; McClure 1991; Festinger 1962 [1957]), rather than the culture itself (Swidler 2001; Archer 1988).

A property of networks in general, is that some aspects of them are more central and some more peripheral. As a general rule (and one that does need to be qualified), the more central actors or "nodes" are in a network, the more power and influence they are likely to have (Burt 1982). As interrelated networks of meaning, cultures have these characteristics, and thus can be seen and analyzed, at least in part, in network theoretic terms. A salient characteristic of cultures then, is their tendency to carry within them priorities–some information is more important than others, some values and ways of behaving are preferable to others.

While the specifics of *what* is privileged varies from culture to culture, a universal of *all* cultures, is this implicit prioritization (H. Simon 1990; Bourdieu 1984). These priorities are reflected in artifacts of the culture, most notably the language; so much so that some theorists see language and culture as virtually the same thing. While I am not necessarily arguing that here, there is no doubt that some of the specific ways in which language is used serve to define and reinforce cultures and subcultures.

Culture and Modernity

Over the last several centuries society has seen dramatic changes, particularly of the sort associated with "modernity" (Dilthey 1976; White

1973; Kohn 2004) World population is now over six and one-half billion and rising, having passed the one billion mark only in the mid Nineteenth Century (United Nations 1992; Cohen 1995). The transformation has been accompanied by rises in literacy and formal education, increasingly concentrated modes of production and urbanization, a steady supplanting of agrarian with industrial ways of living (Chase-Dunn and Hall 1997; Inglehart 1990, 1997; National Research Council 1999).

Modern society, as Snow (1998) famously remarked, has experienced a dehision between two "cultures"–one aligned with the humanities and the other more closely associated with scientism. What is modernity itself then? There are, of course, a number of ideas about that. In very broad brush terms, we might see *modernization* as a process that involves *increasing complexity* on a number of dimensions, particularly those involving technological changes and their attendant social forms.

A number of social observers, most notably Emile Durkheim (1964 [1893]; Durkheim and Mauss 1961 [1902]) and Georg Simmel (1955 [1908]) have suggested that the emergence of the individual as a social form has run parallel to the rise of modernity. Durkheim's famous dictum from The *Division of Labor in Society,* in which he challenges the utilitarianism that had become popular in his time, is indicative here: "Society is not born of individuals, rather, the individual is born of society."

How then does the modern individual, and her post-modern offspring, react to the increasing challenges of modernity? In the face of overwhelming complexity, there is a tendency to shut down and simply ignore mounting evidence of large scale challenges, even while opting for ever more elaborate escape strategies. As Marcuse (1964) saw by the mid Twentieth Century, there appears to be a shift away from critical engagement of pressing social problems, particularly by individuals feeling overpowered by highly technocratic cultures. In Marcuse's dystopian scenario, reality seems so overwhelming that, at some point, the alienated modern/post-modern individual turns away from the complexity and simply seeks to escape into mind-numbing activities.

A number of ideas from epistemology and from empirical studies in cognitive and neural psychology are important to consider here, particularly in cultures where, as is increasingly the case, the technological and social conditions are ripe for what is sometimes referred to as "input overload" (Burns and LeMoyne 2003). Cultures and the institutions embedded in them become characterized by increasingly desperate and narrowly focused "fixes" to the overwhelming complexity of modernity/post-modernity. Politicians are taken seriously when they, for example, propose "abstinence education" for runaway population problems (Ehrlich and Ehrlich 1990), and global warming is dismissed as merely a fantasy created by some other(s) perhaps vaguely characterized as "elitists" or "liberals" (see Burns and LeMoyne 2001).

Even "local" environmental problems often tend to involve large and complex ecosystems in a number of ways. Global warming and deforestation are, in a very real sense, macro-level problems. These in many ways are exacerbated by the structure of the only real global institution—the global economy (Burns et al. 1997, 2003; Ehrhardt-Martinez 1998). There is a mismatch between the scope of a number of environmental problems and the human institutions that, even under the best of circumstances, stand to address them (e.g. Shandra 2007; Smith and Wiest 2005; Robinson 2004).

Institutions and Cultural Lag

Institutions are embedded in the cultures of which they are part, and yet we are in a somewhat unique situation worldwide. Although international trade goes back hundreds of years (Wallerstein 1974), what is new now is the *extent* to which a truly global set of markets have emerged (Chase-Dunn et al. 2000; Chase-Dunn and Hall 1997). Beyond a mere cliché, the rise in "Globalization" has made it increasingly likely to have commodity chains that involve many nations and even continents on a regular basis (Robinson 2004; Hornborg 1998). This has profound environmental consequences (Lofdahl 2002; Burns et al. 1997, 2003; Bell 1973, 1976).

Early in the Twentieth Century, William Ogburn (1961 [1932]) noted that while it is typical for culture to reflect material conditions such as the natural geography and the constraints people encounter to make their livings, there tends to be a time lag between the conditions people face and the cultural adaptations to them. Sometimes that lag could be even as long as several centuries. There is evidence that even in the face of overwhelming modernization there is a persistence of traditional values, and this is true across geographic regions as well as levels of development (Inglehart and Baker 2000).

Yet while constrained by tradition, institutions and ideas still tend to diffuse. John Meyer and his students and associates (e.g. Meyer 1977; Meyer et al. 1997), for example, have documented the increasing uniformity of institutions such as education systems and types of governance in the nation-state. Because of the associated decrease in opportunity costs, the rise of this "Institutional isomorphism" (DiMaggio and Powell 1991) facilitates increasingly larger scale bureaucratic practices. These systems throughout the world have, in many ways, come to be so similar to one another that Meyer argues they can be characterized as a "world polity" (Meyer 1980).

Recent work by some of Meyer's students has attempted to extend the world polity analytic framework to environmental action on the nation-state level. Frank et al. (2000a, 2000b; also see Schofer and Hironaka 2005) find worldwide rises in a number of indicators, including the establishment of environmental impact assessment laws, entry into environmental treaties and the establishment of national parks and environmental ministries, as well as the number of chapters of environmental associations over the course of the Twentieth Century.

There tends to be institutional diffusion within a given society as well. The economy can be seen as a "lead institution" (see Rudel and Roper 1997) and one that, for hegemonic powers such as the U.S. and England, is followed closely by military adventure. In addition to being embedded in a dominant culture, institutions themselves can be said to have their own "Institutional culture". Particularly in global markets, ideas of classical economics such as the "law of comparative advantage" and the related principles of *"laissez-faire"* and "economies of scale" increasingly inform the way global business is conducted. These ideas go back centuries, at least to the work of Adam Smith (1999 [1776]) and David Ricardo (2006 [1817]). Despite critiques (most notably Marx, e.g. 1967 [1867]), they have progressively come into their own over the last two centuries (Bell 1976; Hornborg 2001).

Different institutions change across an array of time trajectories. The economy tends to adapt to external circumstances such as changes in supply or demand for a critical resource quite rapidly. The polity adjusts a bit more slowly, with the culture tending to transform more slowly than either (Parsons 1951, 1966). If in fact there is an emerging world polity, it is firmly embedded in a culture that has had over two centuries to absorb the values of the *laissez-faire* capitalism; to the extent there is an emerging ecological consciousness, it has only had a fraction of that time, and likely has nowhere near as deep a level of cultural immersion.

Culture and the Environment

Ever since the dawn of civilization, there has been environmental degradation, and since that time some cultures have in fact had values of sustainability and others less so (Ponting 1991). And yet the complexities of modern times and their aftermath really are characterized by environmental problems of unprecedented magnitude (McNeill 2000). We now have more people on the earth using more resources with technology increasingly capable of making more profound incursions than ever before in history.

Cultural lag then is crucial to consider, particularly when seeking to understand the trajectory of environmental problems. Yet for reasons discussed, cultures tend to change very slowly, even glacially. With the advent of the industrial revolution, there has been a continuing and steadily increasing pace of change in the material culture. Yet our adaptive culture has not caught up. Rather, the gap arguably is widening.

As societies modernize, cultures increasingly embrace values of consumption (Schnaiberg and Gould 1994; Jorgenson and Burns 2007). While some theorists see signs of pro-environmental cultural values (Dunlap and Catton 1994, 2002; Catton and Dunlap 1978; Inglehart 1990, 1997) in some segments of societies, these appear to be increasingly swamped by a logic of global trade in which economies of scale–scales of extraction and production, as well as consumption–have increased to unsustainable sizes (Rice 2007; Roberts and Parks 2007; Hornborg et al. 2007).

There is an increase over time, particularly in the most developed countries, but also in developing countries, of the ecological "footprint" of consumption (see York et al. 2003; Jorgenson 2003, 2004; Jorgenson and Burns 2007). In fact, arguably the most powerful institution ever witnessed in human history has come into its own with the global market. To be sure, this is met with some cultural countertrends (Commoner 1992; also see Ritzer 1993), themselves not without their own set of problems. Yet these global markets have in some real and powerful ways served to forge a worldwide culture of consumption. This is not to imply that history does not hold a number of valuable lessons for us if we are able to heed them. Jared Diamond's (2005) haunting work *Collapse* details a number of such lessons, as does much of the work of J.R. McNeill (2000) and others (e.g. Chew 2001; Ponting 1991).

Why Understanding Cultural Lag Is Crucial to Environmental Consciousness

A culture typically adapts over time to conform to demographic, technological and material conditions. The fact that scrimshaw was a part of some of the Inuit cultures reflected a presence of whalebone from the hunting and fishing that was part of the everyday reality of the people. Likewise, in industrialized societies, the "sexual revolution" could be seen as a function of a congeries of demographic and technological conditions coming together at once—the post World War II baby boom coming of age at a time when technological advances and their diffusion had led to relatively reliable birth control becoming more widely available; these were at a time coinciding with larger modernization processes, such as the move from agrarian to urban industrial societies with largely age-segregated sub-populations, among other factors.

There are other such examples as well, and an ethic of environmental stewardship, reflected partly in term of "attitudes" toward the environment can be seen in a similar framework. The inverse relationship between population density and the probability of adopting the "New Environmental Paradigm" or NEP (e.g. Dunlap and Catton 1994) is an instance of this principle in action.

The complexity of these issues needs to be appreciated. There is necessarily an implied *ceteris paribus* clause always in effect (that is to say, holding all else equal, which of course is seldom actually the case in complex ecological systems), and any discussion can always be nuanced with the addition of pertinent variables. One of many possible examples of how the picture may be further complicated can be found in the propensity of attitudes to become less flexible through the life course, yet more so with higher levels of education. Many aspects of a social ethos, including norms, values, beliefs and basic ways of seeing the world and in parsing new information are passed on from one generation to the next. Thus, ethics that were adaptive to a prior time are likely to persist, even though they may be maladaptive in the current and future times.

Put another way, rapid environmental change renders many aspects of the natural environment appreciably different from what was the case a generation or two ago—and yet many of the ideas and ways of relating to the environment have become part of the ethos of a socially constructed rugged individualism based in a time when environmental resources were seen as virtually endless.

In North America, for example, the myths of "explorers" such as Paul Bunyan and Davey Crockett have given way to the culture of the four-wheeler, snowmobile and power boat. Even as the environments needed to sustain these are being whittled away, sometimes by these very practices, unbridled selfishness may bask in the frame of rugged individualism, thereby lending it an air of legitimacy. The individualism trope, particularly though not exclusively in North America, may in turn be embedded in larger master frames such as "manifest destiny" and the idea of the "frontier" as articulated by historians such as Frederick Jackson Turner. There may as well be aspects of a religious or eschatological "promised land." Ironically, there is evidence that in modern cultures, many of the attitudes that diffuse most profoundly and enjoy general adoption are born in urban environments where unsustainable consumption patterns become entrenched; it is from there that they tend to diffuse to more rural areas (Fischer 1978; also see Grubler 1991).

Further adding to the problem is the phenomenon of willful ignorance, combined with the very real possibility of learned helplessness in the face of what may seem like overwhelming problems (Seligman and Maier 1967). Social psychologists have long known of the propensity to simply try to escape from seemingly intractable problems, often through vapid, meaningless and mind-numbing activities (Marcuse 1964).

Perhaps ways in which researchers and journalists go about framing questions and discussing their findings may inadvertently feed the very public ethos that ultimately is part of the destruction of the environment. A not insignificant aspect of a society of mass consumerism, particularly one characterized by chronic information overload, is the rise of news as entertainment. The latest reportage on celebrities such as Michael Jackson, Martha Stewart and O.J. Simpson tend to take priority over news of deforestation in the Amazon. Serious environmental problems are regularly reduced to a punch line in a joke (e.g."…it's cooler today in the valley—so much for Al Gore and his global warming…and now a word from our sponsor"). In a time when the special effects of Power Point presentations and the like capture attention, environmental messages become part of a bad news background that can be ignored—life is already depressing enough. This of course feeds into the willful ignorance that characterizes much of human attitudes toward the natural environment.

Enantiadromia

Enantiadromia can be defined as the tendency for an action to set up a reaction to it. In social systems, it is not uncommon for the reaction to have a greater impact than the action itself. This idea was first discussed by the pre-Socratic social observer, Heraclitus; it is a concept familiar to an array of disciplines from history (particularly see W.I. Thompson 1971, 1981) to Jungian psychology, and yet is largely ignored in ecological work.

As a general rule, the larger the action or incursion, the greater the reaction will be—and the reactions typically are only partially predictable at best. A property of virtually all systems, whether they are social or environmental systems, is that they tend to operate on a number of levels (Duncan 1964). The scientific method itself, and the technological processes that stem from it, are based in analysis (breaking a problem down into smaller, presumably understandable, parts). As a consequence of this, even the most sophisticated of studies tends to focus on part of the system and to ignore others (Dietz and Rosa 1994).

Science has engaged in rear-guard action, justifying a materialist perspective, for example, in juxtaposition to that proposed by advocates of an "intelligent design" perspective, and it has amassed a literature to this end (e.g. Gould 1992, 1999; Kitcher 1983, 2007; for an historical overview, see Clark et al. 2007). Yet as modern science basks in the glow of its "triumph" over creationist perspectives of the past (e.g. Eldredge 2000), it tends to ignore large issues about how its methods may need some radical rethinking as society moves into the future.

This is not necessarily to argue against scientific analysis. Rather, it is to point out a limitation of the scientific method on which much of industrial and even some of post-industrial social development and the contemporary culture deriving from them are based. Small and, on some level seemingly ignorable, problems have—now aggregated over the two or more centuries since the beginning of the industrial revolution—brought the world to some staggering problems (Hornborg 2001).

While science has made wonderful breakthroughs, the method itself has not kept a pace in development. Rather, it has reached something akin to a sacrosanct status in modern culture. Consider for example one of the truisms in statistical analysis and education about the probability of Type I error and the convention of the .05 or .01 Alpha level. This certainly has worked well in countless instances. Yet particularly where it is difficult if not impossible to isolate factors in an ecosystem, the slippage or "noise" in that system may cause important factors not to be discernible without years of measurement.

It may well be that the precautionary principle, rather than the traditional .05 level, would lead to healthier outcomes (Steingraber 1998). One of the few examples where this *has* been the case was when DDT was outlawed in the United States in the wake of Rachel Carson's (1962) prophetic work, *Silent Spring*. Based on a preponderance of what looked to be at the time (and did in fact prove later to be) significant dangers of DDT, it was banned. It is an unfortunate fact that it is still used in many developing countries. Further, insecticides, herbicides and fertilizers are on the market now that may be even more dangerous than DDT, and yet because there may not enough "proof" at a given time, they remain on the market. Even more ominous is that the pace of technology is such that new products can be developed faster than a thorough vetting can take place.

The combined force of global business and law has given rise to a culture in which burdens of proof lie not with those who would pollute, but with those who would slow its pace. Adoption of the precautionary principle would shift the burden of proof, at least in part, to those who would seek to introduce and market potentially toxic substances—to prove that they are safe, rather than having advocates for the public good have to prove that they are not. As legal scholars have demonstrated, the time and expense associated with who has the burden of proof in many ways preordains the outcome (Gaskins 1992).

This ethical dilemma tends to be approached from the standpoint of what is good for business in the short run—what is "cost effective". It may be that a cultural transformation would need to take place, where interests of global business are weighed against other values such as environmental stewardship and well being. In this as in other issues, a big part of what we think we know comes from how we make assumptions to fill in blanks about parts of the overall situation that are ambiguous or unknown. Cultural schemata provide the symbols and linkages we use to fill them.

Some Lessons and Questions from a Utilitarian Perspective

The foregoing discussion highlights several salient problems associated with linkages between culture and the environment. Some of those can be seen more clearly through the lens of individual self interest, which has become a central organizing principle of some cultures. It arguably is becoming more widespread in parallel with the globalization of markets, which depend on stoking desires in order to maintain a steady growth of markets and revenues. This situation of unquenchable growth in production, marketing and consumption of growth—characterized as a "treadmill of production" (see Schnaiberg and Gould 1994 for an in-depth discussion)—has profound environmental consequences.

A mismatch between individual and collective interests leads to phenomena such as the Tragedy of the Commons (Hardin 1968, 1993). In this situation, people maximize their own short-sighted individual self-interest by taking all that they can get away with, while the costs are borne by the collective and/or the environment on which they depend.

High levels of individualism have been associated with conservative voting patterns (Burns 1992), and this often is manifested not only in placing a low priority on environmental values, but also in not viewing it as unethical to advance one's individual interests even though it may cause environmental damage. Taken together, we witness a condition where some significant swaths of "mainstream" culture have the effect of legitimizing environmentally unsustainable practices.

Polities are organized on the nation-state or local levels. As such, they typically feed into some variant of the tragedy of the commons problem. In his book *Earth in the Balance*, for example, Al Gore (1993) points out a situation where the hydrological balance around the Aral Sea has been permanently altered. What was once a region with plentiful fishing and farming, saw farmland converted to grow crops that are more lucrative in the short run such as cotton, but which required more water than could be sustained. The problem was exacerbated by the fact that more than one country—in this case, Kazakhstan and Uzbekistan—acting in what they saw as their own national interests, sought to take more from the common waters of the Aral Sea basin than did their neighbors. This led to an acceleration of incursion into the natural cycles such that much of the region is now desertified and unable to grow even the crops it had been producing for centuries.

So the tragedy of the commons operates not only on the individual plane, but on a number of aggregate levels as well. What moves action toward or away from a tragedy of the commons scenario? A key variable here appears to be the relative openness or closedness of networks (Coleman 1986; Granovetter 1985; Burt 1982). People tend to have a sense of responsibility to the collective to the extent they are in a closed network, such that results of their actions come back to them, albeit through a chain of linkages to others. In small, relatively closed social groups, there tend to be strong norms of altruism, and severe sanctions for their violation. In large, relatively open groups, norms tend to be less stringent.

How do these insights apply then to the current situation, in which we have more people in the world than ever, and the groupings of people are, on some level (because of, *inter alia*, the internet and other vehicles of mass communication), less geographically based than ever before? We will consider institutions more explicitly below, yet it bears noting that even in social aggregates, individuals still attempt to maximize their own self interests; and this is every bit as true for leaders as for non-leaders (Fain et al. 1994).

As new communities arise, so do new normative systems. There is, of course, cultural lag in this process, and there is no guarantee that either the old norms or the new will adequately address environmental concerns. In fact, given the pace of technological change, there is the very real possibility that, even though norms also may be changing to be more environmentally friendly (and I am not necessarily implying here that they are), they still may be *increasingly* inadequate to the conditions created in the growing climate of hyper production and consumption. This is precisely because the material conditions such as those associated with technology are changing even faster than the cultural adaptations to them.

Why Institutional Fixes by Themselves Do Not Work

Culture has meta-institutional properties. It serves as a common backdrop against which institutions have meaning. To the extent there are institutional linkages (e.g. between the school system and the family on one hand and labor markets on the other), they take place within a cultural context.

It is almost never the case that institutional "fixes" ever actually remedy a problem in any kind of meaningful way. Changing one institution in isolation typically catalyzes enantiadromic processes with a chain of consequences. Put anther way, since institutions are embedded within cultures, an isolated change in one tends to set a reaction in the rest of the culture to restore the equilibrium that was in place before the change. History is replete with examples. Recent attempts to impose a Western style "democracy" in a nation where there is widespread tribalism, low formal education and a dearth of free markets, *inter alia*, have tended to be adventures in frustration at best.

Advances in technology increase in pace. There is now greater ability than ever in history to make incursions into the earth to extract her resources. As the philosopher Martin Heidegger (1966 [1932], 1999, 2006) observed, technology becomes "enframed" in culture-advances in technology such as the invention and use of the internal combustion automobile, the computer, the chainsaw and the snowmobile go far beyond the uses for which they originally were designed. Ideas, vocabularies and sentiments about them become embedded in the popular imagination, thoughts and discourse. In short, over time, they become inextricably part of the culture and become integrated into the very being of the people.

While it is tempting as well to think that many or all environmental problems will be taken care of once the cultural lag process has allowed the adaptive culture to catch up with environmental problems, this approach is fraught with problems. Buttel (2000), for example, offers cogent criticism of world polity theory on a number of accounts. The fact that developing countries have begun to establish environmental ministries and have entered into treaties does not necessarily translate into environmentally friendly *action* (for an alternative view, see Schofer and Hironaka 2005); this may be particularly true in places where the perception was one of having been coerced by more powerful nations or institutions such as the World Bank and/or the International Monetary Fund. There tends to be a "top down" bias as well in discounting the role of scientists, environmental organizations, and other actors *within* a given nation-state.

Even if the world polity *is* moving in the general direction of environmental stewardship (which as Buttel points out is problematic at best), it still may be the case that environmental degradation is occurring in so many ways at such a great rate that rather than catching up with the change, the lag between the creation of environmental problems and the stewardship that would address them gets larger with time. More broadly, in an age when there are more people in the world than ever in history, with technology to make more profound incursions into the natural ecosystem than ever before, and with more wealth concentrated in relatively fewer hands than at practically any other time in history, the natural world is changing rapidly. Yet culture, albeit adapting in response, does so on a slower trajectory. The outcome then may well be a significant and increasing gap between ecological conditions and the culture that stems from them.

It also is tempting to assume that environmental problems will be "solved" at some point by technological fixes (J. Simon 1983). Yet throughout history, when there has been an increase in efficiency of some technology, the culture tends to adapt in a number of ways, including wider use of the resource. While this tendency goes by various names, perhaps the best know is the "Jevons paradox", named after one of its observers (for a recent in-depth discussion, see Polmieni et al. 2008). Generally speaking, the more profound a technological advance, the more likely it is to threaten the balance of the ecosystems of which it is part if there is not a concomitant adaptation in cultural complexity.

There are numerous examples of this. The introduction of greater fuel efficiency in internal combustion automobiles, while laudable in many ways, did redound in some negative environmental outcomes. More automobiles were produced and bought, and more miles were driven. This led to more infrastructure building to keep pace with the automobiles, including more intense need for gasoline, more roadbuilding and the rise of suburbs and ex-urbs which in turn often stimulated other aspects of ecological imbalance and degradation.

There is now an unprecedented juggernaut of global markets organized around mass production and consumption, economies of scale and the concentration of capital (Foster 1999). This is supported by increasingly extensive and intensive technologies which deplete the world's resources and disturb the natural balance ever more profoundly. These processes are juxtaposed with the absence of a global normative system to provide a check, or even to make sustained meaningful ethical judgements about them.

Concerns we have been discussing pose huge ethical questions that redound in the political sphere, and so society is left with a number of key questions about culture and the environment that relate directly and indirectly to political considerations. From a praxiological perspective, what leads polities to promulgate ecologically healthy and sustainable policies, and what increases the likelihood that, even when those policies are adopted, they will be carried out in good faith? How do polities then evaluate how well policies are working, and assess and reevaluate, given the relative certainty that, even with the best motives, there will be unintended consequences with each change?

Sustainable solutions necessarily involve institutions. Yet they must go well beyond an institutional framework. Ultimately, they must be undergirded by a culture with norms of sustainability, central to its core ethos, and internalized and practised by a critical mass of individuals of good will.

Conclusions

Culture can be seen as a large and interrelated system of ways of thinking, communicating and organizing. Culture forms a large part of what people take for granted, and orients how societies perceive and act upon the world in which we live. A major aspect of the culture of "developed" societies is a scientistic orientation in which there is an implicit belief that nature is something that can be controlled. This is largely an artifact of a congeries of interrelated events including the Newtonian and industrial revolutions and the rise of hyper rational analytical ways of thinking. While these brought a number of advances, they also have in many cases worked against cultures perceiving and acting toward the world in ecological terms.

Because science and technology tend to work through isolating some processes, and to do so on one level of analysis, there is a tendency to disregard other important processes and levels of analysis such that there are likely to be enantiadromic outcomes. This tendency is not just in the natural sciences but the social sciences as well. Consider the widespread tendency in economics to treat environmental effects as "externalities" barely worth even measuring. Add to this the largely unquestioned truisms of classical and neo-classical economics, such as the "law of comparative advantage" and the principle of economies of scale (McCloskey 1985), and the potential impacts to the environment are staggering.

It bears noting, of course, that each of the perspectives we are considering is distinct only in an analytical sense. In a natural system, all of them operate, serving at times to constrain and at others to facilitate one another. The purpose of pointing out problems is not to undercut any particular discipline, but rather to issue a call for humility in making an honest assessment of how thinking in general (and academic thinking is no exception) is embedded in the cultures from which it issues (Nelson and McCloskey 1987).

While societies have always faced ecological constraints, many of the problems such as the rise in greenhouse gases coupled with the depletion of carbon sinks through deforestation, now are already of proportions unprecedented in human history. Given the pace of technological change, cultural adaptations appear to be falling dangerously far behind. The arrogance of the scientistic world view leads to ever greater incursions into the natural environment, coupled with a denial that serious problems (such as global warming) even exist (Taylor 1997), or a sophomoric faith that some future technological breakthrough will necessarily serve as a *Deus ex machina* sort of fix.

For there to be truly sustainable practices, the human cultures that undergird them necessarily would have schemas that reflect the earth's ecological complexity embedded in their ways of perceiving, and thus would have increased chances of acting in harmony with the natural world. The best hope is for the culture to develop sustainable ways of seeing and acting upon the world through the long process of education and developing consciousness and ecological ways of perceiving (Freudenburg and Frickel 1995; Eisenberg 1998).

There is some hope on the horizon as well, as traditional institutions slowly adapt to changing conditions and the crises caused by inadequacies of the past. Nobel Prize winner Muhammad Yunus (1999, 2007), for example, articulates a vision for socially and environmentally conscious business. Yunus advocates the adoption of the "triple bottom line" in business, in which the quest for profits is balanced with environmental stewardship and human dignity. Such a change from the current tendency to have the single bottom line of profits as the *raison d'etre* of business would necessarily require new ways of thinking—new cultural schemas.

Put another way, any changes, institutional or otherwise, necessarily will be embedded in broader cultural practices. Integrating ecological consciousness in culture will be crucial here, as will developing a sense of humility and respect for the larger process of natural ecosystems.

Endnote:

[1]A variety of other vocabularies has been used as well. Theorists with a communication studies background typically use terms such as "motif" (Ibarra and Kitsuse 1993), "ideograph" (McGee 1975, 1980), or "terministic screen" (Burke 1966, 1969a, 1969b) to refer to these strategies people use to organize information and to communicate within a cultural milieu. For one final example, as a way of living, making sense of, and acting within one's life world, Bourdieu (1977, 1984) uses the terms "habitus" and "field". So a number of different ways of conceptualizing and studying these crucial processes have been used, with some overlap, and yet with each emphasizing its own salient properties (for reviews, see Burns 1999; Burns and LeMoyne 2003).

Section II

Knowledge and Management

CHAPTER FOUR

Applications of Fishers' Local Ecological Knowledge to Better Understand and Manage Tropical Fisheries

RENATO A. M. SILVANO, MARIA A. GASALLA AND SHIRLEY P. SOUZA

Small-scale and local fisheries are difficult to manage. This is due to the diversity of fishing resources exploited by fishing communities spread over a large region, coupled with the scarcity of funding and people to conduct fisheries surveys. However, local fishers usually posses detailed local ecological knowledge (LEK), which is a potentially useful, but undervalued, tool to improve fisheries management in tropical developing countries. The goal of this chapter is to provide a review of current literature, addressing three major approaches. First, fishers have provided useful and original data about fish biology and ecology, including aspects less known by scientists, such as reproduction, migration and trophic interactions. Such local knowledge may provide guidelines and hypotheses to advise biological studies. Second, although most studies have addressed fishers' LEK about fish or other exploited resources, fishers also often interact with other aquatic animals, such as cetaceans (whales and dolphins), which prey on fish, sometimes resulting in cetaceans becoming entangled in fishing nets. Recent surveys show that fishers' knowledge about these aquatic mammals may be useful, not only to provide scientific data, but also to support dialogue between managers and fishers that may reduce the impacts of fisheries on cetaceans. Third, in many tropical fisheries, statistics on fish landings are either absent or from a short time scale and data are often incomplete. Therefore, scientists and managers usually have limited information about the dynamics and abundance trends of fish stocks. However, fishers, especially the experienced ones, are usually knowledgeable about fish abundance and catch trends from well before scientific data started to be collected. Such local knowledge may thus provide an invaluable (and perhaps unique) opportunity to reconstruct fish dynamics over the long term. These research areas indicate promising ways to apply fishers' LEK in fisheries and natural resources management before both local fishers and their exploited resources disappear.

Introduction

Tropical coastal and freshwater local fisheries are challenging to manage. Reasons include the diversity of fishing resources exploited by numerous fishing communities, the scarcity of funding and trained staff to conduct scientific surveys and the socio-economic importance of such fisheries (Johannes 1998; Pauly et al. 2002). However, notwithstanding the small scale of local artisanal fisheries, inadequate management (or lack of management) can lead to severe depletion of the fishing resources (de Boer et al., 2001), which ultimately has adverse consequences for the fishing communities who rely heavily on these resources.

Most Brazilian freshwater and coastal fishing operations are conducted by small-scale local fishing communities, scattered around reservoirs, along large rivers (mainly in the Amazon Basin), estuaries and along the extensive coastline. Such fishing communities usually have little or no political power, although some communities have local traditional rules to control the use of and to assess fishing resources (McGrath et al. 1993; Begossi, 1995, 1998; Diegues 1999). Brazilian artisanal fisheries show several characteristics typical of tropical fisheries in developing countries: fishers use multiple kinds of fishing gear to exploit a large variety of fish and invertebrate species, fishery management measures are usually absent or insufficient and there is scant scientific data about fish landings or fish biology (Bayley and Petrere 1989; Queiroz and Crampton 1999; Seixas and Begossi 2001; Silvano and Begossi 2001; Begossi et al. 2000, 2004).

The body of knowledge acquired and transmitted at a local (or regional) scale may receive many names; for the purposes of this chapter we will consider it as local ecological knowledge (LEK). This knowledge has been accumulated and transmitted through generations by the reproduction of "memes" (or "cultural variants"), which are fragments of information, transmitted among people by conversation or other behavioural forms (Dawkins 1979). According to several recent studies, local fishers may have detailed local ecological knowledge (LEK) about exploited fishing resources (Johannes, 1981; Poizat and Baran 1997; Valbo-Jørgensen and Poulsen 2000; Aswani and Hamilton 2004). Until the last decade, fishers' LEK was largely neglected by social and natural sciences, but a growing interest in the nature and importance of LEK has recently been detected in both fisheries and conservation science (Neis et al. 1999; Haggan et al.

2007). For example, LEK may be a useful and undervalued tool to improve knowledge of fish biology and fisheries management (Johannes et al. 2000; Drew 2005), especially in tropical developing countries (Johannes 1998). Brazilian fishers also demonstrate detailed LEK about fish (Begossi and Figueiredo 1995; Paz and Begossi 1996; Silvano and Begossi 2002, 2005; Gerhardinger et al. 2006; Silvano et al. 2006, 2008), aquatic ecosystems (Gasalla 2004a, b) and cetaceans (Souza and Begossi 2006). Furthermore, Brazilian fisheries share the same characteristics and face similar problems as tropical and subtropical fisheries in general. The goal of this chapter is to provide a review of current literature, focused on Brazil, of major potential contributions of fishers' LEK to fisheries management. We discuss three main approaches, or lines of research: first, the usefulness of fishers' LEK to advance biological knowledge (and thus management) of fish; second, application of fishers' knowledge of cetaceans, not only to add to scientific data, but also to involve fishers in management decisions aimed to reduce incidental catches of aquatic mammals; and third, ways in which fishers' knowledge may be an important tool to reconstruct dynamics of exploited fish stocks over the long term.

Fishers' Knowledge about Fish Ecology and Biology

Since a classic study of Morril (1967) inaugurated ethnoichthyology as a scientific discipline, several studies (mainly during the last two decades) have addressed local ecological knowledge that fishers have about fish. Among other applications, fishers' LEK may be an invaluable source of insights and hypotheses (Silvano and Valbo-Jørgensen 2008), which, through comparative biological studies, may provide new data for fisheries management (Aswani and Hamilton 2004) or improve biological studies (Poizat and Baran 1997). Fisher's LEK may be also a source of urgently needed biological data about rare and endangered fish, such as the mero (*Epinephelus itajara*) off the south Brazilian coast (Gerhardinger et al. 2006). In order to complement already available ethnoichthyological surveys (Johannes et al. 2000; Dyer and McGoodwin 1994; Haggan et al. 2003; Begossi et al. 2004; Drew 2005), we review some recent methods and analytical tools to improve the application of fishers' LEK in fisheries management.

Methodological Approaches and Sampling Issues

The science of ethnoecology has experienced improvement in methodological and analytical approaches (Huntington 2000a; Davis and Wagner 2003; Silvano et al. 2008). While there is no single ideal method in ethnobiological surveys, the proper research methodology (according to the objectives of each survey) will influence the quality of the results, and thus, the usefulness of these results to fisheries management. Although a detailed methodological review is beyond the scope of this chapter, there are some major and new issues that deserve a brief discussion.

There are two major approaches regarding sampling of people included in the survey (hereafter referred to as informants) and the kind of questions asked in order to gather LEK data (quantitative or qualitative). The first consists of interviewing as many people as possible (or a minimum number of selected informants) and asking the same standardized and usually closed-ended (direct) questions to all informants (structured interviews) (Begossi and Figueiredo 1995; Paz and Begossi 1996; Silvano and Begossi 2002, 2005; Silvano et al. 2006, 2008). The second involves interviewing few (sometimes even one) informants, usually through open-ended questions or even informal conversations (Marques 1991, 1995; Huntington 1998, 2000a). Both methods are useful, depending on the issues being investigated, and they can even be combined within a single study. However, several ethnobiological studies have neglected a critical sampling phase: the selection of informants (Davis and Wagner 2003). Informants should be selected according to clear, reproducible and well stated criteria (Davis and Wagner 2003). We therefore emphasize the need for carefully selecting informants, especially in more qualitative surveys. The quality of data is linked to the quality of the informant and to the representativeness of that informant's LEK compared to the whole community. Although not mandatory, it is interesting if the sampling design allows for statistical analyses by including a minimum number of informants at each locality studied (Davis and Wagner 2003; Silvano et al. 2006). Besides reinforcing the results, statistical testing renders the whole ethnoichthyological survey more acceptable to scientists and managers, therefore increasing the chances that they will consider LEK in management measures.

Reliability and Accuracy

In ethnoecological studies, it is important that the researcher asks the right questions in the right way, using an adequate interview technique

(Huntington 2000a). Answers should also be correctly interpreted. Another issue recently discussed in the literature regards the reliability and accuracy of LEK data. According to Maurstad et al. (2007), reliability is the degree to which informants are transmitting what they really know and believe, while accuracy is the degree to which the information provided by interviewees (even if reliable) corresponds to real biological phenomena. A robust sampling method and well-conducted interviews should circumvent the problems of low reliable and less accurate information. For example, interviewing many informants, emphasizing those answers more frequently cited (Silvano and Begossi 2002, 2005; Silvano et al. 2006, 2008), reduces the risks of unreliable or inaccurate information (unusual information is discarded or treated as an exception). Alternatively, the reliability of LEK data can be assessed by checking if LEK data is grounded in the informant's daily activities, while the accuracy can be checked by comparing LEK data to information from conventional biological studies (Maurstad et al. 2007). If it is suspected that at least some of the LEK data may be inaccurate or unreliable, this can be added to the discussion of the results. For example, in a highly conflicted scenario at Newfoundland, Canada, two distinct categories of cod fishers held different perceptions about ecological factors influencing the decline of cod populations, but such perceptions are strongly influenced by political and social constraints (Palmer and Wadley 2007).

Ecological Patterns and LEK: from Local to Global, from Simple to Complex

On one hand, the detailed LEK that fishers have regarding their local (or regional) environment may complement conventional scientific biological knowledge, which usually deals with broader and more general patterns (Huntington et al. 2004). On the other hand, the local nature of fishers' knowledge may limit its applicability to broader ecological patterns (such as fish migrations), which may involve large-scale movements across distinct regions and even countries. One approach to assessing ecological patterns over larger geographical scales using LEK consists of using the same methods to study several fishing communities in distinct regions, and analyze the comparable results together to reveal the broad picture (Valbo-Jørgensen and Poulsen 2000; Silvano and Begossi 2005). For example, Silvano and Begossi (2005) observed possible general feeding and migratory patterns of bluefish (*Pomatomus saltatrix*, Pomatomidae, Figure 4-1), based on the LEK of Brazilian and Australian fishers. Similarly, LEK data gathered among 355 expert fishermen from 51 areas along 2,400 km of the Mekong River in Laos, Thailand, Cambodia and Vietnam, provide new and invaluable information about the reproductive behaviour and migration of 50 fish species (Valbo-Jørgensen and Poulsen 2000).

Figure 4-1. The bluefish (*Pomatomus saltatrix*, Pomatomidae), which has worldwide distribution and is an important marine commercial fish exploited by several local fishing communities along the Brazilian coast (Photo: R.A.M. Silvano).

Knowledge about trophic relationships among fish species and between fish and other organisms is important to fisheries management (Polunin 1996; Vasconcellos and Gasalla 2001). Among other reasons, this knowledge allows us to better understand the process of fish contamination by chemicals, such as mercury (Boischio and Henshel 2000). Fishers' LEK may be useful to assess complex ecological patterns, sometimes based on simpler pieces of information. Tropical marine and freshwater food chains are difficult to elaborate based on fish diet data, as such data are usually not available for most species, especially in complex ecosystems (e.g., reefs or tropical lakes). Fishers usually possess detailed knowledge about fish diet and feeding behaviour and such LEK can be useful to elaborate at least simplified food chains, following two general approaches. First, a few experienced fishers may be asked to depict the feeding interactions themselves, usually through drawings (Marques 1991, 1995). Second, many fishers may be individually interviewed about the diet and predators of selected fish species, and then such information may be organized by the researcher in the form of food chains (Silvano and Begossi 2002; Gasalla 2004a).

Formulating and Testing Biological Hypotheses through LEK

One of the most promising outcomes of studies about fishers' LEK would be to provide new hypotheses about fish ecology, to be further examined through biological studies. In a brief review of recent ethnoichthyological studies, Silvano and Valbo-Jørgensen (2008) present 29 hypotheses based on fishers' LEK about feeding habits, migration, reproduction and abundance patterns of marine and freshwater fish from Brazil and Southeast Asia. Many more hypotheses could be devised from the several existing studies about fishers' LEK. However, a more robust approach would be to both elaborate hypotheses based on fishers' LEK and to test such hypotheses through biological surveys (Marques 1991; Poizat and Baran 1997; Aswani and Hamilton 2004). Such double-checking of fishers' LEK accuracy may be essential to reveal new local ecological patterns and to correct mistakes made by both fishers and biologists. For example, in the Solomon Islands, South Pacific, Aswani and Hamilton (2004) proposed four hypotheses regarding the ecology and fisheries of the reef fish (*Bolpometodon muricatum*) based on fishermen's LEK. After examining conventional scientific surveys, the only rejected hypothesis referred to lunar periodicity and fish catches: contrary to fishers' assertions, this fish was not caught most during the new moon (Aswani and Hamilton, 2004). Nevertheless, in other instances, the testing of hypotheses based on fishers' LEK has provided new and unexpected information, sometimes even contradicting biological knowledge. For example, Marques (1991) confirmed the unusual hypothesis, based on fishers' knowledge, that an estuarine catfish eats terrestrial insects on a seasonal basis. Fishers' LEK may also reinforce current hypotheses and concepts in the biological literature. For example, several Brazilian fishermen from both southeastern and northeastern coasts mention that some fish, such as bluefish and mullet (*Mugil platanus*, Mugilidae) migrate from South to North, which matches the (limited) scientific evidence (Silvano and Begossi 2005; Silvano et al. 2006).

Fishers' LEK on Cetaceans

Overview of Studies on Fishers' LEK about Cetaceans

Surveys assessing cetaceans' ecology through fishers' local ecological knowledge (LEK) are relatively recent. Richard and Pike (1993) analyzed experiences with co-management of beluga (*Delphinapterus leucas*) and narwhal (*Monodon monoceros*) hunting in the Eastern Canadian Arctic in the 80's and 90's, concluding that integrating LEK and scientific knowledge is imperative for the effective management and conservation of these species. Huntington (1998, 2000a, 2000b) and Huntington et al. (1999)

interviewed hunters in Alaska and obtained information on the beluga's distribution, abundance, migration patterns, local movements, feeding behaviour, diet, avoidance of predators, calving, bathymetry, ecological interactions and human influences. Mymrin et al. (1999) compared information obtained from young and older hunters about beluga's ecology, and found several areas of agreement, as well as new points that deserve investigation. However, they highlighted the fact that LEK is disappearing with the older hunters.

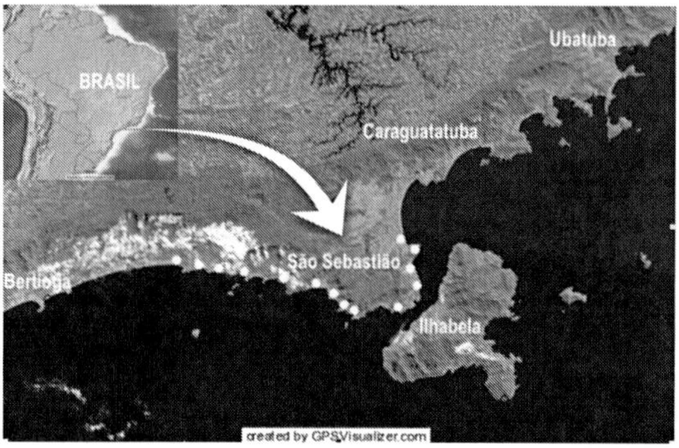

Figure 4-2. Northern coast of São Paulo State, showing fishers' communities studied (white dots) at São Sebastião (Map and satellite image created by GPS Visualizer.com) (Modified from Souza and Begossi 2006).

In Brazil, ethnobiological surveys on cetaceans began in the last decade (Przbylski and Monteiro-Filho 2001; Pinheiro and Cremer 2003; Alarcon and Schiavetti 2005; Peterson 2005; Souza and Begossi 2006) and are mostly related to ecological aspects, such as interactions with fisheries, preferential habitats, feeding and social behaviour. We present more detail on fishers' LEK regarding cetaceans in a case study on the southeastern Brazilian coast (Souza and Begossi 2006).

Ethnobiology of Cetaceans by Fishers from São Sebastião, Southeastern Brazilian Coast

São Sebastião (23°42'18" to 23°45'38"S – 45°25'41" to 45°53'49"W) is situated on the northern coast of São Paulo State (Southeastern Brazilian coast) and marked by the presence of the Atlantic Rain Forest (Figure 4-2). Small communities of artisanal fishers, called "caiçaras", interact with whales and dolphins there, especially with coastal species, which are incidentally caught by gillnets (Souza and Winck 2005). Local researchers confirmed the occurrence of whales and dolphins in that coastal region during the last 14 years, through records of sightings and stranded or incidentally captured animals. According to reports of the "*Projeto SOS Mamíferos Marinhos*", from September 1994 to September 2006, 151 cetaceans of 16 species[1] have been recorded alive or dead (S.S. unpublished data). These reports have also confirmed that encounters between fishers and cetaceans are frequent, involving coastal dolphins at the main fishing grounds. Incidental catches in gillnets are an occasional result of these interactions and involve two species in particular: *Pontoporia blainvillei* (Pontoporiidae) (Figure 4-3) and *Sotalia guianensis* (Delphinidae) (Figure 4-4). The former one is considered vulnerable, according to the International Union for Conservation of Nature (IUCN) and the Brazilian Environmental Agency (IBAMA) red lists (Reeves et al. 2003, IBAMA 2001). The International Whaling Commission (IWC) has recognized the incidental capture of cetaceans as a threat to the populations of small species, especially from the families Phocoenidae, Pontoporiidae and Delphinidae (Perrin et al. 1994).

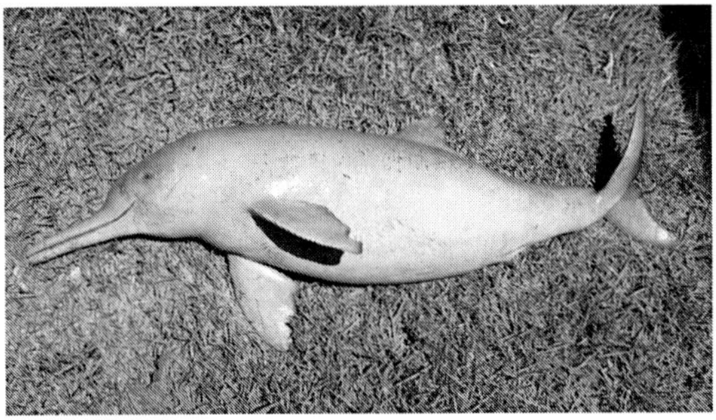

Figure 4-3. Franciscana dolphin (*P. blainvillei*) caught by gillnet at Enseada, São Sebastião, Brazil (Photo by SOS Mamíferos Marinhos).

Accessing Fishers' Knowledge

In order to record fishers' knowledge about cetaceans, we interviewed 70 fishers (mean age of 59 years) from 14 fishing communities in São Sebastião. We used semi-structured questionnaires and showed unlabelled figures and photos of 11 cetaceans species that occur in this area: Family Balaenidae: *Eubalaena australis* (right whale), Family Balaenopteridae: *Megaptera novaeangliae* (humpback whale), *Balaenoptera edeni* (Bryde's whale) and *B. acutorostrata* (minke whale) from the suborder Mysticeti and Family Delphinidae: *Orcinus orca* (killer whale), *Tursiops truncatus* (bottlenose dolphin), *Steno bredanensis* (rough-toothed dolphin), *Stenella frontalis* (Atlantic spotted dolphin), *Delphinus capensis* (common dolphin), *Sotalia guianensis* (marine tucuxi) and Family Pontoporiidae: *Pontoporia blainvillei* (franciscana) from the suborder Odontoceti. Each fisher was individually interviewed after his agreement to participate in this survey. We asked questions about how fishers name, classify and group the species shown in the unlabelled figures. We also inquired about ecological aspects such as areas of occurrence, preferential habitats, seasonality, group size, feeding habits, predators, reproduction and interactions between the species and local fisheries.

Figure 4-4. Marine tucuxis (*S. guianensis*) at São Sebastião Channel, Brazil (Photo by SOS Mamíferos Marinhos).

Naming and Classifying Whales and Dolphins

Cetacean species and genera show high morphological similarity, making identification in nature difficult, generally only performed successfully by cetacean experts. Fishers' perception about cetaceans in the studied region was highly influenced by the phenotypic (or morphological) salience and cultural importance of the whales and dolphins, since these fishers did not use cetaceans for food or sale. Historically in the studied area, incidentally captured cetaceans have not been utilized, according to interviewed fishers (n = 30). Whales and dolphins were classified by most fishers into the category "fish", but with little difference from the other categories ("Mammals" and "Not fish") (Figure 4-5). According to the folk taxonomy proposed by Berlin (1992) the category "fish" represents a life-form, characterized by different groups of animals that live in aquatic habitats, including fish, aquatic invertebrates, turtles, crocodiles, dugongs, whales and dolphins (Anderson 1969, Hunn 1982, Berkes et al. 2000).

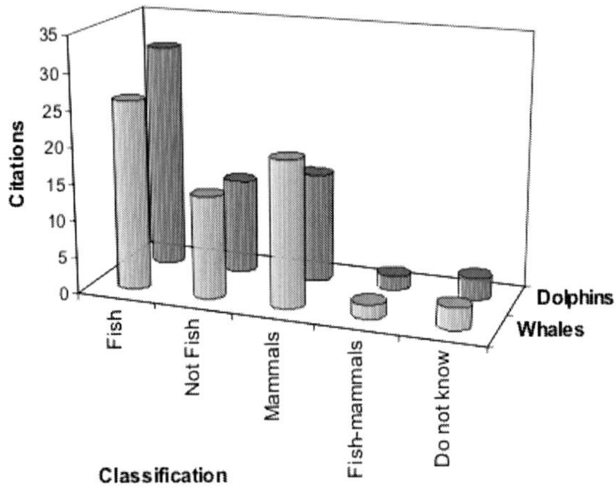

Figure 4-5. Whales' and dolphins' classification according to fishers (n=70) from São Sebastião, Brazil (Modified from Souza and Begossi 2006).

In relation to nomenclature, following Berlin (1992), the studied fishers provide three generic names and 14 folk species of cetaceans, nine of them corresponding to binomials (Figure 4-6). Fishers' answers indicated that the most conspicuous cetaceans were not necessarily the most abundant, but the most caught (*P. blainvillei* and *S. guianensis*) and those of greater size (although they are rare), such as *T. tursiops* and *E. australis*. These species, together with those featured by media (*M. novaeangliae* and *O. orca*), were the most recognized and important in the process of classification and nomenclature by the interviewed fishers, being named to the folk species level. The most caught species *P. blainvillei* and *S. guianensis* happen to be the most threatened in the surveyed area. These species have not been sufficiently studied, thus potentially making the fishers' knowledge valuable for conservation.

Fishers' Perception of Cetacean Ecology

The information provided by fishers from São Sebastião confirmed that they obtain LEK of cetaceans both during their fishing activities and by cultural transmission. Fishers' LEK thus is dynamic and susceptible to changes. There was a high degree of agreement among fishers' answers, data from local research on cetaceans and data from scientific literature on southeastern Brazilian cetaceans. However, for most of the studied species, there are relatively few studies on cetaceans in the region of the northern coast of São Paulo. Three vulnerable species, *Eubalaena australis*, *Megaptera novaeangliae* and *Pontoporia blainvillei* occur in this area, besides other poorly known species.

Ecological topics relating to coastal dolphins (e.g., franciscanas and marine tucuxis), and bottlenose dolphins were better known by the fishers. Occurrence areas and feeding habits were the topics most detailed by the interviewed fishers (Figure 4-7).

Fishers provided 46 different locations of occurrence in the studied area, especially for the species *P. blainvillei*, *S. guianensis*, *T. truncates*, *M. novaeangliae* and *E. australis*. Most fishers (n= 46) affirmed that dolphin movements are conditioned by foraging activities, although some dolphin species could be "residents", while others "are just passing through" in specific seasons of the year. Group sizes for dolphins varied according to the species. Some dolphin species, such as *D. capensis* and *S. frontalis,* are known to form large groups with hundreds of animals, while others, such as *P. blainvillei,* form small groups of less than 10 individuals (Jefferson et al. 1993, Hetzel and Lodi 1993, Whitehead and Mann 2000, Siciliano et al. 2006). Fishers' reports of dolphin group sizes agreed with information from literature and research data for the studied area.

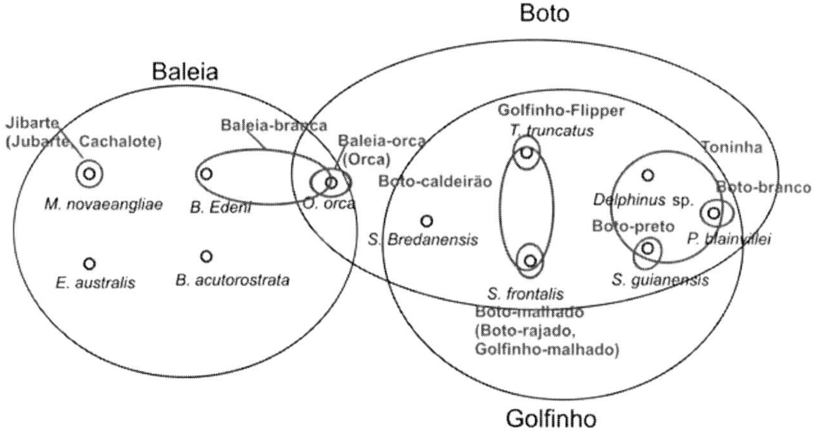

Figure 4-6. Cetaceans' classification, according to fishers from São Sebastião, Brazil. Dash circles correspond to biological species, dot circles / ellipses are folk species and solid ellipses correspond to generic rank (Modified from Souza and Begossi 2006).

Among whales, the migratory species (humpback and right whales passing during their reproductive season in the winter) were especially noted by the fishers (n = 37), confirming the northern coast of São Paulo as part of their migratory routes. Fishers' knowledge of cetaceans' reproduction was mainly related to where and when mother-calf pairs were observed. Nearly half of the interviewed fishers (47%) mentioned the strong bonds linking mother and calf. Whales tend to form smaller groups than dolphins (Whitehead and Mann 2000).

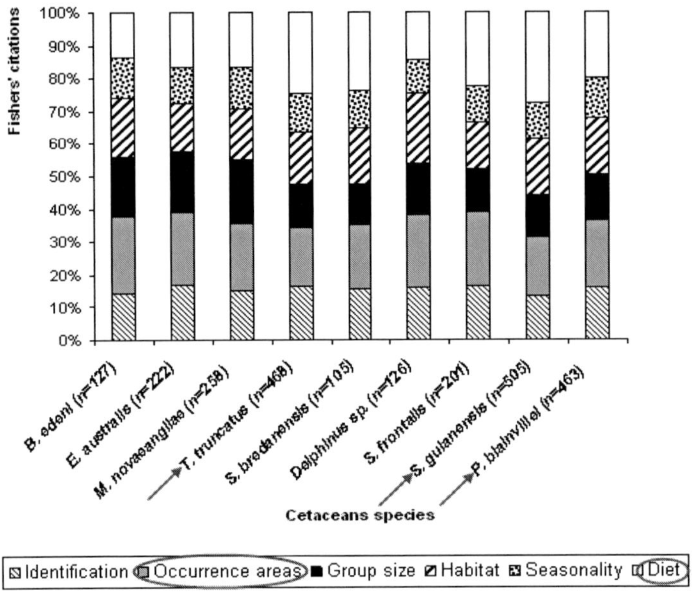

Figure 4-7. Proportion of fishers citations for 6 ecological aspects of nine species of cetaceans (the most mentioned ecological aspects and cetaceans' species are marked by circles and arrows).

Interviewed fishers easily identified fish schools chased by dolphins. Another clue to dolphins diet is obtained by fishers every time a dolphin is caught by gillnet used to catch a specific fish. From these "*in loco*" observations, fishers provided a list including fishes, crustaceans and molluscs as possible prey items for the cetaceans studied (Figure 4-8, Table 4-1). According to fishers (n = 39), the main animals preyed on by whales are anchovies (*Anchoa* sp) and sardines (*Sardinella brasiliensis*), although most fishers confessed that they had not seen whales feeding in nature. However, a "meme" was mentioned by some fishers (n = 11): "the elders used to tell that the whale stops near coastal rocks, open its mouth to let the anchovies in, then it swallows the fishes slowly, because the whale's throat is very narrow". In general, according to the interviewees, dolphins prey on a greater number of species (2 to 18 items) than whales (3 to 6 items).

Table 4-1. Prey items consumed by cetaceans at São Sebastião, according to interviewed fishers.

Biological Family	**Biological species***	**English common names**	**Folk names**	**Citations**
Clupeidae	*Sardinella brasiliensis*	Sardine	Sardinha	67
	Brevoortia pectinata	Menhaden	Savelha	4
Engraulidae	*Anchoa* sp.	Anchovy	Manjuba	73
Hemiramphidae	*Hyporhamphus* sp.	Halfbeak	Panaguaiú	10
Belonidae	*Tylosurus acus*	Needlefish	Timbale	1
Serranidae	*Epinephelus marginatus*	Grouper	Garoupa	1
Pomatomidae	*Pomatomus saltatrix*	Bluefish	Enchova	5
Carangidae	*Caranx* sp.	Jack	Xaréu	2
	Oligoplites sp.	Leatherjack	Guaivira	3
	Selene sp.	Moonfish	Peixe-Galo	1
	Caranx crysos	Blue runner	Carapau	3
Sciaenidae	*Cynoscion* sp.	Shortfin corvina	Pescadinha	5
	Micropogonias furnieri	Croaker	Corvina	1
	Paralonchurus brasiliensis	Banded-croaker	Maria-Luíza	1
	Menticirrhus sp.	Kingcroaker	Betara	1
	Larimus breviceps	Drum	Oveva	2
Mugilidae	*Mugil platanus*	Mullet	Tainha	125
	Mugil curema	White mullet	Parati	47
Sphyraenidae	*Sphyraena* sp.	Barracuda	Bicuda	1
Trichiuridae	*Trichiurus lepturus*	Cutlassfish	Espada	4
Scombridae	*Scomberomorus brasiliensis*	Mackerel	Sororoca	2
Stromateidae	*Peprilus paru*	Butterfish	Gordinho	1
Carcharhinidae	*Carcharhinus* sp.	Shark	Cação	1
Loliginidae	*Loligo* sp.	Squid	Lula	35

Biological Family	Biological species*	English common names	Folk names	Citations
Argonautidae	*Argonauta argonauta*	Argonauta	Argonauta	1
Penaeidae	*Xiphopenaeus kroyeri, Litopenaeus schimitti*	Shrimp	Camarão	5
Total citations				402

*Species identification followed Ávila-da-Silva (2005) and Fishbase (Froese and Pauly 2000).

Accounts provided by the fishers during the interviews regarding dolphin feeding behaviour were very detailed, including information about feeding preferences, movements, chasing strategies and interactions with seabirds. The main prey for dolphins were fish species from the families Engraulidae (anchovies), Clupeidae (sardines), Mugilidae (mullets) and squid (Loliginidae) (Table 4-1). The prey items mentioned by the fishers agree with local research on feeding habits of small cetaceans, which prey on the families Clupeidae, Engraulidae, Sciaenidae (croakers) and Loliginidae (Souza et al. 2008). They also agree with LEK information given by fishers from other points on the southeastern and southern Brazilian coast (Di Beneditto et al. 2001, Pinheiro and Cremer 2003, Peterson 2005, Hassel 2006).

Fishers mentioned six examples of interactions with cetaceans, most of which (60% of answers) were considered as positive interactions, such as "dolphins follow the boat", "dolphins concentrate the school of fishes" or "they perform incredible acrobatics". They also mentioned negative interactions (40% of answers), such as 'dolphins eat from gillnets', "dolphins disperse the schools" and "dolphins tear up gillnets". Among the positive interactions, some of the interviewed fishers (13%) pointed out that dolphins can help by concentrating the fishes near the fishing nets. Fishers mentioned the example of the interaction between *Tursiops truncatus* (the bottlenose dolphin) and fishers from the southern coast of Brazil, who rely on the dolphins to bring schools of mullets near the beach (Simões-Lopes et al. 1998). However, this same dolphin species has negative interactions with fishers in Florida, U.S.A., where dolphins steal fish from gillnets (Zollett and Read 2006). The outcome of interactions between fishers and dolphins seems to be at least partially related to fishing technique and the kind of fish exploited by fishers: coastal fishers that catch mullets and other near-shore estuarine fishes usually show positive mutualistic associations with dolphins

(Simões-Lopes et al. 1998), while offshore fishers using gillnets to catch pelagic fishes tend to be in conflict with dolphins (Zollett and Read 2006).

Among the answers to questions asked of the fishers, incidental catches of cetaceans were the least cited topic. Nevertheless, when available, fishers' information agreed with the literature, that gillnets were responsible for most of these catches (Figure 4-9) and that *P. blainvillei* and *S. guianensis* were the species most often caught. According to the fishers, such incidental catches occur because dolphins get their beaks or flippers entangled in the net when pursuing fish, do not manage to escape and subsequently drown. Although 75% of the interviewed fishers confirmed that dolphins are incidentally caught, it was obvious this topic is sensitive among fishers, who avoided talking about it due to the governmental ban on catching cetaceans. In fact, incidental catches are considered by the IUCN as one of the main threats to the survival of small cetaceans around the world (Reeves et al. 2003).

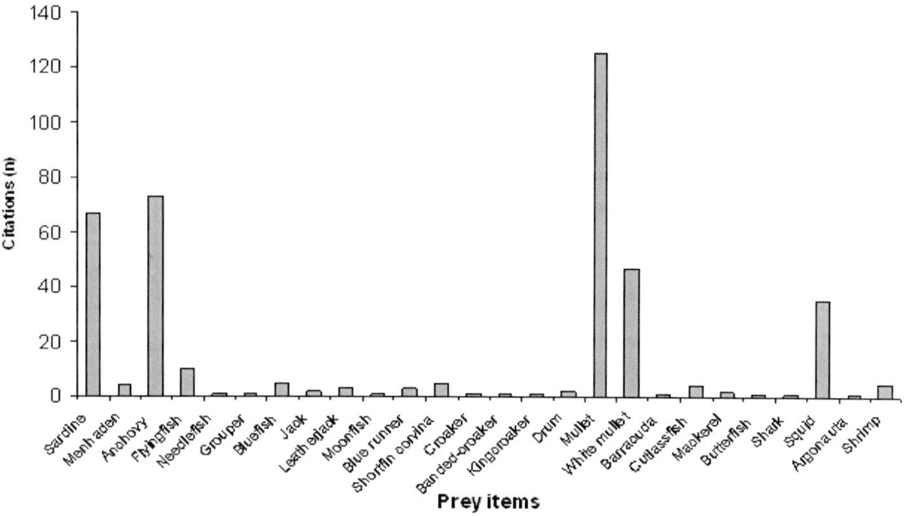

Figure 4-8. Prey items consumed by cetaceans according to fishers from São Sebastião.

Most of the interviewed fishers (73%) believe that cetaceans do not have predators, because they are very intelligent and fast swimming animals. However, 13% of them said that man is the main predator of whales and dolphins, making allusions to dolphin's entanglement in nets or to whale hunting.

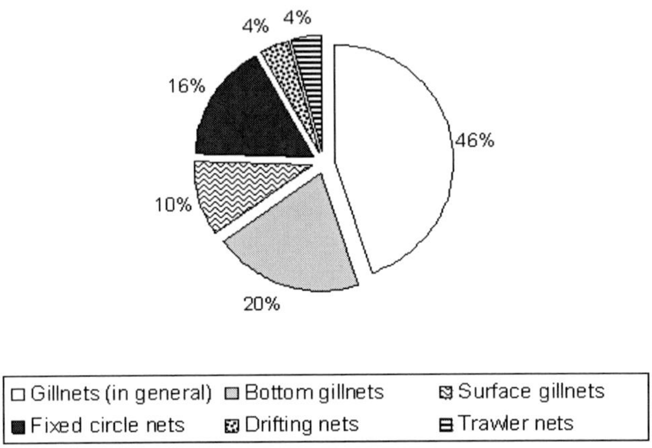

Figure 4-9. Fishing nets that may catch dolphins, according to the fishers' citations (n = 49).

II.6 Fishers' LEK and Cetacean Conservation

The interviewed fishers of the southeastern Brazilian coast demonstrated that they know very well the occurrence areas and feeding habits of the studied cetacean species. Information given by them about seasonality, preferential habitat and reproductive aspects, can contribute to locating protect areas that can be used during reproductive periods by the studied species. According to Parra (2006), knowledge of cetacean spatial dynamics and habitat preferences is essential to the delimitation and effective management of conservation areas to protect threatened species. On the northern coast of São Paulo State, some islands are officially protected marine areas in which fisheries are restricted; however, many fishers do not agree with these restrictions because they have used these areas for many decades. As our study demonstrated, fishers' knowledge is consistent and should be considered in delimitation and management of protected areas. Likewise, Huntington (1998) who studied hunters' LEK of belugas, proposed that hunters' ecological information and insights should be used in addition to other sources of information to make environmental management more biologically and culturally effective. Johannes (2003) also demonstrated that local knowledge can be very useful for locating and to managing protected marine areas and for providing long term data about fisheries. In fact, Johannes et al. (2000) mentioned four examples in which the knowledge of local communities contributed to solve conflicting issues regarding cetacean management in the Arctic.

On the other hand, fishers' LEK seems to be increasingly influenced by popular media, thus becoming globalized and losing important social mechanisms of cultural transmission. Moreover, some change is expected in fishers' knowledge due to their increasing contact with tourism. This is not necessarily bad. As fishers demonstrated empathy for cetaceans, maybe contact with updated information could increase fishers' awareness about cetacean vulnerability. Such awareness could lead to their cooperation in developing alternative management strategies for gillnet fisheries, in order to minimize incidental captures.

Fishers' LEK also provides management alternatives to cope with dynamic changes in ecosystems (Berkes et al. 2000), contributing to the conservation of marine habitats. As research about cetacean biology is of long-term duration and generally expensive, fishers' knowledge could also indicate priority topics for research, especially in those regions of the coast where none has been conducted. Huntington et al. (2004) suggested that LEK could complement scientific knowledge, helping to understand processes that are not completely explained by scientific information.

Fishers' LEK and Marine Population Trends

The interplay of changing marine ecosystems and human societies has often been approached through historic research. This is essential to reconstruct past social and ecological systems (Pitcher 2001; Gasalla 2004a, b) and to reveal "shifting environmental baselines" ("shifted baseline syndrome"), such as when recent trends mask past over-fishing and fish stocks are reduced even before fisheries data start to be collected (Pauly 1995). Comparisons between our tracking standards and reference points from the past can highlight differences important for achieving sustainability. In this sense, fishers' LEK shows potential for reassessing or reconstructing population trends and past abundances, especially in data-poor situations (Neis 1997; Pitcher 2001; Gasalla 2003a, b, 2004a, b). Where long-term data sets are unavailable, older fishers are often the only source of information about historical changes in local marine stocks and marine environmental conditions (Johannes 1998; Johannes et al. 2000,

2001). Fishers' LEK could thus also help to improve management of target stocks and rebuild marine ecosystems, highlighting the importance of "putting fishers' knowledge to work" (Haggan et al. 2003).

The pioneering work of Robert Johannes and collaborators (mostly concentrated in several South Pacific Island countries) highlighted the relevance of fishers' LEK (Johannes 1981, 1998, 2002; Johannes et al. 2000). However, the application of LEK studies is not restricted to ancient traditional fishing communities. It appears that the emblematic cod fishery collapse in Newfoundland could have been the responsible for the particular interest shown in Canada in quantitative research of abundance trends from fishers' LEK (Neis 1997; Neis et al. 1999). In the case of cod, coastal fishermen seemed to be previously aware of the cod stock risk and tried to alert the Canadian Fisheries Department, which relied solely on population models and thus could not reverse overfishing (Wilson 1999). Neis et al. (1999) describe ways to access the large reservoir of information held by fishers, some use of cross-checks to identify consistent patterns, and the usefulness of LEK to interpret quantitative surveys used in fisheries assessment. They propose that local information from resource users can be assembled and used in quantitative fishery stock assessments, thus bringing benefits to fisheries management (Neis 1997, Neis et al. 1999).

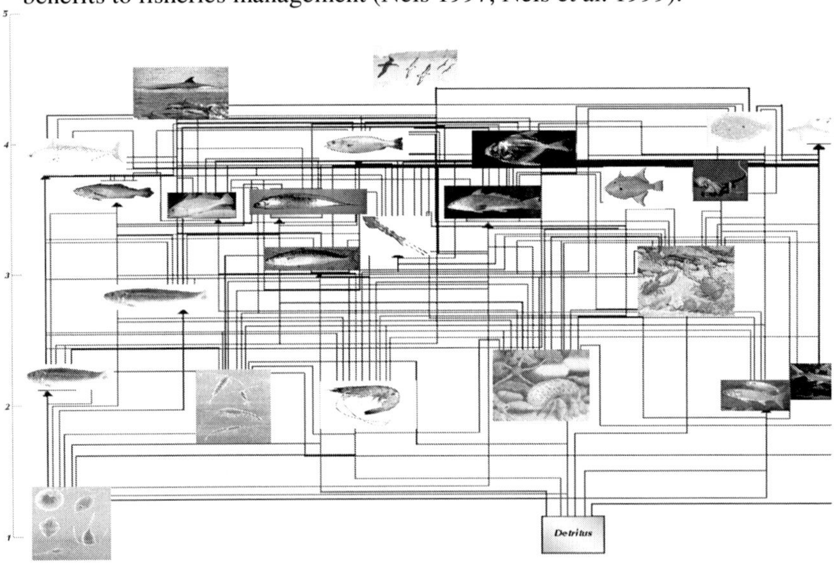

Figure 4-10. An ethno-ecological model of the South Brazil Bight shelf ecosystem based on fishers' perception of fish preys and predators (After Gasalla 2003b).

In Brazil, as well as in many other tropical fisheries, fishery landings statistics are either absent or recent, and data are in several cases incomplete. Therefore, scientists and managers usually have limited information about the temporal dynamics and abundance trends of fish stocks. However, fishers, especially the older ones, usually have knowledge and perceptions about fish abundance and catch trends from well before scientific data started to be collected. Such local knowledge may be a very useful way of reconstructing fish population dynamics over the long term. In fact, historical and archaeological research has shown that many species may have been much more abundant in pre-fishing times than previously appreciated (Pitcher 2001; Gasalla 2004a, b; Pinnegar and Engelhard 2008). In some places, local fishers together with historical and informal literature are the only sources for obtaining information about the situation in the past.

An assessment of fishers' LEK to reconstruct a past ecosystem state was undertaken in Brazil (Gasalla 2004a, b). While most of the studies on LEK focus on artisanal fishing communities, Gasalla (2003a, b, 2004a, b, 2007) took a different approach, investigating ethno-ecological knowledge of fishermen engaged in industrial fisheries along the South Brazil Bight, in the southeastern Brazilian shelf, South Atlantic. The aim of the study was to cover the shelf ecosystem to 100m depth. Surveys concentrated on veteran and retired industrial fishermen who had been fishing in the whole area of study and had kept memories from past catches and about alterations in the marine ecosystem. Gasalla (2004b) investigated the extent of fishers' ecosystem-related knowledge, such as changes in stock abundance and trophic interactions. The ultimate goal was to make that knowledge system available for studying long-term changes in the area and to provide data for ecosystem modelling purposes (Gasalla 2007). Ecosystem models were used to address fishing impacts and required information on the former states of the ecosystem, before the expansion of industrial fishing (Gasalla 2004a).

In fact, the first step was the construction of a trophic model of the shelf ecosystem based on fishers' LEK (Figure 4-10). This information proved to be very useful for modelling purposes, as fishermen were very experienced in recognizing predator-prey relationships (Gasalla 2004b).

Information was obtained by informal face-to-face interviews and questionnaires rooted in progressive contextualization with experienced veteran purse-seiners and trawling fleet's fishers (Gasalla 2003a, b). Industrial fishermen also recognized changes in the marine ecosystem and proposed explanations concerning exploited stocks, reduction in stocks' abundance and trophic relationships (Gasalla 2007). Considering fisher's perception on shifts of fish communities, most of the 85 interviewed fishermen referred to a "disappearance" of several resources from the fishing area (Table 4-2).

In terms of decreasing trends in abundance and species vanished from the South Brazil Bight, the most reported ethno-categories were "sharks", followed by mackerels ("cavalas"), big-size carangids, lutjanids and serranids. Actually, the shark's populations considered most heavily reduced by fishers were *Carcharhinus brevipinna*, *C. limbatus* ("galha-preta"), *C. longimanus* ("marracho"), *Carcharias taurus* ("mangona") and *Sphyrna spp* ("martelo").

An evaluation of trends of change was done for the first time in the South Brazil using an integration of retrospective data from different sources, including historical accounts, anthropological literature, reports and grey literature, as well as scientific surveys and reconstruction and landings statistics (Gasalla 2004a, b). This made the study suitable for cross-validating fishers' LEK. Actually, a possible "pre-fished" South Brazil Bight ecosystem was more populated by big fish, such as jacks (*Seriola* spp, *Caranx* spp., Carangidae), Serranids (*Mycteroperca* spp.), Lutjanids (*Lutjanus* spp., *Rhomboplites aurorubens*) and Scombrids (mackerels). These had disappeared from landing data, but were assessed by fishers' citations. Cross-validation of the decrease in shark populations, sardines, sciaenids and haemulids (among others) and an increase in rays, flatfish and small piscivorous fish were also undertaken (Gasalla, unpublished data). This example shows how quantitative models can be made more robust using LEK.

Table 4-2. Percentage of citation of fishery resources (other than sharks) reported as vanished from the South Brazil Bight coastal ecosystem by veteran industrial fishers (after Gasalla 2004a).

Cited ethno-categories	**Scientific name**	**%**
Cavalas	*Scomberomorus* spp.	47.36
Xerelete	*Caranx latus*	31.57
Olho de Boi	*Seriola dumerili*	28.94
Olhete	*Seriola lalandii*	27.97
Xaréu	*Caranx hippos*	23.68
Miraguaia	*Pogonias cromis*	19.73
Vermelho	*Lutjanus* spp.	17.10
Bagre	Several genders	8.53
Pitangola	*Seriola fasciata*	6.40
Pescada amarela	*Cynoscion acoupa*	6.40
Pescada foguete	*Macrodon ancylodon*	6.40

Cited ethno-categories	Scientific name	%
Pescada cambucu	*Cynoscion virescens*	6.40
Enchova	*Pomatomus saltatrix*	6.40
Lanceta	*Thyrsitops lepidopoides*	4.27
Garoupa	*Epinephelus* spp.	4.27
Cherne	*Epinephelus niveatus*	4.27
Robalo	*Centropomus* spp.	4.27
Galo testudo	*Selene setapinnis*	4.27
Carapau grande	*Caranx chrysus*	2.13
Pescada pan	*Macrodon ancylodon*	2.13
Cioba	*Lutjanus/Rhomboplites* spp.	2.13
Caranha	*Lutjanus griseus*	2.13
Olho de boi jangalengo	*Heteropriacanthus*	2.13
Badejo	*Mycteroperca* spp.	2.13
Bicuda	*Sphyraena* spp.	2.13
Cavala tuiná	*Scomberomorus* spp.	2.13
Cavala fogueira	*Scomberomorus* spp.	2.13
Bonito bacuara	uncertain	2.13
Serrinha	*Sarda sarda*	2.13
Marimbá	*Diplodus argenteus*	2.13
Paraty-guaçu	*Mugil* spp.	2.13
Camburuçu	*Centropomus undecimalis*	2.13
Corcoroca	*Haemulidae*	2.13
Sapinhanguá	Bivalvia	2.13
Piraúna	*Pogonias cromis*	2.13
Leão marinho	*Otaria/Arctophalus* spp.	2.13
Cavalinha	*Scomber japonicas*	1.05
Raias	*Several ray species*	1.05
Carapeva	*Diapterus* spp.	1.05
Escrivão	*Eucinostomus gula*	1.05
Goete da cara preta	Uncertain	0.65
Xixarro	*Trachurus/ Decapterus* spp.	0.65

More recently, studies have been done in other parts of the world, using observations and systematic documentation of fishers' perceptions of trends in abundance. These studies emphasize the usefulness of fisher's knowledge, sometimes called "anecdotal information". A study on the Gulf grouper *Mycteroperca jordani* in the Gulf of California, Mexico, suggested that stocks collapsed in the early 1970s, long before official fishery statistics started to be collected (Sáenz-Arroyo et al. 2005a). These authors noted clear differences in the number and size of fish reported as being caught by three successive generations of fishermen: older fishers recall larger catches in the 1940s and 1950s, but much lower catch rates in the 1960s and especially in the 1990s (Sáenz-Arroyo et al. 2005a). This was similar to Gasalla (unpublished data) findings in Brazil (Figure 4-9), but little of the knowledge about marked declines in fish abundances seemed to have been passed on from the older to the younger generation of fishers, even those in the same families (Sáenz-Arroyo et al. 2005b).

In coral reef ecosystems, Dulvy and Polunin (2004) surveyed fishers' LEK on the giant humphead parrotfish *Bolbometopon muricatum* at 12 lightly exploited islands in the Lau group, Fiji. Fishers reported this parrotfish as being previously abundant, but it had not been caught at six islands since at least 1990 and was considered rare at another four islands. A compilation of giant humphead parrotfish records throughout the Indo-Pacific suggested that this fish has now become globally scarce (Pinnegar and Engelhard 2008).

Although such studies considered the fishers' LEK as just anecdotal, they did not seem to reflect random stories but rather an observational outcome taken from fishers' long-term experience.

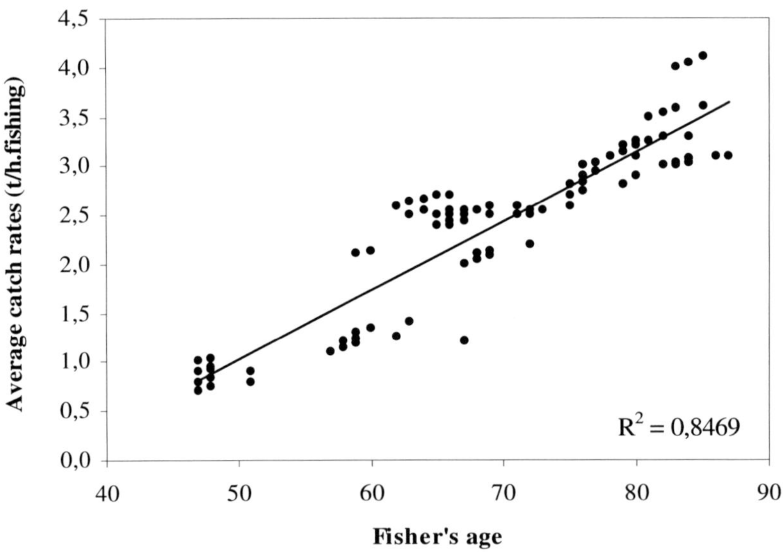

Figure 4-11. A linear relationship of fisher's age and average catch-rates reported by them. (After Gasalla 2003a, 2007).

In fact, some evidence has shown that relying only on modern scientific ecological or fisheries data may result in the aforementioned "shifted baseline syndrome" (Pauly 1995; Saens-Arroyo et al. 2005; Pinnegar and Engelhard 2008). LEK evidence, such as the examples illustrated here contain valuable information for conservation plans in data-poor areas, especially when degradation is rapid and time and funds for scientific research are limited (Johannes et al. 2000). Further research will certainly allow for more exhaustive assessment of historical evidence from LEK and of shifting baselines from different cultures and distinct regions.

Conclusions

A promising approach to sustainably manage local tropical fisheries is co-management, involving local fishers in the research, management and conservation of their own fishing resources. This has been done in the Brazilian Amazon (McGrath et al. 1993; Queiroz and Crampton 1999) and the Solomon Islands in the South Pacific (Aswani and Hamilton 2004). Such co-management can be a way to integrate LEK into management policies, since LEK reflects users' experiences in relation to resource management. This can ultimately make conservation initiatives more effective, as observed for cetaceans on the southeastern Brazilian coast (Souza and Begossi 2006). Fishers' LEK may be especially useful for adaptive management, a dynamic approach in which management measures are constantly monitored and improved according to new data obtained (Berkes et al. 2000). The development and success of co-management approaches depend mainly on the political organization and motivation of the local communities, not only on fishers' LEK (Berkes and Turner 2006). However, fishers' LEK is fundamental to develop co-management or other sustainable management measures (Ruddle and Hickey 2008), as it is usually strongly related to their local fishing activities and rules regulating these activities (Johannes 2002). Furthermore, the propensity of a given local fishing community to develop formal or informal conservation initiatives may be linked to fishers' LEK regarding the exploited fishing resources (Berkes and Turner 2006).

Notwithstanding methodological advances and studies showing the invaluable contribution of fishers' LEK to fisheries-related sciences, the idea that local fishers can manage their resources has sometimes drawn severe criticism (Sheppard 2006). As for other sources of information, LEK should be carefully evaluated and applied to improve local research and management strategies (Huntington 2000a). This chapter brings an up-to-date review of some issues regarding fishers' LEK, aimed at facilitating the integration of LEK into fisheries management strategies. The conventional fisheries management schemes, which are usually imposed on local fishers by governments and fisheries scientists, seem not to be effective at preventing overfishing in several cases (Pauly et al. 2002). This review can contribute to current and future surveys of fishers' LEK, which are an important first step to include fishers in the management and conservation of the exploited fishing resources.

Acknowledgements

We thank the fishers of the studied communities for their kind cooperation, FAPESP/ SP for financial support of research projects and for grants to R.A.M. Silvano; CNPq for a productivity scholarship grant to R.A.M. Silvano and for a post-graduate grant to S.P. Souza; M.A. Gasalla acknowledges support from WWF-Brazil, IFS, CNPq, Fisheries Centre of University of British Columbia, T. Pitcher and D. Pauly; we also thank A. Begossi and P.F. Lopes for useful comments and for the invitation to participate in this issue and an anonymous reviewer for useful comments.

Endnote:

[1]*Megaptera novaeangliae*, *Balaenoptera edeni*, *B. acutorostrata*, *Eubalaena australis*, *Pontoporia blainvillei*, *Sotalia guianensis*, *Stenella frontalis*, *Tursiops truncatus*, *Steno bredanensis*, *Delphinus capensis*, *D. delphis*, *Kogia sima*, *Pseudorca crassidens*, *Orcinus orca*, *Berardius arnouxii and Mesoplodon mirus*.

CHAPTER FIVE

Ethnobotany of Artisanal Fishers

NATALIA HANAZAKI, FLÁVIA CAMARGO OLIVEIRA, TATIANA MOTA MIRANDA AND NIVALDO PERONI

This chapter has its focus on the artisanal fisher's traditional local knowledge on plant resources in particular, and environments in general. Human Ecology and Ethnobotany are closely related approaches, differing in the emphasis on behaviour or on perception, respectively. Artisanal fishers' knowledge and use of natural resources at different scales can be studied through a combination of these approaches. In this chapter the knowledge and use of plants and environments is analysed using case studies of artisanal fishers in the Brazilian coast. Artisanal fishers interact with plant resources through their farming activities, through occasional use of medicinal plants, and through extraction of plant resources for the building of fishing traps, among others. We discuss the general use of plant resources by different island-dwelling communities, contrasting some aspects of their history and environments. We then discuss the distinctive aspects of plant cultivation and plant extraction, focusing on small scale farming activities and on the use of native trees and bamboos to build fishing traps. Fisher communities have been increasingly connected to urban areas and to new economic alternatives, such as tourism exploration. In association to these changes, the knowledge related to the terrestrial environment in general and to the plant resources in particular can persist dynamically among artisanal fishers. Here we illustrate some aspects of such rich knowledge.

Introduction

Artisanal fishers interact with plants in many ways, and with varying degrees of intensity. Despite being characterised by their fishing practices, artisanal fisher activities range from less intensive management actions, such as the extraction of forest products, to intensive management actions associated with the cultivation of the agrobiodiversity (Peroni et al. 2008). These different ways of interacting with the plant world also reflect the heterogeneity of artisanal fishers as a cultural group. Artisanal fishers can occupy different environments, including marine, freshwater and estuarine areas, and can have very different ethnic and cultural origins and influences (Hanazaki 2004). Despite such differences, these communities share a dependence on both terrestrial and aquatic environments. Many communities of artisanal fishers were relatively isolated until recently. They have historically been self-sufficient in meeting many of their needs for food, medicines and raw materials used in fishing activities (e.g. fibres, wood, floaters, hooks, and baits).

This chapter focuses on artisanal fisher's traditional and local knowledge about plant resources, in particular, and environments, in general, through a theoretical framework combining ethnobotany and human ecology. Human ecology and ethnobotany are closely related approaches, differing in their respective emphases on behaviour v.s. perception; in the case of ethnobotany the perception focused on the plant world and on the dynamic relationships between people and plants. The use of both approaches allows artisanal fishers knowledge, and use, of natural resources to be studied at different scales.

Ethnobotany studies among fisher communities have rarely focused on the interplay between fisheries and plant-human interaction. The main focus is usually on the medicinal or edible uses of plants, including the role of native and exotic plants used for medicine (Negrelle and Fornazzari 2007), or general patterns of plant use (e.g. Figueiredo et al. 1993, 1997; Rossato et al. 1999; Hanazaki et al. 2000; Fonseca-Kruel and Peixoto 2004), yet no emphasis was made to the plants related to fishing activities or to the role of the fishing activities for the ethnobotanical knowledge. The land and sea connections were explored in case studies by authors such as Hanazaki et al. (2007), linking four main economic activities related to Caiçara livelihoods: fishing, small-scale agriculture, plant extraction and tourism related activities. The interplays between land and sea were also explored by Jayatissa et al. (2006) in an analysis of two estuarine lagoons of Sri Lanka where fisheries was an important activity and where the recent freshwater inflows into the lagoons declined the fisheries and favoured the spontaneous increase of mangrove vegetation. Liebezeit and Rau (2006) also stressed the potential of mangrove areas which have been used traditionally as sources of non-wood forest products, food and pharmacological agents, in spite of also being essential areas for aquaculture and fishing activities. Ethnomedical plant uses from mangroves were also discussed by Venkatesan et al. (2005). Studies such as these indicate how plant use and fisheries can be balanced to potentially maintain a local economy, especially in mangrove areas economically exploited for timber, non-timber forest products, and fisheries. Mangrove vegetation is more clearly linked to fishing activities, due to the distinctive ecological characteristics of these ecosystems. However, the same does not apply to other coastal vegetation types, such as the Brazilian Atlantic Forest and sand dune vegetation.

Perhaps the most exotic use of plants in fisheries is related to fishing poison. Fishing with the aid of plant poisons is still used in many places in the world, such as Tropical Africa (Neuwinger 2004) and the Amazon (Van-Andel 2000; Prance 1999; Heizer 1987; Moretti and Grenand 1982). In this kind of fisheries, plants with toxic compounds are scraped into small pieces and thrown or pounded into the water, and the fish are killed or stupefied by these toxic compounds, in order to facilitate their capture (Howes 1930; Neuwinger 2004). Ethnobotanical studies dealing with these plants are focused on the plants used and their active principles, such as rotenone-yielding plants (*Lonchocarpus* sp., *Derris* sp., *Tephrosia* sp.) and saponin-yielding plants (Sapindaceae) (Moretti and Grenand 1982).

Another important connection between plants and fishers is related to the human populations who use seaweeds. There are a few authors who addressed this question. Abbott (1996) discussed the extensive uses of fresh and dried seaweeds in different parts of the world, especially with edible purposes, and Turner (2003) studied the ethnobotany of a Rhodophyta among the First Nations on the Pacific Coast of Canada. It is expected that the relationships between artisanal fishers and seaweeds also occur in many other parts of the world and with different degrees of dependence, however this relationships can be covered up by the main extractive activities related to fish resources.

Amazon fishers also have an important ethnobotanical and ethnoecological knowledge related to fishing activities and plants. For example, the Desâna from Alto Rio Negro, Brazil, recognise a variety of plants useful as food for fish, such as the flowers of *Tabebuia obscura* (Bur. and K. Schum.) Sandwith and the fruits of *Genipa* cf. *americana* L., *Cunuria spruceana* Baill., *Byrsonima* sp. and *Astrocaryum jauari* Mart., among others (Ribeiro 1995). The role of fish as seed dispersers is observed by Amazonian Caboclos (Goulding 1980; Smith 1979). For the same Desâna cited above, Ribeiro (1995) also reported other plant species used as raw materials for baskets and fabrics used on fishing and canoe building. At Várzea da Marituba, within the lower São Francisco valley, Brazil, Marques (1995) found 106 plant species directly related to fishing, used in the past or still used currently, such as *Andira nitida* Mart. ex Benth.), *Anacardium occidentale* L., *Ocotea gardneri* Mez and *Manilkara* sp., used for canoe building; *Astrocaryum* sp. used for manufacturing of nets; *Syagrus coronata*

Becc., *Schinus terebinthifolius* Raddi, *Manihot esculenta* Crantz and *Zea mays* L. used for bait.

In order to discuss the knowledge and use of plants and environments we analyse case studies of Brazilian artisanal coastal fishers, especially from the South-eastern part of Brazilian Atlantic coast. Artisanal fishers interact with plant resources through their farming activities, through the occasional use of plant as medicines, and most frequently, through the use of plants for food and handicrafts. They also interact with plant resources through the extraction of trees and a slender class of bamboo for building fish traps and fishing gear. To illustrate this, we discuss the use of plant resources as a whole by different island-dwelling artisanal fisher communities, contrasting their respective histories and local environments. Then, we discuss the distinctive aspects of plant cultivation and plant extraction, focusing on small scale farming activities and on the building of fishing traps using native trees and bamboos. Though the interplay between fishing and farming activities has been discussed elsewhere (see, for example, Hanazaki et al. 2007; Peroni et al. 2008), it is also addressed in this chapter.

Peroni et al. (2008) discussed the artisanal fishers' ethnobotany through a comparison of the different dynamic systems underlying the relationships between fishers and plants. The traditional agricultural systems, such as the swidden cultivation, declined in the last century at the coast, but the fisher-farmer systems are still currently characterised by cultivation of manioc (*Manihot esculenta* Crantz), yams (*Dioscorea* spp.) and sweet potatoes (*Ipomoea batatas* Poir.) as the main crops. The traditional fisher-farmers have maintained a high poly-specific and poly-intraspecific agrobiodiversity managed in restricted areas. In this scenario there is interplay between the farming and fishing calendars, characterised by a multiple use of resources and by complementarities between the production and local economy history (Adams 2003).

The Ethnobotany of Fishers from Cardoso Island

The Cardoso Island, an estuarine island, is located in the coastal Southern Sao Paulo State with an area of over 55,000 acres. This region has tropical climate and ombrophyllous rain forest vegetation on the slopes, and Quaternary forests and sand dune vegetation (or "Restingas") on the coastal plains, with mangrove patches in the estuarine areas. Historical information indicates occupation in this area by ancient groups previous to the Guaiañas and Carijós Indians that lived in the island when the Europeans arrived in

1501 (Barros et al. 1991; Plano de Manejo do PEIC 2000; Sampaio et al. 2005).

The current artisanal fishers in Cardoso Island are known as Caiçaras, or the native rural inhabitants of the Atlantic Forest in the region from the coastal Northern Paraná state to the coastal Southern Rio de Janeiro State. The Caiçara culture is greatly influenced by the miscegenation of Brazilian Indians and Portuguese colonists (Mendonça 2000). The current occupation of the Cardoso Island has a connection with social and economic changes which have taken place in the region. The island was densely populated during the 17th century, when many families migrated to the region (Barros et al. 1991). With time, a gradual population decrease has been noticed, apparently due to factors such as decrease of some farming practices, beginning of the commercial fishing, immigration to urban areas, and creation of a conservation unit in that area (Mendonça 2000). Since 1962 the whole island has been declared a biodiversity conservation State Park. All those facts have influenced the Caiçara's way of life.

The study group is composed of four communities, living in sand dune vegetation areas of the Cardoso Island. Sand dune vegetation is locally known as "Restingas" and is composed of tropical forest species ranging from herbs and bushes to trees, occupying Quaternary sandy soils. The communities of Itacuruçá and Pereirinha are situated on the estuarine side of the island, which is easier to reach from the local town of Cananéia. Foles and Cambriú are located on the open sea side of the island, being more isolated than the two previously mentioned communities, and farther away from the town. Both groups have small family nuclei historically inhabiting these places. These groups are composed of native people who live of fishing, cropping, and resource extraction of plants used for food, medicinal and handicraft work. Recently the Caiçaras have begun exploiting tourism as an economic activity. The main differences between these communities are related to their subsistence practices: while in Itacuruçá-Pererinha tourism activities are continuous throughout the year and rather common, in Foles-Cambriú the fishing activities are the main income activities, because tourism is cyclical, occurring only in the summer months (from December to January).

Semi-structured interviews with adult residents have been conducted in order to obtain information about socioeconomic aspects and about their knowledge and use of local plant resources. We interviewed both men and women; including all the adults who agreed in taking part in this research (methodological details can be found in Miranda and Hanazaki 2008). Botanical samples have also been collected and identified *in loco* and in the laboratory, and were deposited in the reference collection of the Laboratory of Human Ecology and Ethnobotany, at Universidade Federal de Santa Catarina.

The interviewed inhabitants of these communities mentioned a total of 200 folk taxa known and/or used, corresponding to 154 species and 65 botanical families. The most highly represented families were Myrtaceae (19 species) and Asteraceae and Poaceae (12 species). Among the species cited, 105 were native and 61 were exotic (for further details, see Miranda and Hanazaki 2008).

Table 5-1 shows the plants mentioned by at least 20% of the members of the four communities. The most cited ones were: *Schizolobium parahyba* (Vell.) S. F. Blake, known locally as "guapiruvu" and mentioned by 68.6% of the interviewed individuals (n = 35); "hortelã" or *Mentha arvensis* L. and *Mentha* sp., was mentioned by 49% (n = 25); "caxeta," or *Tabebuia cassinoides* (Lam.) DC., was mentioned by 47% (n = 24); "araçá," or *Psidium cattleyanum* Sabine, was mentioned by 43% (n = 22); "boldo," corresponding to *Plectranthus barbatus* Andrews, *Vernonia condensata* Baker and *Salvia* sp., was mentioned by 41% (n = 21); and *Chenopodium ambrosioides* L., or "erva-de-santa-maria", was mentioned by 37% (n = 19) of the interviewees. The species *Cymbopogon citraturs* (DC.), locally known as "capim-cidró" or "erva-cidreira", *Psiduim guajava* L. ("goiaba", "goiaba branca" or "goiaba vermelha") and *Garcinia gardneriana* (Planch and Triana) Zappi or "vacupari" were mentioned by 33% of the interviewees.

The use categories established here were: medicine, food, and handicraft, which includes plants used to make goods, fishing traps, houses and canoes. These categories were found to be the most common ones in other ethnobotany studies (Prance et al. 1987; Figueiredo et al. 1993; Hanazaki 1997; Hanazaki et al. 2000), and they include extracted and/or cultivated plants. Species used for ornamental purposes, for firewood or for feeding animals were placed in a category named "others" and were mentioned only a small number of times.

The most common type of plants cited belongs to the medicine group (35% of the plants; n = 82). Plants used for handicrafts and food were equally common at 28% (n = 66) each. The remaining 9% (n = 22) of plants fell into the "others" category. The medicinal plants most cited were "hortelã" (*Mentha arvensis* L. and *Mentha* sp.), cited by 49% of the interviewees (n = 25) and used especially to treat colds, and "araçá" (*Psidium cattleyanum*), which was cited by 43% (n = 22) and is used to treat diarrhoea and stomachaches. "Hortelã" used as a seasoning, and "araçá" were the most cited food plants, mentioned by 49% and 43% of the interviewees, respectively. These two food plants are consumed "in natura".

The species *P. guajava* and *G. gardneriana* (Planch and Triana) Zappi, which are also used for food, were cited by 33% of the interviewees (n = 17).

The most commonly mentioned plants in the category handicrafts were "guapiruvu" (*S. parahyba*), which was cited by 68% of the interviewees (n = 35); "caxeta" (*T. cassinoides*) cited by 47% (n = 24); "araçá" (*P. cattleyanum*) cited by 43% (n = 22); "vacupari" (*G. gardneriana*) cited by 33% (n = 17); "guanandi" (*Calophyllum brasiliense* Cambess.) cited by 27% (n = 14); and "brejaúva" (*Astrocaryum aculeatissimum* (Schott) Burret) and "canela" (*Inga* sp., likely a mismatch between local and scientific species) cited by 20% (n = 10) of the interviewees. The Cardoso islanders indicated that the species *S. parahyba* (Vell.) S. F. Blake is used to build canoes, and, according to Barros et al. (1991) and Lorenzi and Matos (2002), it is useful because of its soft and smooth consistency, and it has economic potential due to its use in making boxes, paddles and tools. The same characteristics were noted of *T. cassinoides*, used by Cardoso islanders to make small artisanal pieces, such as animal miniatures; musical instruments, such as the local guitar and violin named "viola" and "rabeca", respectively, and fishing accessories, such as net floaters. Several species were indicated as being useful in making fishing traps called "cerco-fixo", including "araçá", "vacupari" and "guanandi". This kind of fishing trap is further detailed in this chapter.

Given the different geographic locations of the four communities, we noticed some differences in the use and knowledge of the environmental resources among the studied groups. A large number of local plant names, botanical species, botanical families and native species were quoted by both communities on the estuarine face of the island (Pereirinha and Itacuruçá) but not by the inhabitants of the ocean face (Foles and Cambriú) (Table 5-2). In the first group, we noted an equal importance between the medicinal and handicraft species, with each category comprising 35% of the cited plants (n = 71). Plants used for food were 23% (n = 46) of the mentioned plants, while the remaining 7% (n = 14) were in the "others" category. In contrast, the use categories most important for the members of the Foles-Cambriu communities were the food and the medicinal categories, each one totalling 37% (n = 41) and 35% (n = 39) of the cited plants, respectively. The plants used as handicrafts were mentioned by 25% (n = 28) of the interviewed people, and the remaining 3% (n = 4) were under the "others" category (Figure 5-1).

Table 5-1. Used and known plants cited for at least 20% of the interviewed people (n = 51) at four communities from Cardoso Island, Brazil. Uses: 1- Handicraft; 2- Medicine; 3- Food; 4- Others; und.- undetermined

Local name	**Scientific Name (Botanical Family)**	**Botanical Family**	**Origin**	**Citations**	**Uses**
Guapiruvu	*Schizolobium parahyba* (Vell.) S. F. Blake	Caesalpiniaceae	native	35	1, 2, 4
Hortelã	*Mentha arvensis* L. *Mentha* sp	Lamiaceae Lamiaceae	exotic	25	2, 3
Caxeta	*Tabebuia cassinoides* (Lam.) DC.	Bignoniaceae	native	24	1
Araçá	*Psidium cattleyanum* Sabine	Myrtaceae	native	22	1, 2, 3
Boldo	*Plectranthus barbatus* Andrews *Salvia* sp *Vernonia condensata* Baker	Lamiaceae Lamiaceae Lamiaceae	exotic	21	2
Erva-de-santa-maria	*Chenopodium ambrosioides* L	Chenopodiaceae	native	19	2
Capim-cidró/erva-cidreira	*Cymbopogon citraturs* (DC.)	Poaceae	exotic	17	2
Goiaba/g. branca/g. vermelha	*Psiduim guajava* L	Myrtaceae	native	17	2, 3
Vacupari	*Garcinia gardneriana* (Planch and Triana)	Clusiaceae	native	17	1, 2
Guanandi	*Calophyllum brasiliense* Cambess.	Clusiaceae	native	14	1
Cataia	*Pimenta pseudocaryophyllus* (Gomes)	Myrtaceae	native	13	2
Laranja	*Citrus aurantium* L	Rutaceae	exotic	13	2, 3
Abacate	*Persea americana* Mill.	Lauraceae	exotic	12	2, 3
Aipim/mandioca	*Manihot esculenta* Crantz	Euphorbiaceae	exotic	12	3
Maracujá	*Passiflora alata* Dryand *Passiflora edulis* Sims	Passifloraceae Passifloraceae	native	11	2, 3
Pitanga	*Eugenia uniflora* L.	Myrtaceae	native	11	2, 3
Araticum	*Annona crassifolia* Mart. *Rollinia sericea* (R. E. Fr.) R. E. Fr.	Annonaceae Annonaceae	native	10	2, 3
Brejaúva	*Astrocaryum aculeatissimum* (Schott) Burret	Arecaceae	native	10	1, 3
Caju/caju branco	*Anacardium occidentale* L.	Anacardiaceae	native	10	2, 3
Canela	*Inga sp.*	Mimosaceae	native	10	1
Salva-vida	*Lippia alba* (Mill.) N. E. Br.	Verbenaceae	Exotic	10	2

Using diversity indices such as Shannon-Wiener and Simpson to analyse the diversity of cited and/or used plants revealed greater plant diversity in the Pereirinha-Itacuruçá (estuarine face) group compared to the Foles-Cambriú (ocean face) group.

It was expected that there would be a greater interaction between plant resources and the communities that were more isolated. That could be reflected in a wide diversity of used and/or known plants, with a more detailed knowledge on their ecosystem. However, the reduced diversity of used and known plants mentioned by the residents from Cambriú and Foles may be a result of social, economic, and environmental factors. Fishing seems to be enough to provide the animal protein needs of the families, who have also been practising subsistence agriculture. In general, the knowledge in these communities is focused on plants used for food. It is important to note that a great number of plants cited in these locations are exotic. The habitants of Cambriú and Foles have a greater dependence upon cultivated plants when compared to the other two communities, combined with the fact that the food category was the most cited by the Cambriú and Foles interviewees. The greater diversity observed in Itacuruçá and Pereirinha communities may reflect a greater interaction between the members of those communities and the local environment, which is indicated by their knowledge related to native local species useful for specific handicraft purposes. This fact could have a connection to tourism as an important economic activity in Itacuruçá and Pereirinha. Some residents of these communities, especially the younger ones, work as tourism guides for the conservation area, which contributes for ethnobotanical knowledge conservation. Ethnobotanical skills are frequently valued by ecotourists and by many park visitors. This practice involves the local inhabitants, strengthens concepts and ecological principles, values the Caiçara knowledge and thus avoids some eventual loss of cultural characteristics.

The consequences of the tourism in small societies are discussed by authors such as Robben (1982), Mansperger (1995), Diegues (2004), Parada (2004), and Calvente et al. (2004). Robben (1982) argued that the practice of tourism activities may be beneficial to such societies as it creates new economic opportunities. On the other hand, negative consequences of disorganised tourism are problems related to land appropriation, higher dependency on urban centres, social stratification, population flow, changes in the subsistence of those communities, and loss of autonomy (Robben 1982; Mansperger 1995). In this sense Parada (2004) believes that the adaptation of local societies to new situations, such as in the case of increasing tourism-related activities, is a quite natural process present in any culture and it does not necessarily imply loss of cultural richness of such communities. The community must understand the value and the importance of their own culture, which has occurred in Itacuruçá and Pereirinha.

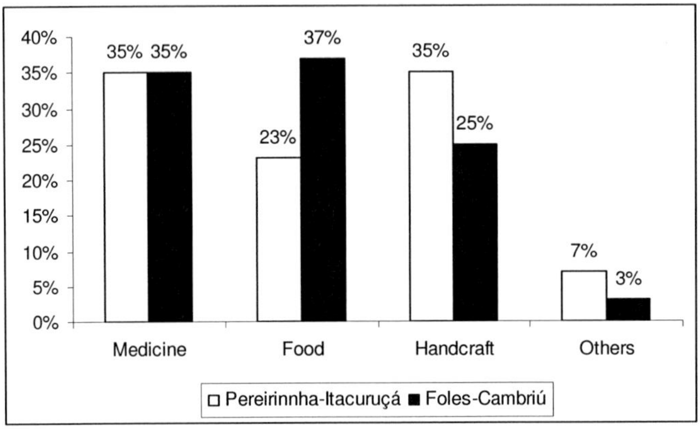

Figure 5-1. Plants cited in each use category, for 51 interviewed residents from Cardoso Island, Brazil.

Regarding the changes experienced by the Caiçaras, the dynamic aspect of the local knowledge is revealed as some concepts are lost or changed, and new elements are introduced to the existing knowledge. Recently, the practice of tourism and the creation of protected areas have been pointed out as the main socio-economic changes contributing to transform the Caiçara culture (Parada 2004; Miranda and Hanazaki 2008). This scenario has been observed especially at Pereirinha and Itacuruçá. As previously stated, the practice of tourism activities can contribute to the maintenance of the general plant knowledge in spite of its potential negative consequences. The maintenance of ethnobotanical knowledge must be considered, as it occurs broadly, not necessarily being shown in a detailed level of analysis. For a local inhabitant, for example, to recognise a species and to know about its use involves a different kind of knowledge necessary to recognise a species and, in fact, to use it for some purpose. The actual use of the botanical species has been limited due to the creation of the conservation area, which limits the extraction of forest products.

Some exceptions occur on knowledge and use of forest species. The ethnobotanical knowledge related to the construction of a fixed fishing trap ("cerco-fixo") by the estuarine communities of Cardoso Island can be an additional factor contributing to the greater diversity of cited plants at the Itacuruçá and Pereirinha communities. The use of fixed fishing traps occurs only at the estuarine communities; at the open sea communities the fishing trap cannot be fixed due to the sea tides and wave exposure. This specific knowledge involves a great number of plant species and is found only among the communities in the estuarine area.

Ethnobotany of a Fishing Trap

In order to investigate the ethnobotanical knowledge related to "cerco-fixo" and held by fishers from the estuarine region of Cardoso Island, a study was undertaken considering all the estuarine communities of the Island: Itacuruçá, Pereirinha, Sítios, Marujá, Enseada da Baleia and Pontal do Leste. Additional information was gathered through interviews with key-informants who live outside of the Island but also fish using the "cerco-fixo". Semi-structured interviews were conducted regarding socio-economic and ethnobotanical information, and plant species were collected with fishers and botanically identified. In addition, participatory observations were conducted during the construction of "cercos-fixos" and during the extractions of plant species used in this process.

Table 5-2. Comparisons of the local knowledge regarding the number of citations of local plant names for communitites from Cardoso Island, Brazil.

Characteristics	Pereirinha-Itacuruçá	Cambriú-Foles
Local plant names	156	101
Scientific species	124	86
Botanical families	57	47
Native plants	88	43
Exotic plants	36	36
Number of interviews	20	31
Cultivated plants	92	175
Not cultivated plants	373	160
Shannon-Wiener index (*H'*; base 10)	2.04	1.83
Simpson index (*1/D*)	8.42	46.95

"Cerco-fixo" (Figure 5-2) is a fixed fishing trap. In Brazil, it was originally used by Brazilian Indians for trapping migratory species of fish, such as mullets (*Mugil* sp.). Nowadays it can be found in many localities along the Brazilian coast from the north to the south and within many non-Indian communities. It is cited as being the principal artisanal fishing practice of the south coast of the Brazilian state of São Paulo (Radasewsky 1976; Mendonça and Katsuragawa 2001), being important as a source of food, income and cultural significance, and as a key element of natural resource use in the region (Santos and Garavello 2003).

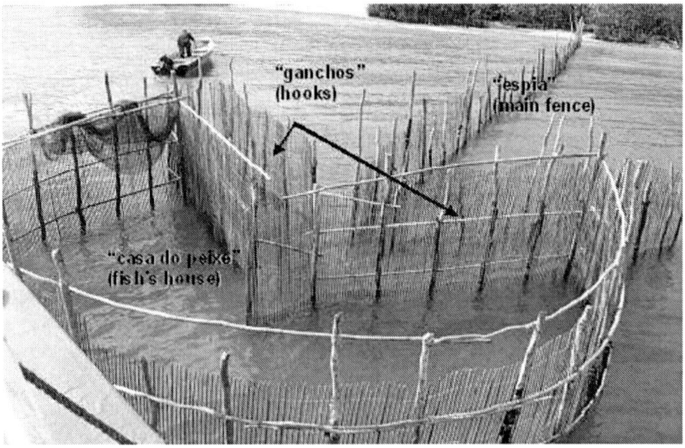

Figure 5-2. The main components of a *cerco-fixo*, a fixed fishing trap (Photo: Flavia C. de Oliveira).

Along the south coast of São Paulo state this fixed fishing trap is built in the estuarine region and is placed perpendicular to the mangrove edge. The protected waters of the estuary are more proper for constructing fixed fishing traps as these permit higher durability due to an absence of waves in the open shoreline. Fishers of the south coast of São Paulo state build the "cerco-fixo" using wood and a slender class of bamboo, called "taquaras", taken from distinct zones of tropical forest in this region. The fishing trap is composed of three main parts called: "espia" (the main fence), "ganchos" (hooks) and "casa-de-peixe" (fish's house) (Figure 5-2).

The building of a "cerco-fixo" begins with by securing structural timber posts ("mourões") (Figure 5-3a) in the mud and sand soil substrate (Figure 5-3b) at the edge of the mangroves making the trap's basic structure.

After fixing the posts in the substrate, fishers wrap the trap with fences comprised of "taquaras" woven together with wire (Figures 5-4a; 5-4b). Some of them claim that, in prior times, "taquaras" were woven with vines instead of wire.

As a final step in the fixed fishing trap construction, structures called "arcos" (arcs) and "varas" (poles) are placed on the trap. Like the structural posts, arcs and poles are made from timber. The poles are tied perpendicularly to the main fence and hook posts, as well as to the "taquaras", making these structures firmer. Meanwhile the arcs (Figure 5-5) are tied perpendicular to the fish house posts and "taquaras". Due to the fact that the fish house is the curved part of the trap, the arcs need to be made of flexible wood that can be bent to an arc shape. For the poles, and also for the main fence posts, some fishers choose to use exotic bamboos instead of trees. The arcs are the only component of the trap that cannot be made of bamboos, as they need to be flexible.

Figure 5-3. (a) sticks; (b) fishermen securing the sticks in the substrate at the beginning of a fixed fishing trap construction (Photos: Flavia C. de Oliveira).

In addition to those materials extracted from the local vegetation (trees, bamboos and "taquaras"), some materials used in fixed fishing trap construction are obtained from the market, including wire and nails.

In this research we generated an estimate of the amount of material used to construct a fixed fishing trap. The estimates for one "cerco-fixo" are: posts made from trees (50-180 units), poles and arcs made from trees (25-35 units), "taquaras" (60-200 dozen), wires (10-50 Kg), nails (½-4 Kg), main fence posts made from bamboo (70-80 units), and main fence poles made from bamboo (4-10 units). If fishers choose to use bamboos to make posts or poles the amount of trees decreases.

The amount of each material used to build a fixed fishing trap depends on the size of the main fence. The main fence size itself depends on two factors. The first factor is the fish species captured during each period of the year. During the cold season (May to September/October) the main fish species captured is mullet (*Mugil* sp.), a species that passes further from the mangrove edge requiring a longer main fence. This is not necessary during the hot season (November to March/April) because the fish species captured during this period pass nearer to the coast mangrove edge.

The second factor determining the main fence size is the length of the flat part of the land where the fishing trap is built. The influence of the flat extension of the substrate on the estuary floor is due to the fact that the hooks and the fish house must be placed in the deepest area occupied by the trap. Therefore, the length of the flat segment of substrate floor will be the length of the main fence.

Nowadays some fishers are choosing to use galvanised screens instead of the woven "taquaras" in the construction of some parts of a trap, particularly the fish house. Plastics can be used for wrapping the posts, poles and arcs and, in some cases, "taquaras", in an effort to increase the trap durability. Along with the galvanised screens, these plastics are modern elements that have appeared amidst this artisanal fishing art in recent years.

According to fishers, the "taquaras" that are not enclosed with plastic can last from 2 ½ to 6 months, with longer durability in the cold season (5-6 months). When enclosed with plastic "taquaras" can last from 6 months to a year. Posts not enclosed with plastic can last from 2 to 5 months, with some kinds of wood able to persist for a year. When enclosed with plastic they can last from 6 months to 4 years, with some species of wood able to last for 8 years.

Figure 5-4. (a) weaving the slender bamboos together with wire to make the fence; (b) the same process in detail (Photos: Flavia C. de Oliveira).

The wrapping of wood with plastic and the use of galvanised screens can contribute to changes in the body of ethnobotanical knowledge held by fishers. As the fishers enclose the wood with plastic they do not have to make differentiations between tree species in terms of resistance because, according to them, enclosed with plastic "any wood can last for a long time". From this perspective, the incorporation of plastic into the building of the fixed fishing trap can narrow the knowledge held by fishers regarding tree species. In the same way, the use of galvanised screens can contribute to the loss of knowledge related to "taquaras". This is due to the fact that when fishers stop extracting "taquaras" to build the trap they no longer practise the dynamic of knowledge construction through the management of plant resources, with this dynamic not being passed to future generations.

An amount of three local plant names of "taquaras" and 112 local names of trees and palm trees were mentioned by the interviewed fishers who work with fixed fishing traps (for further details, see Oliveira 2007).

"Taquara" species were found to be related to two different genera (*Merostachys* and *Chusquea*). The 112 plant names of trees and palm-trees were found to belong to 34 botanical families and 61 genera (89 specimens identified at the species level, nine at the genus level and one undetermined) (Table 5-3). The botanical family with the highest representation was Myrtaceae, (25.8% of species identified), followed by: Melastomataceae (6.7%); Lauraceae (6.7%); Annonaceae (5.6%); Sapotaceae (5.6%); Rubiaceae (4.5%); Fabaceae (4.5%). Within the Myrtaceae family the highest genus representations were *Myrcia* and *Eugenia*. Table 5-3 illustrates the species of trees and palms identified for this study by local fishers as useful for building the "cerco-fixo".

Figure 5-5. Some woods being bent and dried to become arcs (Photo: Flavia C. de Oliveira).

It is possible that the Myrtaceae species are the most sought after for two main reasons: 1) in the type of vegetation where the fishers obtain wood, Myrtaceae is the family that offers species with optimal strength for being buried into the substrate and maintaining the structure of the trap in strong currents (Record and Hess 1972); 2) in this type of vegetation Myrtaceae is the most abundant family, thus having more visible species that can be used in the trap. This latter explanation is consistent with the theory of ecological appearance which has been applied to ethnobotanical questions (Philips and

Gentry 1993a, b; Albuquerque and Lucena 2005). The ecological appearance theory suggests that the probability of a fisher encountering a useful individual of the Myrtaceae family is greater than the probability of encountering a useful species belonging to other botanical families.

Although many different tree and palm-tree species were identified as having the potential of being used as posts and/or poles and arcs, some species more than others are preferred by the fishers because of their strength and rot-resistance and/or because of their suitability as arcs and/or poles. A good arc, for example, has to be flexible without breaking when it is bent, held and dried in a curve shape. Additionally, there are timber species that have the preference of the fishers because they always grow very straight and reach a favourable height and diameter for trap construction.

Fishers of the southern coast of São Paulo demonstrate a detailed knowledge of plant species that are useful in the construction of fixed fishing trap, as well as of the local environment managed by them for the procurement of these species. This detailed knowledge allows the fishers to construct fishing traps that are both effective and durable. It is noteworthy that technological changes, such as the use of galvanised screens and plastics, just as restrictions of natural resource use imposed by environmental legislation in the region can promote changes in the body of knowledge relevant to the construction of the "cerco-fixo". There is also evidence of the dynamic aspect of traditional and local knowledge regarding the fixed fishing trap and of associated practices, such as the adaptability of these elements to new situations.

Agrobiodiversity Connecting Ethnobotany and Plant Domestication in Fisher Communities

Ethnobotanical knowledge regarding plant species reveals different degrees of plant-fishers interactions. An important connection between fishers and plants occurs through the agrobiodiversity cultivation. According to Peroni et al. (2008), despite being characterised by fisheries, Caiçaras use plants with varied degrees of management—including the intensively management actions such as the cultivation of the agrobiodiversity.

The analysis sets focus on the plant management practised by fisher-farmers, mainly in their household, following the approaches of Netting (1993), Brookfield (2001) and Brush (2000).

Table 5-3. Tree species, including palms, utilized in the construction of *cerco-fixos* along the southern coast of the state of São Paulo, Brazil.

Scientific name	Family
Abarema brachystachya (DC.) Barneby and J. W. Grimen	Mimosaceae
Amaioua intermedia Mart.	Rubiaceae
Andira anthelmia (Vell.) J. F. Macbr.	Fabaceae
Andira fraxinifolia Benth.	Fabaceae
Blepharocalyx salicifolius (Kunth) O. Berg.	Myrtaceae
Byrsonima ligustrifolia A. Juss.	Malpighiaceae
Calophyllum brasiliense Cambess.	Clusiaceae
Calyptranthes concina DC.	Myrtaceae
Chrysophyllum flexunosum Mart.	Sapotaceae
Chrysophyllum inornatum Mart.	Sapotaceae
Clethra scabra Pers.	Clethraceae
Cordia sellowiana Cham.	Boraginaceae
Coussarea contracta (Walp.) Müll. Arg.	Rubiaceae
Eclinusa ramiflora Mart.	Sapotaceae
Erythroxylum amplifolium (Mart.) O. E. Schulz	Erythroxylaceae
Esembeckia grandiflora Mart.	Rutaceae
Eugenia cuprea (O. Berg.) Nied.	Myrtaceae
Eugenia multicostata D. Legrand	Myrtaceae
Eugenia stigmatosa DC.	Myrtaceae
Eugenia sulcata Spring	Myrtaceae
Eugenia umbelliflora O. Berg.	Myrtaceae
Euterpe edulis Mart.	Arecaceae
Faramea sp.	Rubiaceae
Garcinia gardneriana (Planch. and Triana) Zappi	Clusiaceae
Geonoma gamiova Barb. Rodr.	Arecaceae
Geonoma schottiana Mart.	Arecaceae
Gomidesia fenzliana O. Berg	Myrtaceae
Gomidesia spectabilis (DC.) O. Berg	Myrtaceae
Gomidesia tijucensis (Kiaersk.) D.Legrand	Myrtaceae
Guarea macrophylla Vahl	Meliaceae
Guatteria australis A. St.-Hil.	Annonaceae
Guatteria hilariana Schltdl.	Annonaceae
Heisteria silvianii Schwacke	Olacaceae
Hirtella sp.	Chrysobalanaceae
Ilex amara (Vell.) Loes.	Aquifoliaceae
Ilex pseudobuxus Reissek	Aquifoliaceae
Ilex sp.	Aquifoliaceae
Ilex theezans Mart. ex. Reissek	Aquifoliaceae

Scientific name	Family
Lacistema hasslerianum Chodat	Lacistemataceae
Laguncularia racemosa (L.) C. F. Gaertn.	Combretaceae
Leandra cf. *acutiflora* (Naudin) Cogn.	Melastomataceae
Leandra cf. *fragilis* Cogn.	Melastomataceae
Licania hoehnei Pilg.	Chrysobalanaceae
Malouetia arborea (Vell.) Miers	Apocynaceae
Manilkara subsericea (Mart.) Dubard	Sapotaceae
Marlieria racemosa (Vell.) Kiaersk.	Myrtaceae
Marlieria tomentosa Cambess.	Myrtaceae
Matayba guianensis Aubl.	Sapindaceae
Maytenus robusta Reiss.	Celastraceae
Metrodorea nigra A. St.-Hil.	Rutaceae
Miconia cf. *cubatanensis* Hoehne	Melastomataceae
Miconia cf. *fasciculata* Gardner	Melastomataceae
Miconia sp.1	Melastomataceae
Miconia sp.2	Melastomataceae
Miconia sp.3	Melastomataceae
Myrcia bicarinata (O. Berg) D. Legrand	Myrtaceae
Myrcia fallax (Rich.) DC.	Myrtaceae
Myrcia glabra (O. Berg) D. Legrand	Myrtaceae
Myrcia insularis Gardner	Myrtaceae
Myrcia macrocarpa DC.	Myrtaceae
Myrcia multiflora (Lam.) DC.	Myrtaceae
Myrcia pubipetala Miq.	Myrtaceae
Myrcia racemosa (Berg.) Kiaersk.	Myrtaceae
Myrcia sp.1	Myrtaceae
Myrcia sp.2	Myrtaceae
Nectandra oppositifolia Nees	Lauraceae
Ocotea aff. *glaziovii* Mez	Lauraceae
Ocotea dispersa (Nees) Mez	Lauraceae
Ocotea elegans Mez	Lauraceae
Ocotea puberula (Rich.) Nees	Lauraceae
Ocotea pulchella (Nees) Mez	Lauraceae
Ocotea sp.	Lauraceae
Ormosia arborea (Vell.) Harmas	Fabaceae
Pera glabrata (Schott) Poepp. ex Baill.	Euphorbiaceae
Pimenta cf. *pseudocaryophyllus* (Gomes) Landrum	Myrtaceae
Podocarpus sellowii Klotzch ex Endl	Podocarpaceae
Posoqueria latifolia (Rudge) Roem. and Schult.	Rubiaceae
Pouteria beauripairei (Glaziou and Raunk.) Baehni	Sapotaceae
Protium heptaphyllum (Aubl.) Marchand	Burseraceae

Scientific name	Family
Psidium cattleyanum Sabine	Myrtaceae
Rapanea ferruginea (Ruiz and Pav.) Mez	Myrsinaceae
Rapanea guianensis Aubl.	Myrsinaceae
Rapanea venosa (A. DC.) Mez	Myrsinaceae
Roulinea sericea (R. E. Fr.) R. E. Fr.	Annonaceae
Roupala cf. *consimilis* Mez	Proteaceae
Rudgea recurva Müll. Arg.	Rubiaceae
Seguieria cf. *guaranitica* Spegazzini	Phytolacaceae
Siparuma brasilienses DC.	Monimiaceae
Siphoneugena guilfoyleiana C. Proença	Myrtaceae
Swartzia acutifolia Benth.	Fabaceae
Tapirira guianensis Aubl.	Anacardiaceae
Ternstroemia brasiliensis Cambess	Theaceae
Tibouchina cf. *pulchra* (Cham.) Cogn.	Melastomataceae
Tibouchina multiceps (Naud.) Cogn. Mart.	Melastomataceae
Trichilia lepdota Mart.	Meliaceae
Virola bicuhyba (Schott ex Spreng.) Warb.	Myristicaceae
Xylopia brasiliensis Spreng	Annonaceae
Xylopia langsdorffiana A. St.-Hil and Tul.	Annonaceae

These approaches have created a basic framework to analyse the use and management of local resources by local farmers. In the context of the agrobiodiversity use and management some authors have argued about the necessity of including approaches which can connect the household unit to different spatial and temporal scales, such as the community-level and the regional level (Zimmerer 2004, Brookfiled and Padoch 2007). This is particularly important in the fisher-farmer context because of the long life time using the areas and the profound dependence on the plant local resources. Zimmerer (2004) argued that the seed flow is a factor that bridges the household and community levels of analysis on resource management in order to address the sustainability policies, programs and projects that have been targeting the agriculture, environment, and conservation. This has amplified inter-scale framework, combining both local and spatial scales to regional and temporal scales, though less attention have been given to the consequences of management under the plant domestication point of view. In the Atlantic Forest, information about species use and management by fisher-farmers using this inter-scale framework is scant.

In order to analyse the agrobiodiversity used and managed by fisher-farmers in different and complementary scales we included other communities from the South of São Paulo State coast: Pedrinhas, São Paulo

Bagre, Icapara, Aquários, Vila Nova, Praia do Leste, Sorocabinha, Ilha Grande, Subaúma, Porto Cubatão, Itapitangui, Prainha, Agrossolar, Papagaio, Juruvaúva, and Ubatuba (municipalities of Iguape, Cananéia, and Ilha Comprida, see also Peroni and Hanazaki 2002 for further details). The studied group in this case is composed by 34 fisher-farmer households. Data was collected through interviews with the householders and visits to their cultivation areas.

Along this extended study area the use and management of biodiversity, including agrobiodiversity, could be analysed based on its use frequency and management intensity. This includes the management of different local categories of land-use types and landscapes. The details of these categories, based on the use, management intensity and local classification are summarised in the Table 5-4. Age and history of use were two important criteria used by farmers to make distinction of field types. The types have been identified using ethnographic data combined with ethnobotany and remote sensing data (Peroni 2004; Peroni et al. 2008). The use of diverse land use types to sustain diverse use categories of plants is very explicit, such as medicinal plants, extractive (income generating) plants and crop species.

The land uses exemplify degrees of use and management resulting in various influences on the species population compositions (Table 5-4). The manipulation of species populations, sometimes isolated, occur in a lower management degree, while intensive actions and activities can happen in a higher level of species communities. Thus, management degree increase is directly related to different organizational levels, from populations to communities. This is very apparent especially in the swidden areas. Following Clement (1999), the species in this field type can be considered semi-domesticated or domesticated ones (Peroni et al. 2008). The land preparation phase of a swidden cultivation area illustrates an activity that starts a dynamic disturbance factor in the surrounding plant community. Regarding agrobiodiversity cultivated within these swiddens, the management level is high and the influences of such actions are quite complex (Martins 2001; Peroni and Martins 2000). Peroni and Hanazaki (2002) found 261 varieties of the 53 crop species among 33 Caiçara farmers, illustrating a poly-specific and poly-varietal characteristic. Some household units can manage 14 varieties of manioc (Peroni et al. 2007). The manioc agrobiodiversity cultivated by these Atlantic Forest farmers is mid-high when compared to Amazonian sites where the centre of origin of this species is probably located (Emperaire and Peroni 2007).

The analysis of agrobiodiversity allows the formulation of questions which may explicit the relations between fisher-farmers, plant species, and the environment. Questions related to the maintenance of varieties have been widely discussed, emphasising Amazonian people (see, for example, Kerr and Clement 1980, Martins 2001, Pinton and Emperaire 2001). These studies have shown the maintenance of a high number of species and varieties under cultivation, as well as the wide use of the environment to its reproduction. On Atlantic Forest coastal region, studies focusing on the agriculture of fisher-farmers such as Caiçaras people were done by Adams (2003), Peroni and Hanazaki (2002), Peroni and Martins (2000), Sanches (2001) and Peroni et al. (2007).

According to Peroni and Martins (2000), Martins (2001) and Peroni et al. (2007), the management based on swidden cultivation connects the species to the environment manipulation, emphasising the inter-relations between cultivation activities and the evolutionary dynamics of plant species, such as manioc. These authors showed that fisher-farmers have cultivated this species as a staple food, and have maintained essential processes of evolutionary dynamics, such as the use of fire in the swidden cultivation systems. As pointed out by other authors, the local folk variety diversity has been amplified by conscious and unconscious processes, such as the incorporation of recruited seed plants of manioc after the management of swiddens (Elias et al. 2001; Emperaire and Peroni 2007). This example shows the ability of farmers to modify the ecological and genetic structures of the species populations.

The intense use of such systems in the past (Adams 2003) has created a mosaic of land use stages, or swidden-fallow areas, reflecting in a mosaic of secondary succession stages of the vegetation. Along the Brazilian coast, extensive areas correspond to regenerating fallows that have been used more or less intensively in the past. Fisher-farmers rarely use plants or manage them from only one field type or only one land use stage within this mosaic of possibilities.

Conclusions

In a world gradually connecting fishing communities to urban areas and transforming semi-isolated communities into semi-urban, activities and knowledge related to both the terrestrial environment and resources do persist among artisanal fishers. This knowledge is related to different types of plants, from cultivated and highly managed to extracted plants used infrequently for medicinal purposes, and plants extracted in considerable amounts to build fishing traps.

Table 5-4. Description of land-use types and folk categories in south coast of São Paulo State of Atlantic Forest.

Land use	**Field types**	**Folk name(1)**	**Use category2**	**Description**
Forest	"Pristine" or mature forest	*Coivara*	Medicinal Plants and extractive plants (economic income). **(-/-)**	Inaccessible places; advanced stage of succession; inside protected areas; less disturbed in recent years; slopes with more than 30%.
Fallows	Regenerating fallows	*Tigüera* (1 year)	Medicinal plants; collect of tuber crops (manioc, yams). **(+/-)**	Initial stage of regeneration; herbs; ferns; and rapid growing species
	Regenerating fallows	*Capoeirinha* (>2, <5 years)	Medicinal and extractive plants (economic income), edible fruits. **(+/-)**	Scrubs; ferns; and rapid growing species
	Regenerating fallows	*Capoeira* (>5, < 15 years)	Medicinal and extractive plants (economic income), edible fruits, plants for handicrafts. **(+/-)**	Advanced stages of regeneration; tree species
	Regenerating fallows	*Capoeirão* (> 15 years)	Extractive plants (economic income), plants for handicrafts. **(-/-)**	The most advanced stage of regeneration; tree species
Swidden	Cultivated plot	*Roça*	Native (manioc, sweet potato and yams) and introduced species (subsistence) **(+/+)**	Swidden agriculture systems (4 or more years of use; 2 years of fallow)

Land use	Field types	Folk name([1])	Use category[2]	Description
Agroforestry	Home gardens	*Quintal*	Short cycle plants and fruit trees (subsistence) **(+/+)**	Near the house with introduced and native species.
Specific human historical landscapes	Former local occupations of abandoned homes (regenerating fallows)	*Tapera*	Long cycle plants, fruit trees; manioc **(+/-)**	Soils with high fertility resulted of log term domestic deposits remains
	Anthropogenitic soils; ancient sites	*Sambaqui*	Germoplasm reservoir of crop species; local varieties of manioc **(-/-)**	Anthropogenic soils on areas of ancient deposits of shells; High soil fertility
Human (Community)	Anthropogenic areas	*Caminhos, terrenos*	Medicinal plants **(-/+)**	Exotic and cosmopolitan species, weeds

[1] Time of regeneration

[2] Use frequency and management intensity: (-/-) less frequency and low intensity; (-/+) less frequency and high management; (+/-) high intensity and low management; (+/+) high frequency and high intensity (Source: Modified from Peroni et. al. 2008).

For several local people, fishing and the use of plants are two deeply connected activities. These connections can be more or less explicit, depending on the research focus. It is important to note the interaction between socio-economic factors and the use and knowledge of plantsIn many situations, socio-economic factors influence the existence or lack of plant resource use patterns in different human groups. In the situations analysed here, with different communities on the same region, the existence of communities with different kinds of interactions with the environment has been noticed, resulting in differential knowledge on nature, more specifically on plants. The intensity of use of extracted plants for building fixed fishing traps is also different when considering the location of the communities in the island, whether estuarine or open sea. The maintenance of fishing and farming activities is also different in each situation, yet most of the agrobiodiversity maintained by these fishers and farmers is in jeopardy since the farming activity itself is being threatened by the recent economic changes.

Acknowledgements

The authors thank all the interviewed fishers and their families from Cardoso Island and Cananéia region for allowing this research. We also thank Jeffrey Stoike for helping on English review, the Brazilian agency CAPES for T. M. Miranda and F. C. Oliveira master scholarships, to FAPESC (Santa Catarina Research Foundation) for financial support to fieldwork, and to FAPESP 03/13688-9 for N. Peroni post-doctoral grant.

CHAPTER SIX

Adaptation and Indigenous Knowledge as a Bridge to Sustainability

KEITH D. MORRISON AND SIMRON J. SINGH

Introduction

The concern for sustainability is a phenomenon sweeping across the globe. It is developing a momentum that makes it difficult to be considered to be a mere passing enthusiasm. Furthermore it is a movement that is operative at multiple levels; from local community groups to professional research teams, and politically at all levels right up to the United Nations. Moreover it is a phenomenon that is however not overtly ideological, as there is a multiplicity of ways in which it is being developed. The term sustainability appears to arise in the attempt to define what is an already existing intuitive concern. Indeed, attempts to define sustainability have foundered on the many rocks of the multiplicity of approaches and activities (Atkinson et al. 2007). What can perhaps be stated safely is that there is a phenomenon able to be termed sustainability which inspires multiple ideals, guiding principles and regulative concepts providing meaning, purpose and hope. The phenomenon is associated with horizons for the future that are believed in.

Reflections on the phenomenon of the sustainability movement need to deal with the relatively recent strength of the movement, yet that it also appears to be an innate and universal phenomenon among cultures. It is possibly this tension that has led to sustainability conceptually to become linked to adaptation within a flux of socio-ecological systems. Such adaptation can be recognised as innate and universal, while exhibiting at particular periods relatively greater significance, depending on the intensity of the need for change. So whilst sustainability has and will always be a concern for all cultures in all periods, the uniqueness of our present global situation has made it into a powerfully visible movement. But does this mean that it is only now in the present sustainability movement that the innate and universal phenomenon has become reflexively recognised for the innate and universal role it has? In this chapter we argue that this is not the case. Indeed we argue something closer to the reverse, namely that the present dominant global culture based on a capitalist economic system emphasizing capital accumulation of resources and the consumption of them, is the substantive cause for the dire need to adapt, and that this has arisen due to loss of awareness of the need for sustainability. Moreover the cause of the loss is due to a presumption of superiority, borne of colonial expansion by the cultures now forming the dominant global culture.

We argue that the present dominant global cultural influence has lost its awareness of the need to co-evolve with the ecosystems lived within, and that it is this loss, which we define as the loss of indigenous knowledge, that is producing both the presumption of superiority and the dire need to change. We argue that the dire need for change is essentially the need to recover an indigenous cultural outlook and praxis held prior to participating in colonial expansion, and that when this is achieved, the need for the sustainability movement will diminish as the necessary adaptation will have occurred.

We start by reviewing how the present sustainability movement is a phenomenon within the dominant global culture. In doing so we explore what has been learnt by the sustainability movement about the universal nature of the concern for sustainability. Starting with these very valuable contributions to our knowledge by the sustainability movement, we pick up on an obvious but sometimes ignored implication, namely that the present dominant global culture giving birth to the movement has to change to overcome the problems it has caused. This is well recognised in the sustainability science literature, for example in the fields of Industrial Ecology and Ecological Economics concerning themselves with material and energy flows (e.g., Ayres and Simonis 1994; Fischer-Kowalski 1997a; Fischer-Kowalski and Haberl 2007; Matthews et al. 2000; Martinez-Alier 1990) and a need for "de-growth" in both capital accumulation and consumption that recently brought hundreds of scholars from various disciplines together in Paris[1]. We complement these perspectives with ones concerned with the effect of dominant global culture on the cultures it has dominated through colonisation. We explore how literature on the postcolonial situation provides insights into what needs to change within the dominant global culture.

From the two complementary perspectives of the literature on sustainability and the postcolonial situation we then proffer a trans-disciplinary sketch of how an innate and universal concern for sustainability is manifest differently through socio-ecological transitions. We then develop a critical discussion of indigenous knowledge and sustainability by arguing that the contemporary recovery of an adequate understanding of sustainability within dominant global culture requires a recovery of lost indigenous knowledge; knowledge lost when Western cultures became a dominant global influence through the process of colonisation. We note that indigenous knowledge has been and continues to be lost within the former Western colonial countries. We also recognise that terms analogous to indigenous knowledge are often utilised for example, traditional knowledge and local ecological knowledge, as explored by Silvano, Gasalla and Souza in Chapter Four of this volume.

We then explore generic features of indigenous knowledge to reflect on whether or not the notion of sustainability is already present within indigenous cultures. We also reflect on how the generic features of indigenous knowledge can open doors for dominant global culture to learn from indigenous knowledge about how to recover sustainability. Next we explore how explicit study of the understanding of sustainability held by indigenous knowledge, especially those continuing to exist as living traditions, provide potentially an efficient and effective way forward for the sustainability movement. But we finish with a warning; to take this approach is politically fraught because of how indigenous knowledge is presently still considered by many within the dominant global culture. Great care is needed to avoid the assimilation of extant indigenous knowledge into the dominant global culture. The adaptation to remain sustainable shown by extant indigenous cultures is still largely one of resistance to dominant global culture. The necessary change required within dominant global culture, which indigenous culture can help reveal, is deeper than popularly thought. If this is not recognised, there is the risk that not only will dominant global culture fail to adapt, but by a misdirected look toward indigenous knowledge, it will also destroy a bridge left to cross over to the promised land of sustainable development.

The Measuring of Sustainability

Major global institutions have produced impressive libraries of literature about sustainability. These include United Nation organisations, for example UNESCO, UNDP, and the pre-eminent conservation organisation, the IUCN, but they are not restricted to these. A simple search with google scholar gives us about 780,000 publications which has "sustainable development" in their titles. But even from these sources the literature is vast, and any attempt to review the literature extremely difficult. Nevertheless there are themes that are discernible, along with their history. Much of the literature can be found to have had their genesis in the definition of sustainable development in what is affectionately termed the Brundtland Report, "Sustainable development is development that meets the needs of the present without compromising the ability of future generations to meet their own needs" (WCED 1987). The two parts to this archetypal definition, one about the present and one about the future, are arguably the source for many of the themes found proliferating in literature about sustainability. Not wishing to limit the value of other points and the value of other definitions by focusing on this one definition, but it by itself provides ample scope with which to explore what has been discovered about sustainability by the global movement over the last two decades.

The definition by the Brundtland Report throws the challenge to ensure that the present needs of people are met, and to do so in a way that does not undermine the possibility for future generations to do so. Often the first part is associated with development, and the second with sustainability. To paraphrase, we need to develop to ensure we fulfil the needs of the present but not so future generations are not able to sustain what we develop. Sustainability therefore focuses on the constraints to development, in the quest of seeking to ensure we avoid over-reaching them. So we find an inherent tension where we seek to increase resource use to fulfil the needs of the present whilst recognising that there are constraints beyond which any increases cannot be sustained. In summary there are two sets of goals defined: one associated with development, for example the Millenium Development Goals of United Nations (UN Millenium Project 2005) organisations to fulfil our present needs; and the other associated with sustainability, for example the Millennium Ecosystem Assessment (MEA 2005). The challenge taken up by those guided by this definition of sustainable development has been to attempt to integrate the two sets of goals. This has evolved over the decades. The Agenda 21 resolution of the 1992 Rio Earth Summit symbolised the integration as the intersection of three circles representing the environment, the economy and society (Agenda 21 1993). More recently culture, ethics and spirituality have been recommended to form a quadruple bottom line (Inayatullah 2005). The need to configure what is required to be integrated has not however often been the explicit topic of discussion, but has been addressed by Bossel (1998; 2001) who argued from the perspective of what is needed for adaptation. From this perspective, he pointed that there is a well recognised "law of requisite variety" proposed by Ashby (1956), which implies that the indicators utilised within an ecological system to ensure adaptation is not arbitrary and must represent the significant environmental impacts on necessary system functions. He continues that there is a generic set of "orientors" or types of indicators necessary for all living systems, namely:

existence, effectiveness, freedom for action, security, adaptability, coexistence and psychological needs. In apparent agreement from independent research, Max-Neef et al. (1991) argued for a similar typology in their "human-scale development" model for human well-being, namely: subsistence, affection, understanding, participation, leisure, creation, identity, and freedom. Many integrative frameworks for indicators have been proposed, and belie in part this common structure. Reed et al. (2006) review some of them relevant to community development, and make a summary distinction between "top-down" approaches focussing on biophysical concerns and "bottom-up" approaches focussing on socio-cultural concerns, pointing out the need to integrate the two approaches into an adaptive process whereby local communities participate in the development of indicators.

The evolution of understanding about sustainability points to how sustainability has come to be considered goal-seeking behaviour at two levels. One deals with biophysical processes and the other with symbolic/cultural processes (Fischer-Kowalski and Weisz 1999). Different terminology is utilised by different authors, but to utilise one approach, there are 1st order objectives dealing with biophysical processes and 2nd order goals dealing with symbolic/cultural processes. 2nd order goals are inspired by ideals and are horizons of possibility, and their quest has been termed second order cybernetics (Scott 1996). The horizons of the future constituting 2nd order goals are associated with development to fulfil the needs of the present. They are universal and may also be innate, necessary orientations to fulfil the "law of requisite variety" as an adaptive process. 1st order objectives by contrast need to adapt to environmental changes to ensure that 2nd order goals associated with development of needs are fulfilled. Shorter term 1st order objectives refer to specific concrete objectives and the constraints and opportunities the specific environment provides. They are associated with sustainability and the realistic ways in which to contextually fulfil needs. The two levels co-exist dynamically to describe society-nature interactions in which there are two sets of causations; one natural and the other cultural (Fischer-Kowalski and Weisz 1999). Max-Neef et al. (1991) modelled the interaction in terms of "fundamental human needs"–which are 2nd order goals–and contextually specific "satisfiers" or ways to fulfil "fundamental human needs", which are 1st order objectives. The challenge this model makes is to ensure that "synergistic satisfiers" that fulfil multiple needs are achieved, and "pseudo-satisfiers" compensating for unmet needs and "violators" (of satisfying other needs) are avoided. This points out how socio-ecological transitions to relatively greater or less sustainable development are potentially always underway, depending on what "satisfiers" are sought to fulfil universal needs for development (Ekins and Max-Neef 1992). Moreover, satisfiers are institutional in structure; they are socio-cultural systems and so implicate social structures defining social status and definitions of affluence. To attempt to capture the dynamic of the resulting possible socio-ecological transitions a sustainability triangle has been proposed (Figure 6-1).

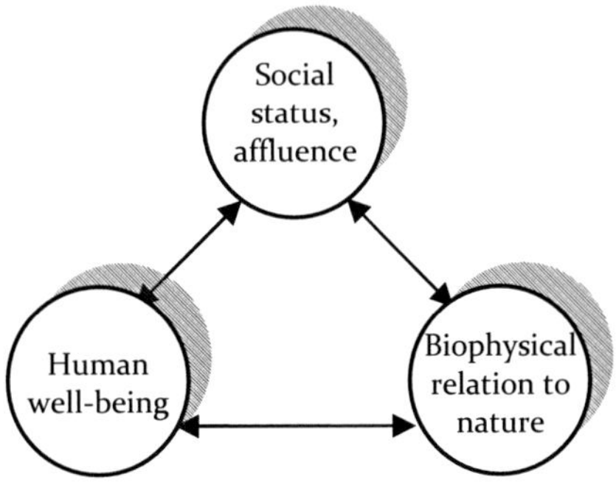

Figure 6-1. The sustainability triangle (after Fischer-Kowalski, 1997b, and Haas, per. comm)

Every society has its own understanding of human well-being or quality of life which varies from indigenous to industrial; e.g. in indigenous cultures, preferences may be for more leisure time to reproduce social relations, while in industrial (global) cultures preferences may be for a certain diet, living space, etc. Human well-being is also related to how a given society views affluence, in other words, what determines social status in a society. For example, industrial societies tend to view status in terms of income or capital, or the consumption of certain brand names, while in indigenous cultures social status tends to be determined by the ability to provide for others and wisdom to resolve conflicts. The definition of affluence and social status in any given society would have a direct impact on the biophysical relation to nature (in terms of material and energy flows) via a chosen means of production to fulfil their quest for well-being and affluence.

To intervene for sustainable development into the adaptive process of socio-ecological transitions requires establishing adequate monitoring regimes to provide feedback about the mismatches between: 2nd order goals defining human well-being and 1st order objectives to satisfy social status and determinants of affluence; between 1st order "satisfiers" and metabolic natural requirements, and between 2nd order goals defining human well-being and metabolic natural requirements. There remains however an open question in relation to this model. It is whether or not direct monitoring between human well-being and natural metabolic processes can occur without intervening social system satisfiers? This is addressed later on in relation to the socio-ecological transitions that emerge due to evolutionary selection. What is clear is that the complexity of the dynamic implicates all four bottom lines. What is not so clear is how to integrate the multiple goals, and hence multiple objectives of the four bottom lines to guide intervention and hence sustainable development of socio-ecological systems. Integration of the separate relative fulfilment of objectives is required to be able to provide some measure of the singular terms, sustainability and sustainable development, even in the short term. Determining the appropriate metric for the integration of multiple objectives has a vast literature. The issues revolve around how the objectives are related and the relative commensurability of the indicators used (Costanza and Daly 1992; Gustavson et al. 1999; Hediger 2000; Munda 2005; Reed et al. 2005).

Adaptation is therefore a real-world ever-changing contextual dynamic of attempting to continue to develop in the right direction. Adaptation at the 1st order level refers to how, once objectives are defined, sustainability and sustainable development is actually sought to be achieved. But adaptation also occurs at the 2nd order level to ensure "satisfiers" fulfil needs, and that the defined needs are appropriate. Much of the systems literature terms consideration of 1st order adaptation as "hard" system analyses. This literature also recognises the 2nd order definition of appropriate "satisfiers" as also an adaptive process even though it is not able to be measured, and terms it "soft" system analysis. Moreover the development of appropriate needs is recognised as a learning process, if not strictly adaptation to a changing context. Attempts to integrate these three analyses together have come to be termed "soft systems methodologies", and are now commonly incorporated into planning and management for sustainability and sustainable development (Checkland 1981; Checkland and Scholes 1990).

The reference to indigenous cultures and indigenous knowledge by United Nations organisations and also the IUCN with regard to sustainability, are mainly in relation to "soft" issues. Even though occasionally local, including indigenous, monitoring regimes are explored as possible "hard" systems that can be learnt from, usually indigenous knowledge is intuited to bear upon what are appropriate definitions of needs and how to ensure appropriate "satisfiers" or ways to fulfil needs. This is illustrated well by the science and technology policies for the sustainable development programme of UNESCO (2008), where the general conference has authorised the director general to implement nine actions including, "To reinforce community capacities to record, manage and mobilize local and indigenous knowledge in order to shape sustainable development and natural resource management to local requirements and needs". Indigenous cultures and knowledge are considered to be particularly vulnerable yet also symbolising the horizons of sustainability and how to ensure they are maintained (Macchi 2008).

The Postcolonial Situation

Dominant global culture is the product of European colonial expansion. Colonial and neo-colonial expansion was and is respectively an attempt to sustain certain objectives by imposing it onto other cultures. In terms of systems theory, "hard" system analyses were and are carried out with a relative absence of "soft" systems analyses. The expression of this involves a presumption of superiority on the part of the European cultures and the now dominant global culture towards the cultures they are imposing their objectives for development upon. Dominant global culture is presumed to be the norm through which all other cultures are to be evaluated. The presumption is false and arises only because of a relative loss of "soft" system analyses. This is understood by some sectors and there are now efforts to seek communities to define their own indicators and monitoring regimes for sustainability and to respect indigenous cultural practices and their knowledge (Reed et al. 2006). But it is still not the norm. This is not surprising, and is well attested to by postcolonial literature dealing with the interaction of dominant global culture with indigenous cultures. It is still necessary to consider literature produced by the dominant global culture on sustainability as being produced through a lens that is colonial in its heritage. What we need to do is ask how does this affect the understanding of sustainability obtained, and in particular does it limit it?

The limiting effect is most clear in the types of genre that are privileged in the literature on sustainability and sustainable development produced by the dominant global culture. Cultures generally comprise of a wide range of genre, from dance and poetry to science and technical skills, to economic institutions of exchange. Sustainability discourse produced by the dominant global culture however usually limits itself to a very narrow range of genre. Systematically manageable quantifiable measurement is the privileged discourse. "Hard" system approaches that are not nested within "soft" system context belie the colonial heritage. The "soft" context is not absent but the emphasis on "hard" systematic practices using an "all purpose-currency" or exchange value with which to evaluate everything into a single "hard" system means that remote or exogenous control and management of activities is possible and commonplace.

It is widely recognised that global environmental problems, and thus the threat to sustainability, is a consequence of the ways humans interact with their environment. In more concrete terms, the problem lays in the biophysical relation between man and nature in terms of material and energy exchanges, as well as changes in ecosystem parameters to meet their basic and extended needs. The environmental crisis owes much to the global dominant culture (based on the capitalist logic of high production and consumption) that requires a constant throughput of materials and energy from the environment. Thus, we are constrained from two sides: input and output. On the input side we are confronted with the limits to resources that are available for sustainable exploitation and scarcity in terms of finite geological reserves. The other problem is on the output side in terms of wastes and pollution that we dump on to the environment, especially those from geological reserves that are in effect a dump onto the biosphere (Fischer-Kowalski and Haberl, 2007; Adriaanse et al. 1997; Matthews et al. 2000).

Sustainability was and is not a crisis in indigenous cultures to the extent their metabolism is directly linked to their resource base they are entitled to exploit spatially as well as to seasonal variability. Overuse of their environment immediately has effects on food availability for the next season. There are therefore clear "hard" systems indicators able to be monitored by indigenous cultures. In order to effectively regulate the use of resources, and avoid extinction, indigenous cultures devised social institutions and cultural rituals that are followed in order to satisfy the need for sustainability within their own means. Homeostatic understanding is intrinsic to indigenous knowledge.

It was the introduction of an "all-purpose-currency" for exchange value, facilitating a type of market and trade (due to the colonial legacy), that has led to a breakdown of the delicate and intimate relationship between society and nature in indigenous cultures (Polanyi 1944; Polanyi 1977; Hornborg 2001). Means of production changed, and over exploitation of resources for the market/trade began to take place. The "all purpose currency" ensures a supply of food from the outside, and thus the dependency on local resources for food diminished gradually. It is no longer as necessary to regulate resource use for basic survival, since the market assures food and other needs from other systems through trade, as long as they conformed to supplying specific raw materials for the market. This leads to the gradual destruction of their environment. Such a process had an effect on the local culture, which is gradually engulfed by the dominant global culture, even though some of the social and cultural traditions still remain as a cultural hang-over of the past, but in effect, they have lost their functional meaning to regulate resources and maintain a sustainable lifestyle. Of course, this loss varies from culture to culture depending on the historical, political and economic context. In summary, colonial contact has been an important driver for a socio-ecological transition to take place from indigenous regimes to an agrarian and even industrial socio-ecological regime, leading to unsustainable trends across the globe (Singh 2003).

This process of imposing systematic processes from without is also recognised as potentially psychologically disturbing. Bateson (1972) characterises it as an imposition to cause an effect, and how power can be defined. Bateson argues that the imposition is carried out by the "gate-keeping" of information pertaining to the 2nd order level so that people are put into a "double-bind" whereby they are invited to participate in a system but not allowed to be creative agents in dialogue with those making the invitation (Harries-Jones 1995). The invitation is a mask obscuring manipulative control and imposition. This unfortunately is played out in the fraught politics within post-colonial situations, irrespective of whether there are actually intentions to disturb indigenous people or not. The consequence is a vacillation between submission and belligerence on the part of those within colonised cultures towards those within the dominant global culture. The phenomenon is also well recognised by psychology (Fromm 1991) and postcolonial literature (Said 1993). This is of course not only an emotional reality but also a very real breakdown in the metabolic relations with nature, as access to resources is lost. This is still occurring, for example in how indigenous cultures are those most vulnerable to climate change (Macchi 2008).

The external management and control by those within exogenously controlled institutions towards those within indigenous communities appears to conveniently allow the avoidance of personal relationships. This not surprisingly appears to disturb those within indigenous communities because it leaves no room for them to voice their views and to participate on their own terms into the process of sustainable development. For indigenous communities to participate on their own terms, exogenous systematic or "hard" processes that influence their livelihoods need to be topics that they can dialogue about within personal relationships, and can challenge within 2nd order "soft" adaptation by changing the institutions of status and determinants of affluence that are being imposed to supposedly satisfy their fundamental human needs. One way in which this can be approached is through the learning theory initially proposed by Argyris and Schon (1978), whereby three nested learning loops are proposed: single loop learning refers to adjustment of indicator values; double loop learning refers to beliefs about what the indicators should be, and triple loop learning refers to the processes followed (Keen and Mahanty 2006). According to this framework, what is missing is "triple-loop" learning to question the processes implicated by socio-cultural structures. Failure to recognise this "soft" dimension of adaptation is what appears to restrict the genre of the dominant global culture to exogenously controlled systematic data collection, so as to unreflexively seek 1st order hard system adaptation within unchallenged institutional structures of social status and determinants of affluence. The consequence of this, perhaps conveniently, is to establish and maintain a one-sided power relationship of domination by the exogenous dominant global culture over indigenous cultures, even when sustainability and sustainable development is being sought. A consequence of this fraught postcolonial situation is the obscuring of how indigenous cultures themselves achieve sustainability. In particular, how indigenous culture carries out 2nd order adaptation. This not well perceived because their activity is sought to be controlled and managed in terms of existing institutional structures and the use of an "all-purpose-currency". This applies also in a subtler way to how indigenous communities' awareness of the sacred or divine is analysed. Instead of seeking how to perceive how indigenous cultures operate in relation to what they call the sacred or divine, there is often a systematisation of their activity in terms of belief systems (eg. Berkes et al. 2000): "double-loop" learning is recognised, but not "triple-loop" learning. Indigenous knowledge is interpreted in terms of belief systems, and contrasted to scientific knowledge systems held by the dominant global culture. This can be seen played out in how it is still highly problematic to consider that the quadruple bottom line includes spirituality as a reality rather than as merely a belief. When such systemisation of indigenous knowledge occurs, 2nd order and 1st order adaptation is confounded, so potential insights into the relation of awareness of the sacred and the divine to 2nd order adaptation are obscured. To avoid this requires authentic dialogue and open-mindedness to the possibility that the indigenous narratives about the sacred and divine tell of experienced realities that really influence processes. We take this open minded approach in our chapter to see what indigenous knowledge can potentially offer our common quest for sustainable development. Such an approach brings appreciation of how indigenous cultures actually adapt, and potentially what indigenous knowledge has to offer to assist dominant global culture recover its own sustainable development. We claim that it can potentially free us from the unsustainable logic of capitalism to increase capital accumulation and consumption.

A Trans-Disciplinary Framework for the Sustainability Movement

The dominant global culture is not only critically challenged by indigenous postcolonial perspectives, it is also critiqued from within. The critiques from within Western cultures are wide ranging, with major themes emerging from Marxian influenced literature; psychotherapy and phenomenology; systems and cybernetics; and ecological and post-modern perspectives. The questions that need to be asked of these endogenous critiques are threefold: whether or not these critiques are adequate to understand what is wrong; whether or not they are able to point to what needs to change; and whether or not they are able to herald the required change? To summarise: are the critiques sufficient to ensure that postcolonial situations are healed not only in the former colonies but also in the former colonial homelands? If the problems the sustainability movement are trying to correct are due to dominant global culture, and that the dominant global culture is the continuation of the former European colonial cultures, then correction of the root causes in former European cultures that have given rise to a dominant global culture are at least a facet of what the sustainability movement needs to address. But it would be naive to assume that this is all that needs to change. Even though such idealism does address the "soft" 2nd order issues, by itself it fails to address the real economic, social and ecological processes and their momentum, which 1st order "hard" adaptation deals with. 1st order "hard" adaptation is necessary to implement 2nd order "soft" correction of the cultural sources of error. The real forces unleashed irreversibly by colonisation will not change merely by the recovery of lost ideas or creation of new ones. But if the influences of more relevant ideas within cultural processes are to even nudge the forces within economic social and ecological systems towards increased sustainability, some real progress toward sustainable development will have been achieved. "Hard" system processes can be directed in more appropriate directions through correction of the cultural influences on human motivation and intentionality. Transition management literature outlines the challenge similarly, by defining three scales of relative inertia to adaptation by use of ecological metaphors, namely "niche", "regime" and "landscape" (Geels 2002; Foxon et al. 2008). "Niche" refers to local innovation opportunities; "regimes" to broader institutional maintained practices; and "landscapes" to general cultural norms underlying institutional practices. Innovative ideas and practices emerge in "niche" and have the task to transform "regimes" and ultimately "landscapes". The process of transformation is guided by vision providing long-term societal goals, with shorter term practical objectives met along the transition.

If the cultural causes of the unsustainability of dominant global culture are addressed through internal critique, the reality of indigenous cultures and their self-determination becomes of prime importance. Addressing the cause of colonisation recovers the reality that indigenous cultures are all intrinsically equal partners with the dominant global culture and each other. There is serious need to consider how further development can incorporate authentic dialogue and partnerships between dominant global culture and extant indigenous cultures. This is of course not a novel view. It has been a view, albeit a minority one, in Western cultures ever since the Romantics of the 19th Century response to industrialization and associated colonization. Indeed a type of human ecological irony can perhaps be observed in how Western colonization has brought about contact with what is needed to correct the cause of the colonization it carried out. But early Romanticism did not stem the tide of colonization and industrialisation, including militarization. Other attempts to try to deal with the negative feedback from those colonized, and the classes formed by industrialization emerged in waves: Marxism, based on a reading of phenomenology, attempted to address the economic and to an extent ecological processes; Darwinian theories challenged not only the privileging of Western culture but also presumptions about other species and nature generally, and was complemented by psychoanalysis addressing the psychological disturbance caused by the loss of appreciation of the material and biological reality. Later, ecology and systems disciplines emerged as scientific attempts to appreciate the inherent interrelationships between all agents, human and otherwise, built in large part on groundwork by phenomenology and the Darwinian tradition. These have been applied in the fields of Industrial Ecology, Ecological Economics and Ecological Engineering. Finally, contemporary populist sustainability movements have emerged, for example the concern for "sustainable consumption" influenced strongly by the 1992 Earth Summit and the hundreds of Agenda 21 groups that emerged afterwards to seek ways to implement the recommendations (Geer Ken et al. 2008; Bilharza et al. 2008). Also of course local indigenous initiatives have likewise emerged in response to the postcolonial situation, for example the Chipko movement in Nepal (Weber 1987) but are found in every continent (Clark 2002).

The historical emergence of endogenous Western critiques has been a bit like the Olympic flame being passed on from one runner to another and is now finally reaching the Olympic stadium where all runners from all around the world come together to meet. Even though Romantic idealism was and is insufficient, it is conceivable that if the history of responses to the cultural cause of what the sustainability movement is now addressing can all be rolled up together into the present along with indigenous cultural knowledge and indigenous initiatives, a better grasp of what to do may be able to be gained. This however brings its own challenges. To large degree they are the challenges of the Society for Human Ecology with their annual international conference and interdisciplinary journal "Human Ecological Review". Even though metaphorically speaking there is a flame being passed on, the various responses are nevertheless distinct in historical context and also in the disciplines and paradigms they are based and in which they are developed, and of course indigenous cultures all have their own unique heritage. The challenge is how to integrate such a diverse set of texts that exist in separate, if to an extent overlapping, contexts? It is naive to attempt to seek an integration that will result in a total solution. Such total system attempts conflate “soft” 2nd order and “hard” 1st order adaptation. Once again, “soft” 2nd order idealism is not sufficient by itself even if conceptualised as dealing with economic, social and ecological processes. The real-world processes have to be lived as praxis within “hard” 1st order systems continually adapted under guidance of “soft” 2nd order learning. But neither need it then become an Olympian competition to determine the best way forward. Rather what is necessary are a diversity of trans-disciplinary frameworks that attempt to accommodate significant disciplines and cultural traditions, along with appreciation of the adaptive learning processes involved. This maximises opportunities for authentic dialogue and partnerships for productive collaboration. A place to start is the creation of a trans-discipline that accommodates Western disciplines. If trans-disciplinary frameworks can be synthesised to assist integration of a significant endogenous Western self critiques, they may also be helpful to open pathways to appreciate indigenous cultural traditions as well, to facilitate and initiate authentic dialogue.

Attempting to find an optimal trans-disciplinary framework is however naive. Relatively more adequate trans-disciplinary frameworks are nevertheless likely to be able to be constructed in the concern to facilitate sustainable development in a postcolonial world, in its multiple contexts. The important principle is to base their construction on the ability to accommodate the various disciplines that have emerged along with indigenous cultural traditions. Given this, some parsimony is then necessary. But what the most effective and efficient frameworks are will change and co-evolve with the changes in the situation in which they emerge and need to be applied. Appropriate trans-disciplinary frameworks will themselves undergo both 2nd order and 1st order adaptation. We proffer below just one possibility, to be illustrative of the necessary task.

Our proposal is based on the "adaptive cycle" metaphor stemming from within the ecology discipline and now utilised by the fields of adaptive environmental management (Holling 2001; Gunderson et al. 1995). The metaphor relates capital accumulation and system organisation in a four stage continuous process of transition; from increase in both capital and organisation; to a decrease in both capital and organisation; to an increase in capital while organisation is still decreasing, to finally a decrease in capital while organisation is beginning to increase again, before both starting to increase again and so on. We however only use the "adaptive cycle" as a seed metaphor, which we extend and critique. The critique and extensions are informed from evolutionary biology, phenomenology, the theological heritage of Western cultures, and contemporary sustainability science. Only a brief sketch can be made here and it will serve to focus on what aspects help appreciate what indigenous cultures have to offer sustainable development.

The "adaptive cycle" metaphor can symbolize much of what is observed as systemically happening within living systems at multiple levels, including socio-ecological transitions (Weeks et al. 2004). One of the valuable contributions of the metaphor is how it is nested to describe the complex dynamics of living systems at multiple scales, spatial and temporal, within the biosphere. The term "panarchy" has been coined to describe nested "adaptive cycles" making up the biosphere (Gunderson and Holling 2002). The "adaptive cycle" metaphor needs however to be complemented by other metaphors, including those utilised by transition management.

Adaptive management, especially as it is being developed by the "Resilience Alliance" in their journal "Ecology and Society", but not restricted to this forum, is critically developing the role of adaptation within sustainable development, but seldom is there explicit critique of existing dominant global cultural patterns of capital accumulation and consumption and the associated social status and determinants of affluence. Adaptation of social structures is generally interpreted as "adaptive governance" (Folke et al. 2005; Gunderson 2000) to ensure that existing patterns of capital accumulation are able to be maintained through discovering new opportunities for resource exploitation. Even though this does not contradict the need to also critique existing patterns of capital accumulation, it does point to the need to complement the adaptive management approach with vision for socio-ecological transitions involving different capital accumulation patterns. The need to complement the adaptive management approach with transitional management approaches has already been mooted (Pahl-Wostl 2007; Foxon et al. 2008). Through considering the human ecology of emergent socio-ecological systems we propose further to this that metaphors utilised by indigenous knowledge, including indigenous and traditional theological Western knowledge along with transition management, fulfil a role to guide the developmental of "soft" 2nd order dimensions.

Evolutionary biology utilises a hierarchical model where levels in the hierarchy are formed by "holons" (Michod 1988; Okasha 2003; Wilson et al. 2008; Godfrey-Smith 2006). "Holons" are qualitatively different levels of natural organisation that have emerged through the evolutionary process. Such a view accords with the thermodynamic view of evolution that argues that increasing order and complexity emerges through the evolutionary process so as to more efficiently dissipate energy surpluses (Swenson 1997; 2000; Swenson and Turvey 1991; Salthe 2007; Barab et al. 1999; Andrew 2000). From this point of view, all human social systems emerge within the same "holon" and so hierarchy cannot essentially and need not be found within social systems. This reveals starkly the limitation of adaptive management utilising only the "adaptive cycle" metaphor. In adaptive management, social systems are considered to be hierarchical, with smaller social systems governed by broader and more powerful social systems (Gunderson and Holling 2002). Whereas there are conceivably different inertias for adaptive change operating at different scales of "niche", "regime" and "landscape", what this involves is a complex network, not a hierarchy in an evolutionary sense. Moreover, innovative adaptations occurring in "niche" are found to be inspired by the presence of a "holon" emerging above social systems. It has been discovered that evolutionary interaction between social groups leads to the emergence of a "new, higher-level type organism" (Wilson et al. 2008). Clues as to the nature of the "new, higher-level type organism" "holon" above social systems can also be found from evolutionary biology, which recognised that at least two types of reproduction occur at each "holon", and hence at each "adaptive cycle". They are germ-line reproduction and somatic reproduction. Germ-line reproduction involves differences, whereas somatic reproduction involves exact replication. Taken together they can be symbolised by the "adaptive cycle", but if each of the two types of reproduction are considered separately a more sophisticated dynamic is required and points to what the "holon" above social systems may be. Germ line reproduction is associated with the need to increase diversity. Coextensive with this is the death of individuals. Death defines individuals; individuals are that which die, as death of old forms is necessary if new forms are to come into existence (Buss 1987). The nested socio-ecological systems forming the co-evolving "panarchy" of the biosphere utilise death to ensure diversity. Diversity and hence death of individuals are intrinsic to adaptation. But, even though what is defined as an individual depends on the scale at which adaptation is occurring within the nested "adaptive cycles" of the biosphere, only human individuals appear capable of being self-aware of individuality, death and our role within the "panarchy" of "adaptive cycles" (Morris 2007). A prominent psychoanalytical view points out that humans have to face the reality of death to affirm their individuality (Becker 1973). Furthermore, the denial of death, because it also threatens individuality, leads to a loss of contextual awareness of the biological roots of our nature, including the role of sexual reproduction. The psychoanalytic tradition generally suggests that a deep seated denial of death leads to a commonplace failure to adequately adapt biologically. Phenomenological tradition adds to this by pointing out that our biological nature is ecological, which psychoanalysis would say includes our socialisation by socio-cultural influences (Morris 2007; Fromm 1962). Namely, the face of all living beings is a control panel through which we engage with our environment. Accordingly the face of a living being is an image of the world the living being lives in. In other words, the face of a living being reflects the significant influences that need to be adapted with so as to fulfil the "law of requisite variety". Generally for animalian beings, this occurs within a social system and so is a collaborative socio-ecological venture (Morris 2005). However for the human individual who is reflexively facing the reality of mortality, their biology and hence ecology includes a different dimension: their face also reveals the image of another dimension to their world. This is the realm of the whole biosphere, often termed the sacred or divine known in "I-Thou" personal relationships (Buber 1970; Zizioulas 1985, 2006). Their face, which is controlling their actions, is responding to perception of the role of the individual and death to allow diversity for the enhancement of new life. This is what is meant by each individual human being in their essential "wildness" being an "image of God", within traditional Western culture (Morrison 2002). The evolutionary biological meaning of the life of any individual being is to give one's life to nurture new life. Actions by individuated human beings to respond intentionally to nurture new life are truly innovative, and qualitatively beyond those of animalian social systems. This is the emergence of a "holon" above that of social systems that selects for ultruism (Godfrey-Smith 2006). Adaptation within sustainable development is inspired by the "holon" of community life, nurtured paradoxically by individuated human beings, and not controlled by higher levels of governance and social system structures of power, though they do provide influence. This is in agreement with paradoxes found within systems and social learning literature, where "communities of practice" constrain individual expression to facilitate individual critical creativity, out of which emerges greater complexity than that provided by the sum of unconstrained individual expression (Wilden 1980; Senge 1990; Wenger 1998; Davidson-Hunt 2006; Pahl-Wostl 2006; Fernandez-Gimenez et al. 2008). Moreover, nested natural systems are found to a have a fractal attractor which is scale independent (Liebovitch and Scheurle 2000). It can be considered to be analogous to the "image of God" seen in the wholeness of each part of nature (Morrison 2002). The various scales of social system, described metaphorically by "niche", "regime" and "landscape", can potentially be developed to resonate together, along with actual ecological niche, regimes and landscapes, as all are essentially driven by the same natural fractal attractor. This is the essence of our proposed trans-disciplinary framework. It enables the problematic interaction between human well-being and metabolic relation to nature to be understood, as well as answering the question of why and how institutions or social systems dealing with social status and determinants of affluence can and should incorporate ethics. It explains how socio-ecological transitions can nurture sustainable community development.

Potentially, human well-being can interact directly with the metabolic processes of nature in a way that other animalian life cannot, because humans can exist as individuals in a way that other animalian life does not. The relationship to nature is potentially one of nurturing of the life of nature as a whole, rather than through instinctive exploitation. Accordingly, institutions of social status and the determinants of affluence are not necessarily driven by the desire to satisfy the acquisition and consumption of capital, as they are in other animalian social systems. They are potentially also ways in which to facilitate learning to become an individual to face the reality of death, and to transcend it by nurturing all life to enter into the "holon" of community above socio-ecological systems. Given that this is the case, a culture can and should choose to limit capital accumulation and consumption if it to benefit the life of nature as a whole. A community will then emerge to ensure that resources are used sustainably for the benefit of all. But paradoxical and complex dynamics controlling contradictory movements are known to arise in such homeostatic systems of resonance (Gorsky et al. 2000). The need to understand these complex homeostatic dynamics to ensure sustainable resource use is nevertheless key to the task of sustainable science as it is being developed, including Industrial Ecology, Ecological Engineering and Ecological Economics. It is our claim that this task is already achieved in indigenous cultures and communities. It is an intentional, well understood and articulated process. It emerges out of dialogue between individuals who know the meaning of death and their individuality, and hence are able to reflexively transcend the nested "adaptive cycles" making up their life within the "panarchy" of the biosphere, by seeing how social systems at all scales can potentially resonate together. They can perceive horizons of meaningful activity to nurture all life within the socio-ecological systems they participate within by perceiving the sacred or divine whole in each of them, because they are themselves seeing with a face that is an "image of God" to perceive the one natural fractal attracter of them all. Cultural traditions generally describe these as sacred or divine horizons; as transcendental dimensions of beauty and goodness which are known when truth is sought without denial, in other words in humility, and where the vision by individuals of this potential and future homeostasis inspires the possibility of working together toward achieving it. This is already known at the heart of the theological heritage of Western cultures and reproduced and embodied through metaphors and rituals (Zizioulas 1985, 2006; Schmemann 1974). It is also at the heart of the reconstructive post-modernism agenda (Dombrowski 1997). This agenda explicitly recognises the need to re-integrate the genre of Western culture which nurture awareness of the sacred or divine into dominant global culture. Whereas deconstructive post-modernism seeks to unmask the meaninglessness and perversity of claims to superiority by Western cultures, reconstructive post-modernism focuses on how indigenous knowledge, including traditional theological knowledge, of Western cultures is capable of discerning between what needs to change culturally and what does not need to change. In other words, it seeks to communicate how to ensure "soft" 2nd order growth, learning and adaptation responds to the transcendental presence of the sacred or divine to guide "hard" 1st order adaptation utilising scientific knowledge to respond to the particular socio-ecological systems participated within.

Such a trans-disciplinary framework also allows consideration of what gives rise to the loss of sustainable development. It arises when there is denial of death; when horizons of meaning for individual life are not able to be perceived. This results in a regression back to instinctual 1st order adaptation within socio-ecological systems. To use Max-Neef et al.'s terminology, "pseudo-satisfiers" and "violator satisfiers" arise due to the denial of death. "Pseudo-satisfiers" and "violator satisfiers" establish systems of social status and affluence which destroy homeostasis. Ellul (1980) for example critiques the emergence of the "technological system" of consumerism and militarization as a collective failure by individuals to respond to the "call" to individuation, which is in effect a collective denial of death, and can be expected to lead to increasing social and ecological mayhem due to fixation onto conflicting "limiting situations" rather than horizons of resonating opportunities provided by the fractal attractor (Kearney 1995). Fromm (1991) similarly argues that it is what gives rise to totalitarian regimes. Ironically, even though this failure involves a regression back to instinctual expression of 1st order adaptation in loss of 2nd order growth, learning and adaptation, it is also associated with a denial of socio-ecological reality and accordingly a loss of concern to ensure the sustainable use of resources. The consequence of this is a confused loss of appreciation of the meaning of biological processes, including sexuality, leading often to emotional frustration. The postcolonial trauma is played out here.

The Structure of Indigenous Knowledge

Indigenous culture comprises of multiple genre within the multi-faceted praxis of community life operating at multiple scales; from familial "niche" to political "regimes" and cultural "landscapes". The range of genre challenges attempts to systematise indigenous knowledge. The reference however to sacred or divine realms and the use of ritual incorporating mythical language and motifs is not essentially different to Western theological genre and so lends itself to being interpreted along similar lines to the traditional cultural heritage of Western cultures. It is probable that considering how traditional Western culture understands itself will provide insights into how to understand indigenous knowledge on its own terms. Taking this approach we are invited to make a fundamental distinction between positive (defining what it is) and negative (defining what it is absent but potentially present) knowledge, and between systematic and pre-systematic aspects of indigenous knowledge. These are the distinctions found in Christian theology, where negative knowledge is considered to be the highest form as it includes the least distortion (Palamas 1983). Positive knowledge is always distorted in the sense that it is always relative to the intention of the definer. Negative knowledge can overcome the distortions of positive knowledge by not having any intention, though it is at the expense of not being able to provide any substantive content. Nevertheless non-substantive content, which we can call context, is still meaningful; indeed it is what provides meaning. For example, awareness of the infinite and eternal dimensions of the beauty and goodness of sacred or divine horizons is doubly negative knowledge, as both infinite and eternal are negative definitions. They arise by negating finitude and time, but they are what provide meaning to all activities within time and space, and provide the long-term vision for sustainable development. Similarly, creative openness by innovative "niche" is inspired by it. Negative knowledge refers to real knowledge of divinity or sacredness as the source of as yet undefined natural possibilities, and is termed "noetic" insight, whereas positive knowledge refers to already experienced real necessities and constraints, and definable possibilities based on these. Negative knowledge looks to the evolution into community, and positive knowledge comprises of knowledge about socio-ecological systems. Negative knowledge is 2nd order knowledge and positive knowledge 1st order knowledge. The union of the two is a complex adaptive process. Soja (1996) suggests it is a "trilectic" involving radically innovative "thirdspace", in agreement with McGowan's (1991) view that post-modern philosophy recognises inherent "freeplay" within every process and system. In social learning theory the "freeplay" within "thirdspace" is termed "shadow spaces" within socio-ecological networks (Pelling et al. 2008), and is of course where transition management sees "niche" emerging. Interestingly, Aulin (1987) also provides a mathematical proof for such indeterminacy wherever there is individual intentionality.

Systematic approaches are constituted by positive definitions and so belie a particular set of utilitarian genre, for example, the classification of medicinal plants. Pre-systematic approaches by contrast are couched in negative definitions to allude to as yet unrealised potential utilitarian genre in the attempt to open up potentially new possibilities, for example, through healing and exorcism rituals. Pre-systematic features of indigenous knowledge cannot be described or understood in a purely positive way. Pre-systematic features refer to the role of wisdom. It is only achieved by every person anew in every context by becoming aware of values which are not being fulfilled, through a manifest lack of community well-being, and hence recognition of the need for and potential for improvement, along with appreciation of what is working to nurture life. It is to discover new wisdom for themselves, but guided of elders who know how to do so. Pre-systematic features of indigenous knowledge encourage a person to be innovative. Heidegger in his attempt to recover the indigenous roots of Western philosophy described it thus:

> It is not a direct possession like everyday knowledge of things and of ourselves. The πρώτη φιλοσοφία is the έπιστήμη ξητονμένη: the science sought after, the science that can never become a fixed possession and that, as such, would just have to be passed on. It is rather the knowledge that can be obtained only if it is each time sought anew. It is precisely a venture, an "invented world". That is, genuine understanding of being must itself always be first achieved (Heidegger 1992: 11).

It is arguable that the essence of indigenous knowledge is precisely this ability to continually reinvent a world of possibilities, and also the ability to guide others to learn how to do so. The ability is gained from the experience of "noetic" insight gained through meditative thought (Heidegger 1956). Homeostasis requires catalysts, which in social systems are leaders mentoring growth of life into community (Gorsky et al. 2000). Homeostasis requires social systems which maintain social status exhibiting leadership to mentor growth of life and determinants of affluence that exhibit altruistic behaviour.

Negative and positive knowledge, and pre-systematic and systematic approaches, are both complementary pairs of the "trilectical" process of adaptation, learning and growth. Using these distinctions, which accord with the trans-disciplinary framework outlined above, we find that indigenous knowledge includes negative and pre-systematic knowledge and integrates them with positive and systematic knowledge to a degree that is largely missing in dominant global culture. The integration is achieved through "noetic" insight and mentoring relationships. The integration is explicitly recognised by indigenous knowledge, just as it is in traditional Western theology. In Western theology it is termed "theandric" knowledge as it is the union of the awareness of the sacred or divine horizons with human socio-ecological reality and facilitated by interpersonal relationships (Zizioulas 1985, 2006). Indigenous cultures generally provide a wide range of metaphors to communicate this, which include the union of light and dark, and the union of male and female. Moreover the very notion of a horizon where the sky meets the earth is alluded to as the union of Earth mother and Sky father. Levi-Strauss (1966) in his structural anthropology developed through empirical study of indigenous cultures recognised this essentially binary nature of cultural structure, though he did not develop its theological reconciliation, transcendence and integration into community. Rappaport (1967, 1979, 1999) in his environmental anthropology did begin to do so. In particular, he explores the role of rituals to facilitate awareness of the natural whole as the context for decision-making.

The reliance on "noetic" insight and inter-personal dialogue frees decision-making from being constrained by systematic thought and institutional power structures, whilst utilising all available systematic thought and other utilitarian knowledge and socio-cultural structures. "Noetic" insight inspires innovative "niche" to emerge to transform the "regimes" and "landscapes" of socio-cultural structures. It is reliance on "noetic" insight that allows indigenous knowledge to remain appropriate for ever-changing contexts, and why it is also termed "localising knowledge" (Raffles 2002: 331-332). But it is also why it always takes time and effort to gain and cannot be simply passed down, or learnt, theoretically. It is only learnt through elders spending time to guide youth who can even so only ever learn it through their own striving to come up with their own "intimate knowledge" (Raffles 2002: 332). The creative innovation involved yields three main features of indigenous knowledge: figurative thought, narratives and substantive knowledge.

Figurative thought comprises of symbols that attempt to communicate the need for openness and "noetic" insight to ensure wise innovation, in particular insight into what sustains natural capital and social well-being. Within indigenous knowledge generally this is recognised as involving poetry, including mythical genealogies and legends, dance and ritual (Carrithers 1992; Haami and Roberts 2002; Roue and Nakashima 2002). Implicit in the expression of figurative knowledge is "relational knowledge" (Raffles 2002: 329-331). To be effective in nurturing wise innovation, rituals have to be freely participated within, involving trust and respect.

Samoa provides an example of the dimensions of figurative thought and how it is integrated within practical adaptation to seek resonance between all systems. Figurative thought is commonly expressed by Samoan elders, "matai". Indeed when becoming a "matai" one is expected to learn a high esoteric version of Samoan language that comprises solely of the use of metaphor, as a series of proverbs that are freely added to. This was the case generally among South Pacific cultures, and even though the high esoteric language is lost among most now, the use of proverbs to guide decisions is still commonly maintained (Paterson 1992, 1994). In Samoa, "matai" are still a source of independence and cultural innovation, and also charged culturally with enabling others to also become independent and innovative. This is why in Samoa, "matai" are expected to be their extended family's entrepreneurs and why the community council comprising of family "matai", the "fono", discusses entrepreneurial opportunities (Morrison 2008). A "matai" works as an entrepreneur with their extended family by creating a collective vision that works for the whole family. Similarly the community council of "matai" works to ensure larger business plans work for the well-being of all families and their natural capital, namely land, and their community natural capital, namely beaches and lagoons. The "fono" involves free ranging dialogue involving figurative language to explore opportunities in an open way. At the conclusion of a "fono", the rules of the village are adapted to suit changing situations, with decisions made by consensus. This occurs every week for several hours after the first religious rituals of the holy day of the week and before the second religious rituals of the day. There is explicit and facilitated integration of "bottom-up" ("niche") and "top-down" ("regime") processes by the cultural "landscape".

This is very significant as the need for explicit and appropriate facilitation and structuring of participatory processes has begun to become clear (Gardner and Lewis 1996; Cooke and Kothari 2001; Reed 2008). This is a clear example of what indigenous knowledge can offer the quest for sustainable development.

Narratives provide horizons to integrate the multiple aspects of development (Carrithers 1992). Narratives use figurative language to encourage openness to "noetic" insight into issues of ecological sustainability and social well-being, and to also integrate specific utilitarian and systematic knowledge that is relevant to the case the narrative is being related to. Narratives fulfil the same role as trans-disciplinary frameworks. They provide a vehicle for the integration of the various perspectives which have been expressed. Using the Samoan example again, within a "fono" the process of starting a speech is to first of all summarise and synthesise what everyone previously has said, before making one's own contribution. The speech is told as a narrative to seek comprehensive integration of multiple perspectives in the quest for a consensus that works for everyone, rather than to debate and argue for a singular truth to win a battle by superior power. The narrative feature of indigenous knowledge enables adaptation to incorporate all views within the community, for the community's development. This can and does of course extend to integrate insider and outsider systematic and utilitarian knowledge, but only as long as there is authentic dialogue based on trust and personal relationships.

Storytellers have a significant role in indigenous cultures. In Samoan culture, they are usually either "matai" or senior women healers and are recognised by titles that refer to them as being "servants of God". It is considered that their ability to narrate a vision for others is a divine act of revelation. It is in other words an expression of "theandric" knowledge integrating 2nd order and 1st order adaptation. In Samoa this includes communion with those who have died, which opens up the eternal and infinite horizons transcending death, and hence the divine experience of the 2nd order level. Awareness of these "wild essences" of things as they reveal themselves of their own accord as personal I-Thou relationships of communion is what unites 1st order practical knowledge with the 2nd order horizons, and forms the content of narratives (Buber 1970; Zizioulas 1985; Morrison 2002; Morris 2007). Moreover the source of the inspiration and perception of this communion with "wildness" is told by the storytellers to be from the divine presence in the places they live, traditionally from birds (Turner 1989). The relational aspect of the figurative knowledge utilised extends to include non-humans, to embrace the socio-ecological system context in which adaptation has to be carried out. Such totemic relationships are commonly recognised as motifs of indigenous cultures. In Samoa two concepts have traditionally been used to summarise the process, “va” and “mauli” (Aiono-Le Tagaloa 2003). “Va” refers to the space in which life occurs, and includes material, biological and ecological dimensions, but also interpersonal relational dimensions of community opening up to eternal infinity. “Va” refers to the experience of the potential resonance of all natural systems. The whole task of a “matai” is to maintain “va” and to make it visible to others, which is the same thing as nurturing life into relationships within community. Traditional architecture in Samoa is designed to emphasise the extension of relational space to include all things, by avoiding walls in building, to utilise round building where possible, and to have villages structured as rings with the meeting ground in the middle beside the building where the “fono” meets. The “mauli” is the centre of one’s being that enables all the dimensions of “va” to be integrated together. So a “matai” has to sit to be a presence that unites heaven to earth. They are metaphorically described as trees that provide shelter, and as poles that hold up buildings.

Substantive knowledge is implicit in skills and practices, and is developed experientially, for example, by learning to carve (Fischer 2004: 25). Intrinsic to this type of experiential learning is the role of a mentor. A mentor is a role model who is more fully individuated in relation to the skill and practice than the person learning from them. That is, they are freer and more creative and able to apply their knowledge and skills to situations as they need to, in sum, they have greater adaptive capacity. The mentor-apprentice relationship is however relative, with everyone being a mentor and apprentice to different people depending on the skill and practices involved. The mentor-apprentice relationship does not involve an authoritarian ethic but rather a rational one (Fromm 1947). This is quite explicitly recognised in Samoan cultures, where even the “matai” role is able to be temporarily given to someone else to take over if the resident “matai” has to go somewhere else for a while. Everyone has the potential to become mentors, and will do so as they learn from their own mentors. One way it has been described in the South Pacific context is to say that everyone has a “mauli” which needs to be shined by others; that is by those who are their mentors. The shining of someone's “mauli” by their mentors has of course to also be sought and accepted. A way this is put in Samoa is to say that to become a “matai” one has to learn to serve the “matai”. The role of a mentor is the generic reality that is encapsulated and developed most fully by elders, storytellers, healers and pastors.

The role of substantive knowledge has been described by an indigenous Chinese classic, the “Tao de Ching” by Lao Tzu, as the hole in the middle of a wheel (Judge 1994). It is what makes the wheel useful. The hub of the wheel refers to narratives and figurative knowledge. Figurative knowledge and narratives are of no more use by themselves than trans-disciplinary frameworks and all other 2nd order idealism. Moreover just as there are multiple possible hubs of trans-disciplinary frameworks, there are also multiple possible narratives and metaphors within figurative knowledge. They all help create the possibility of a hub, which integrates all spokes and a hole in the middle of the wheel so that something useful can be done with the wheel. But, even though the distinctions can usefully be made, figurative knowledge, substantive knowledge and narratives do not in practice exist as separate types of knowledge. "Primal leadership" implicit in mentoring the learning of substantive knowledge uses figurative knowledge and narratives to make explicit the implicit relevance of the substantive knowledge learnt (Goleman et al. 2002). The purpose implicit in the practice of substantive knowledge is made explicit by figurative knowledge and narratives. "Primal leadership" provides insights into the processes of how to gain wisdom to know how to adapt to take up opportunities; to change what needs to and when necessary, and to keep on with what does not need to be changed. "Primal leadership" facilitates communion as participation into the "holon" above socio-ecological systems, providing and facilitating rituals and traditions that do not need to change, and which guide adaptation within changing contexts. Experiential learning couched in meaningful narratives and metaphorical allusions to eternal and infinite horizons, nurtures both "emotional intelligence" (Goleman 1996, 2006) as well as discursive knowledge. The richness of experiential learning sparks "noetic" insight to unite the heart and mind.

One feature of the integrated knowledge expressed by "primal leadership" is the creation of economy that is empowering of others. This is beyond co-evolutionary selection of socio-ecological systems; it is growth into the "holon" of community. Independence is respected and nurtured; for individuals, as well as for families, and also for villages and tribal groups. Associated with this is the commonly held view within indigenous narratives that resources are not owned so as to be able to be traded, but are a gift, and they are in turn gifted to others when required so as to enhance the independence of others. In traditional Samoan culture this is a fundamental rule (Schultz 1953). This shows the interrelationship between the notion of self-determination or sovereignty and that of the stewardship of natural resources. Both are intrinsic to indigenous knowledge. Self-determination or sovereignty is the divine presence creatively worked with to continue to provide the gift of life offered by the divine. These complex "trilectical" dynamics making up the adaptive capacity of indigenous knowledge are however not well recognised. The systemisation of indigenous knowledge is the problem. This can be understood by considering the use of positive knowledge. Care is always required when using positive knowledge. If knowledge is constructed by outsiders, the definitions are relative to intentions towards communities from external agents. The intentions may be supportive, but can also be extractive and /or oppressive. There are two main types of positive knowledge, utilitarian and systematic, and both can be problematic when utilised by outsiders in relation to indigenous knowledge. Utilitarian knowledge by outsiders of indigenous knowledge, for example, knowledge about agricultural and medicinal plants, is problematic because it can be motivated by exploitation of indigenous intellectual property (De Walt 1994; Purcell and Onjoro 2002). Even if the motivation is not exploitation they are typically criticised for failing to recognise cultural and community values (Bebbington 1993; Agrawal 1999; Posey 2002; Purcell and Onjoro 2002). Whereas systematic knowledge by outsiders is constructed upon sets of precepts that attempt to systematically elucidate the cultural values of indigenous knowledge, for example, in terms of metaphysical beliefs, what the precepts should be remains "an on-going ideological debate" (Sillitoe and Bicker 2004, 8). There are multiple competing views and all are between and within multiple outsider disciplines, typically from Western academia (Agrawal 2002: 283-285). They are all criticised for attempting to encapsulate indigenous knowledge into an outsider "cognitive construct" (Clammer 2002: 58; Posey 2002: 27).

The systemisation of indigenous knowledge often results in it becoming property within a Western style state bureaucracy. There is not surprisingly a reaction to this state of affairs. The political reaction is a moment in the postcolonial situation which Said (1993) termed "nativism". He defines "nativism" as the attempt to express cultural forms to recover a lost identity. The losing of identity is not primarily due to the systemisation of indigenous knowledge. Indeed the interest in indigenous knowledge that leads to its systemisation is a sign that overt rejection of indigenous culture is on the wane. However the systemisation of indigenous knowledge is not sufficient to overcome "nativism". "Nativism" is associated with fraught nationalist movements that are often at odds with traditional authority within indigenous cultures. Moreover they do not have a history of being successful modes of political practice. As Said has indicated, the challenge is to achieve the socio-ecological transition after "nativism". We proffer that it may require the recovery of the full adaptive capacity of indigenous culture, and this would require more than the mere systemisation of indigenous knowledge. It may require the recovery of pre-systematic knowledge. But paradoxically this does not require traditional forms found within a particular indigenous culture. Other narratives, including trans-disciplinary frameworks incorporating Western scientific disciplines, can be of assistance. For example, seeking and facilitating the union of pre-systematic and systematic knowledge through visualising the "adaptive cycle" metaphor may be helpful. This can aid the recognition of the complex interaction of creative novelty and stable capital growth: negative knowledge and pre-systematic approaches are phases which introduce novelty and diversity; positive and systematic phases are periods of growth in capital and organisation. This understanding may in turn assist in the recognition of what needs to change and what does not need to, so as to fulfil the "law of requisite variety". If changes in practices are recognised as necessary, it is possible that they could be helpfully informed by outsiders. If the implicit ideology or intentionality of outsiders are made explicit, they can be deconstructed, and if helpful, reconstructed by the indigenous community into their indigenous knowledge. Far from being an unusual situation, this process of hybridisation is recognised as the normal process of how indigenous knowledge develops (Dove 2002). If outsiders, including those in bureaucratic organisations recognise this process, authentic collaborative development may be able to occur. Signs are that this may have already begun (Taylor 1998; Reed et al. 2008). It is what agencies like the UN cluster and IUCN are attempting to incorporate and facilitate.

Indigenous Knowledge as a Bridge

Indigenous knowledge can do more however than merely successfully cope with and develop through outsider influences. Collaborative development is more than merely respecting and assisting the facilitation of indigenous cultures' self-determination and development. Indigenous knowledge can assist in transformation of the unsustainable practices of the contemporary global dominant culture. Already the indigenous roots of the global dominant culture are being sought to provide insights into the transformation (McIntosh 2004). There is however a risk of this becoming a Western version of "nativism". It is not sufficient to merely adopt archaic forms. To do so is a conflation of 2nd order and 1st order adaptation, but in this case, the assuming of 2nd order adaptation into 1st order so as to lose 2nd order learning, for example, the literal interpretations of mythical narratives so as to lose their meaning. No less than with colonised indigenous cultures, what is required is reintegration with pre-systematic knowledge. This includes the reintegration of the literature of Western cultures, along with the critical disciplines and emerging trans-disciplinary frameworks, with scientific knowledge. What is commonly still missing however, even if this synthesis is sought, is the inter-personal nurturing role of mentors. It is possible that extant and successfully adapting indigenous cultures can provide clear guidance about how to use mentoring relationships, and hence "primal leadership" and self-determination to put integration into practice, as ongoing praxis.

What is arguably most required to be learnt by dominant global culture is associated with inter-personal relationships. So to the extent there is authentic dialogue and partnerships within collaborative development, the process of seeking, establishing and maintaining of collaborative development may transform dominant global culture into sustainability. But if the successful transformation through collaborative development is to occur, it is necessary to avoid several things. It cannot be a romantic reaction content to take on archaic cultural forms. Neither can it involve the assimilation of indigenous knowledge, including traditional Western cultural knowledge into systematised knowledge. What is required is recovery of adaptive capacity to be able to recognise what systematic knowledge needs to change, and what does not need to change; to continue to develop according to existing paradigms that which does not, and to creatively explore new paradigms where it does need to change. Moreover, there needs to be trans-disciplinary frameworks to coordinate both processes through the various genre contributing to culture, including, rituals recognising the unknown and spiritual realm, and economic structures facilitating access to resources. This has challenging practical consequences.

There is need to explicitly overcome the spirit-matter dichotomy in global dominant culture. Horizons of meaning need to be known to be found in the material world, not above it. Specifically, death and sexuality need to become recognised more explicitly as part of life. Fixated collective ideology, including "nativism" and bureaucratization, need to be recognised as attempts to deny the reality of death and sexuality. Furthermore, it has to become recognised that this can be achieved by valuing "wildness" as the glimmer of transcendental truth and meaning which open up horizons of hope. An explicit recognition of "wildness" as revealing the transcendental light through the praxis of substantive activities is required. There is need to recover awareness of the meaning for life and death in the eternal and infinite context. Acquisition of capital and power is not the meaning. Accordingly, awareness of horizons of meaning can help with decreasing resource consumption. In turn this can allow openness to realistic opportunities for continued development, guided by for example the "adaptive cycle" metaphor. This will allow the development of apt local sustainable practices. Diversity of practices will increase. Consumption is however facilitated by economic structures. So it is also necessary to ensure that economic structures are in accordance with eternal and infinite horizons. In other words, economic structures must be instruments of the mentoring relationship. In particularly, they must nurture economic independence and entrepreneurship. Economic structures need to facilitate access to resources for all individuals, families and communities, and to avoid dependence of individuals, families or communities on others. Economic structures must be socio-ecological systems that are used to facilitate the formation of community; where integration is sought in the "holon" of divine presence in the eternal and infinite horizon of communion rather than in corporate organisation.

Conclusion

Indigenous knowledge does appear to have a degree and type of adaptability missing in dominant global culture. Arguably it is the adaptability needing to be recovered if dominant global culture is to recover sustainability. Indeed it is perhaps what the sustainability movement is seeking as it responds to the negative feedback that the dominant global culture is getting from other cultures, the environment and from within itself. Attempts to incorporate the adaptive capacity of indigenous knowledge into dominant global culture can be guided by trans-disciplinary frameworks that incorporate the range of responses from within itself, including attempts to recover the indigenous knowledge of Western cultures. This is however difficult for two reason. First it requires fundamental changes to dominant global culture. Second, indigenous identities need to be taken up critically, with care to avoid exclusive nationalistic and/or racist expressions. The way forward requires authentic partnerships and critical dialogue. Extant indigenous culture, mentoring second order adaptation as well as acquisition of substantive and systematic knowledge, can be helpful. It is potentially a bridge to sustainability. It is a cooperative development bridge that the sustainability movement has begun to look at. But to do this is problematic as indigenous knowledge has begun to be assimilated into dominant global culture due to a failure to appreciate the need to share power. If the bridge is to be formed it will have to be a political bridge. But it will also require cultural evolution of growth into communion. Slowly dominant global culture will have to be renewed culturally through better ideas and cultural practices to gain authentic hope and meaning.

There is political change required to enhance opportunities for authentic dialogue and partnerships to safely occur. If it can occur it will produce positive feedback so that the bridge to sustainability becomes strengthened. Presently the bridge is looked at suspiciously by both indigenous cultures and the dominant global culture. Possibly the co-evolution of global socio-ecological systems due to present unsustainable practices will produce the socio-political feedbacks that will create the possibility of such a bridge. At the moment there is underway the recovery of traditional Western cultural heritage and their integration with trans-disciplinary frameworks and literature. But there also needs to be engagement with extant indigenous community life.

Indigenous knowledge does appear to exemplify and provide practical guides of how to correct what needs to change to fulfil what the sustainability movement is seeking to change. But it is the adaptive capacity of indigenous knowledge that achieves this, not any particular form it may have. What needs to be understood is the adaptive capacity of indigenous cultures in face of colonisation. To face this and to engage in authentic dialogue and cooperative partnerships with indigenous cultures is to begin to walk across the bridge to the distant shore of sustainability.

Acknowledgements

The second author gratefully acknowledges the support of the Austrian Science Fund (FWF) for making possible the participation to the 2007 Human Ecology conference in Rio de Janeiro, Brazil.

Endnote:

[1]http://events.it-sudparis.eu/degrowthconference/en/themes/

CHAPTER SEVEN

Co-managing commons: Advancing Aquatic Resources Management in Brazil

CRISTIANA S. SEIXAS, CAROLINA V. MINTE-VERA, RENATA G. FERREIRA, RODRIGO L. MOURA, ISABELA B. CURADO, JUAREZ PEZZUTI, ANA PAULA G. THÉ AND RONALDO B. FRANCINI-FILHO.

Contributing authors: Paula Chamy, David Gibbs McGrath, Helio C.L.Rodrigues, Guilherme F. Dutra, Diego Corrêa Alves, Francisco J. B. Souto.

Over the past three decades or so, the development of Common-Property Theory, or simply the "Commons Theory", has contributed to the advancement of Human Ecology, and has also helped to inform—and to certain extent to improve—policy worldwide. The term *co-management* and its variants[1] emerged in the literature to describe collaboration among resource-users and other key stakeholders in dealing with both internal and external forces affecting commons management (e.g., Jentoft 1985, 1989; Pinkerton 1989; Berkes et al. 1991; Borrini-Feyerabend 1996; Wilson et al. 2003). Initially, co-management was used to describe governance systems in which decision-making power and responsibility was shared between government and local resource-users (Berkes et al. 1991; Sen and Nielsen 1996; Pomeroy and Berkes 1997; Singleton 1998), ideally, to combine each one's strengths and mitigate their weaknesses (Singleton 1998). Scholars recognized that there was a hierarchy of co-management arrangements ranging from those in which the local communities (resource-users) simply inform government of their decisions to those in which government simply informs users of legislations issued and/or carries on enforcement (Berkes 1994; Sen and Nielsen 1996).

Over the years, the concept of co-management has evolved as scholars and practitioners noted that managing commons often requires the involvement of parties other than users and government, particularly in complex commons such as coastal zones (Jentoft 2000). More recently, it has been accepted by many authors that "co-management is a collaborative and participatory process of regulatory decision-making between representatives of user-groups, government agencies, research institutions, and other stakeholders. Power sharing and partnership are essential part of this definition" (Jentoft 2003: 3). Carlsson and Berkes (2005) highlight the fact that "most definitions of co-management have problems in capturing the complexity, variation and dynamic nature of contemporary systems of governance." These authors propose approaching and analyzing co-management in terms of "networks of relationships".

Considering the dynamic nature of governance systems, and the need to adapt co-management institutions and improve ecological knowledge in face of social conflicts and environmental uncertainties, another term was coined in the Commons literature: *adaptive co-management* (Folke et al. 2002; Berkes et al. 2003; Olsson et al. 2004a; Armitage et al. 2007). Adaptive co-management can be seen as "a process by which institutional arrangements and ecological knowledge are tested and revised in a dynamic, ongoing, self-organized process of learning-by-doing" (Olsson et al. 2004a: 75).

To broaden our understanding on collective-actions for managing the commons, it is important to note that "democracy, transparency, accountability and sustainability are key defining attributes of co-management" (Jentoft 2003: 3; Bundy et al. 2008). Moreover, co-management may be seen as consisting of "on-going processes of problem solving" (Carlsson and Berkes 2005).

Co-management arrangements are being developed in different regions of the world for managing commons resources as distinct as fish, wildlife and forests. Examples come from places such as Sweden (Olsson et al. 2004b), Brazil (Seixas and Berkes 2003; Glaser and Oliveira 2004; Kalikoski and Satterfield 2004), Chile (Castilla and Fernandez 1998), Bangladesh (Ahmed et al. 1997) and Canada (Kendrick 2003), among many others. Efforts to synthesize lessons learned from cross-case analysis can be found for regions such as Asia (Pomeroy et al. 2001; Wilson et al. 2006), South Africa (Hauck and Sowman 2003), South East Asia and Southern Africa (Nielsen et al. 2004), Europe (Symes et al. 2003), North America (Pinkerton 1989), and Brazil (Kalikoski et al. in press). In some of these places, the legal framework fosters formal co-management, while in other areas, informal co-management arrangements evolve from self-organizing processes, which may become a formal arrangement later on.

Seixas and Davy (2008), while studying factors that contribute to self-organization processes of successful initiatives managing commons, found out that at least six factors foster such processes: (1) involvement and commitment of key players (including communities); (2) funding; (3) strong leadership; (4) capacity building; (5) partnership with supportive organizations and government; and (6) economic incentives (including alternative livelihood options).

In this chapter, we borrow insights from Commons Theory to investigate the following questions: How have policies contributed to or hampered the creation and implementation of co-management processes in Brazil? More specifically, what factors hinder and foster co-management of aquatic resources in Brazil? We investigate five co-management processes that were established formally or informally. The five cases were presented at the XV International Conference of the Society for Human Ecology held in Rio Janeiro in October 2008. First, we introduce the types of co-management arrangements in place in Brazil; second, we describe the five cases; and third, we discuss advancements and challenges in co-management of aquatic resources in Brazil in the light of Commons Theory. Our ultimate goal is to influence policies that may foster co-management arrangements in the country. Such policies should allow for adaptive management measures and simplification of burocratic process in order to shift the focus to local coastal communities' needs and sustainability goals.

Co-Management Arrangements in Brazil

Co-management arrangements in Brazil are both formal and informal and may take place within or outside protected areas. We considered formal co-management arrangements those that evolved from, or were legitimized by, a legal framework provided by the government. Informal co-management arrangements are those that emerged from self-organization processes incorporating other stakeholders in addition to primary resource-users (including government agencies), but which are not yet supported by a legal framework or policy.

The Coastal Management National Plan (Law 7661/1988), the Water Resources National Policy (Law 9433/1997) and the National System of Nature Conservation Units (Law 9985/2000) are all examples of public policies that allow for resource user participation in decision-making processes (i.e., foster formal co-management processes), although decision-making power is hardly ever evenly shared among government and other stakeholders – the former always holding the bulk of power.

Besides increased stakeholders' participation, the Coastal Management National Plan promotes decentralization and provides guidelines for the establishment of State coastal management plans. For instance, in the State of São Paulo the State Coastal Management Plan (São Paulo State Law 10019/1998) establishes a state-level Coordinating Group in addition to four regional coordinating groups in order to elaborate the State Coastal Management Plan and four Regional Ecological-Economic Zoning Plans, respectively. Both the state and regional coordinating groups are formed by one-third of State government representatives, one-third of Municipal Governments' representatives, and one-third of civil society organizations, highlighting the concentration of decision-making power in government agencies. In addition to the disproportionate weight of "civil society organizations", it is also noteworthy that there is no mechanism ensuring a balanced weight of stakeholders' sectors in the one-third non-government seats in the coordinating groups.

The Water Resources National Policy establishes the rule that watersheds shall be managed by committees formed by government agencies and civil society representatives, in which government agencies hold 50% of the seats; that is, each watershed management committee shall function as a deliberative body in charge of decision-making. Apparently, in this co-management arrangement the decision power is more evenly shared but the concern with representation is still valid, especially for historically marginalized groups, such as subsistence and small-scale fishers, as well as other traditional populations that strongly rely on extractivism.

The National System of Nature Conservation Units (SNUC) classifies protected areas (= Nature Conservation Units) into two major groups, no-take conservation units and sustainable-use areas; each group encompassing a range of protection levels and an array of societal participation on management arrangements. For instance, National Parks, Ecological Stations and Biological Reserves are no-take protected areas, while Environmental Protected Area, National Forests, Extractive Reserves, Sustainable Development Reserves, and Fauna Reserves are sustainable-use protected areas. According to the SNUC law, all protected areas shall have a council formed by representatives of government at all levels (municipal, state and federal) and representatives of civil society. All protected areas have Consultative Councils, with a limited decision-making power, with the exception of Extractive Reserves and Sustainable Development Reserves, which bear Deliberative Councils with the majority of seats attributed to resource users. Although management plans shall be elaborated after the Consultative Council is heard, all rules and management structures have to be approved by the government. In sum, all protected areas in Brazil should experience some degree of formal co-management, although many do not have yet a council.

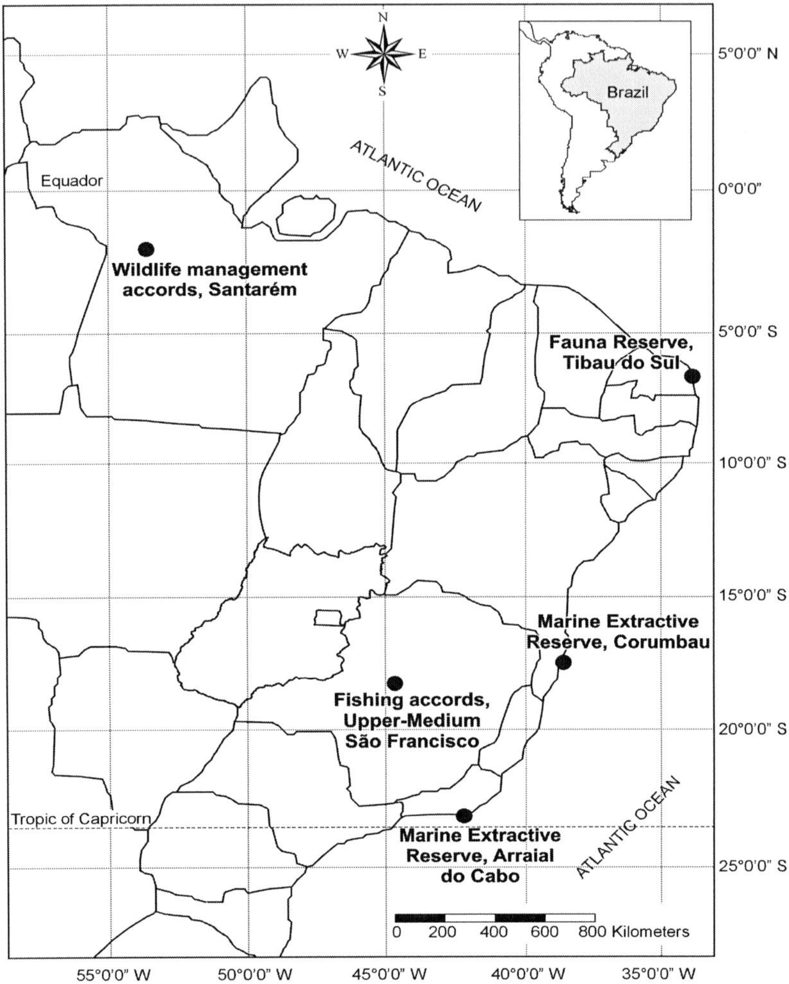

Figure 7-1. Localization of the five Brazilian case studies analyzed (map designed by Pedro Ferraz Cruz).

In the exceptional case of Extractive Reserves and Sustainable Development Reserves the decision-making is more evenly shared between government and society. In some cases, local resource-users are engaged in the planning and management processes even before the formal establishment of the unit[2] and sometimes management plans[3] are preceded by the development of a Utilization Plan, in which users and other interest groups (e.g., NGOs, academia) have a stronger role. This "intermediate" document aims to put in writing customary rules related to resource use and to establish how the unit shall be utilized while additional studies are completed and as the organization evolves.

Guidelines for the elaboration of the Participatory Management Plan were recently issued by the Federal Government Chico Mendes Institute (ICMBio) (Normative Instruction 01/2007). After the completion of the required studies, discussion of the results and long-term planning for sustainability of the unit, a draft Participatory Management Plan is crafted blending traditional and scientific knowledge. The management plan shall be approved by a Deliberative Council "chaired by the agency in charge of its administration and composed by representatives of public agencies, civil society organizations and traditional populations living in the area" (SNUC Art. 18 § 2º and Art. 20 § 4º). Although contemplating a number of mechanisms aiming to promote even distribution of decision-making power, this complex process has not yet been completed in any of the Brazilian Extractive or Sustainable Development Reserves, indicating that its own complexity may be a major hindrance. In addition to the aforementioned legal frameworks—within and outside protected areas—some self-organized informal co-management processes were legitimized by the Brazilian government in the past. For instance, Fishing Accords have been informally developed in the Amazon since the 1980's. The accords were crafted and implemented by local fishers in response to the intensification of commercial fishing in order to control access and increase productivity in local lakes (Castro 2000). In 2002, the Brazilian Environmental Agency (IBAMA) issued legislation (IBAMA Normative Instruction 29/2002) regulating such accords. Fishing Accords shall be made collaboratively by resource-users and government, and shall be monitored and evaluated yearly. Regional Fishing Councils were recognized by the government as legitimate bodies to elaborate such accords.

Another instance, the Forum of Patos Lagoon was set up in 1996; it encompasses 21 institutions which represent fisher organizations, industries, government, NGOs and researchers. The Forum meets on a regular basis to discuss management measures intended to rebuild the capacity of fishery resources in the Patos Lagoon, among other issues. In 1998, a Federal Decree (171/1998) formalized the exhaustively-discussed and agreed upon management measures as a first initiative of this co-management arrangement (Kalikoski et al. 2002). Since then, other regulations have been issued. The Forum of Patos Lagoon as well as other similar fora in place in Brazil acts as consultative body; the government still holds the final decision.

Apart from the on-going co-management processes, such as those of the Forum of Patos Lagoon and the Regional Fishing Councils in the Amazon, provisional, informal co-management agreements have also taken place in Brazil. For example, three regulations were issued in 1981, 1986 and 1993 by the Federal government specifically for the Ibiraquera Lagoon upon request by local fishers (Seixas and Berkes 2003). Finally, some informal co-management processes are carried out on no-take protected areas, such as National Parks and Biological Reserves, although in disagreement with the current Federal legislation (Seixas and Kalikoski, in preparation).

Case Studies: Co-Management Processes in Brazil

In this section we describe five co-management processes involving aquatic resources from five Brazilian states (Figure 7-1): two of them are inland cases and three are coastal marine cases. The first two, at the Amazon varzea (floodplain) and the São Francisco River, emerged as research-action projects aiming to trigger co-management outside protected areas, while the others include co-management initiatives within sustainable-use conservation units.

Participatory Management of Aquatic Wildlife Species at the Amazon Varzea, Santarém, Pará[4]

Turtles and caimans have long been used by Amazonian varzea people and their exploitation has undergone a series of changes since the beginning of Portuguese contact. The appropriation of indigenous labour and knowledge allowed for the establishment of an extractive system for producing turtle oil as well as other products. Intense use—during 300 years of commercial exploitation—led to the decline of these resources. During several decades of the twentieth century, caiman were intensively and illegally hunted for their skins and later for their meat. All these resources continue to play important roles in regional subsistence and trade, but the absence of management measures put their viability at risk. In addition, restrictive legislation and its strict interpretation by local authorities worsened the situation by placing the State and its agents in opposition to the varzea population. This intensified conflicts transformed basic elements of household economy into clandestine activities. Because capturing these species remains illegal, it is challenging to accurately assess their levels of capture.

In the varzea region of Santarém (Pará State) a participatory project[5] engaged communities into mapping the distribution and migratory patterns of caiman and river turtles in community-managed lakes. Communities are interested in developing participatory management systems for local stocks of these species, with the objective of submitting management plans to government authorities for their sustainable use. Workshops, censuses and experimental fishing using local techniques were undertaken in 2006 and 2007 to obtain data on stocks. Consumption was monitored through questionnaires and carapaces and crania were collected in order to obtain precise data on population structure.

Four large turtle nesting beaches are under community-based protection and present great potential for management, including sustainable quotas for commercial egg harvesting and hatchling selling for turtle farmers. Nevertheless, formal management is hindered by the lack of a legal basis for wildlife use in Brazil and by conflicts regarding ownership and benefits sharing. In addition, most locals do not realize that turtle populations are in fact a valuable resource that could be managed collectively and on a legal basis. Moreover, local leaders wonder whether the community is capable of respecting locally-devised rules, since many community members are collecting eggs in protected areas and selling turtles in the winter, when the animals are more vulnerable to being caught in flooded areas.

Improved caiman management was a local demand expressed during our first visits and participatory rural appraisals were undertaken in each community. The arguments presented for management included the frequent caiman attacks on people and domestic animals such as cattle, dogs, ducks and chickens, and the destruction of fishing gear by caimans. To address these concerns, during 2006 and 2007, three workshops were held to train locals on caiman counting, to plan caiman censuses and to monitor and evaluate data. Community groups estimated caimans in the 2006 and 2007 dry season (September to December) and in the 2007 rainy season (January to June). At a fourth workshop, in February 2008, two communities already began discussing preliminary management plans. A fifth meeting is planned to prepare a formal proposal for legal experimental harvesting by these two communities during 2008.

Our project began in 2006 in fourteen communities. Community engagement decreased over time and currently only two communities are still engaged in estimating caiman stocks in order to manage them for sustainability. At least two factors may help explain this trend: (i) the lack of any short-term benefits from management, and (ii) the uncertainty about whether government decision-makers will allow for the implementation of a caiman harvesting program in the area. Both the State and Federal governments have shown interests in caiman management; the Pará State Environmental Agency and the Fisheries and Aquaculture State Agency are both interested on our project—the latter even sent a representative to the last workshop. Two proposals for community-based harvesting quotas for wild caimans were sanctioned by the Brazilian Environmental Agency (IBAMA) for two Sustainable Development Reserves (Mamirauá and Piagaçu-Purus) in the State of Amazonas. This is a clear precedent which favors the future approval of the proposal being crafted with the input of communities in the Santarém varzea.

Fisheries Co-Management at the Upper-Middle São Francisco River, Minas Gerais[6]

The Project "Fish, People and Water" is a participatory project[7] that has been carried out since 2004 and, which aims to facilitate the establishment of a fishery co-management agreement, involving professional artisanal fishermen, among other stakeholders, in the São Francisco River in Minas Gerais State. The idea is to build upon the Fishing Accords, which are a more developed process of fishery co-management initiated in the Amazon region. For this purpose a partnership was established with an Amazonian NGO experienced in co-management processes.

The project has addressed the inefficiency of the current fisheries management model in Minas Gerais State. There, management is centralized in government agencies, such as the Federal Environmental Agency (IBAMA) and the State Forest Institute (IEF), which are both in charge of regulating and supervising fisheries, and the Environmental Military Police, which is in charge of rule enforcement. This centralized management has been ineffective in solving conflicts related to the management and use of fishery resources in the Upper-Middle region of the São Francisco River Basin. These problems include (i) conflicts between local communities and the State arising from distrust between the two parties—some artisanal fishing methods are considered by the State as "predatory" and some local management practices are prohibited; (ii) conflicts between different government institutions, mainly between IBAMA and IEF, resulting in contradictory hard-to-comply-with regulations, which increase the conflicts between the members of the communities and the regulators; and (iii) conflicts among different users of fisheries and water resources, including professional artisanal fishermen, sport fishermen, ranchers, mining industries, hydroelectric companies, and others.

The lack of an agreement on resource management places fish stocks at risk, because (i) it hinders fishers' compliance with fisheries regulations which in turn may result in violent rule enforcement; (ii) it hampers protection of floodplain lakes, which despite being nursery grounds for young fish have been exhaustively exploited and damaged by agriculture; (iii) it allows for degradation of the ecosystem by industrial pollution, threatening certain fish species and possibly jeopardizing the entire fish assemblage (fish kills of surubim *Pseudoplatystoma corruscans* have been detected since December 2004). The risks for fish stocks endanger the sustainability of one of the most important sources of protein and income for local people: artisanal fisheries.

Another management challenge is the lack of information about fish stocks and fishing effort. Periodic participative studies on fish landings are needed for an assessment of the stocks. Fishing effort is also difficult to measure due to a considerable number of "clandestine" fishermen—commercial fishers not registered with fisher organizations ("colônias") and people living along the river who fish periodically for food or leisure.

The participatory project described here involved six municipalities in northern Minas Gerais State and encompassed four fisher organizations that represent about 450 fishing families. Activities such as workshops, courses, and meetings have been carried out in order to promote community empowerment and to develop conditions for governmental decentralization in managing common-pool fisheries resources. As a result, a broader range of individuals from fishing communities are expressing themselves effectively in meetings and participating in constructive discussions; goals and issues of concern are being effectively organized; dialogue with government regulators has been established; relationships with environmental police have been remarkably improved; and, opportunities for women and youth have been developed.

In sum, considerable progress has been made in strengthening fishing communities' organizations and enhancing their voices, mainly through the creation of the "Fisheries Working Group" (*GTPesca*). Since 2005, GTPesca has held seven meetings with broad participation (around 150 attendees, including fishermen, government agents, policemen, researchers, NGO staff members and others interested in fisheries co-management). However, the lack of both the capacity and the ability of government institutions to work in a decentralized and participatory manner, couple with the lack of social, political and economical empowerment of artisanal fishing communities, remain critical issues hindering the implementation of a co-management process in the Upper-Middle São Francisco River region.

The Fauna Reserve for Estuarine Dolphins (*Sotalia guianenis*) at Tibau do Sul, Rio Grande do Norte[8]

International "beach and sun" tourism is rapidly increasing in Brazil. Rio Grande do Norte State (RN) has experienced a 400% increase in this type of tourism over the last decade (Embratur 2007). A parallel increase in demand for leisure activities such as dolphin watching along the coast has taken place. Pipa Bay, at Tibau do Sul municipality (RN), is a famous international tourism hot spot, with an increasing number of hotels and second homes, and increasing tourism boat traffic. From 2002 to 2005, the number of boats offering dolphin-watching services increased fourfold (from two to nine), with boats randomly entering the bay where the dolphins feed and rest. Major changes in dolphins' behavior during the presence of dolphin watching boats were observed over the four-year period (Ferreira et al. 2005). Ferreira and colleagues proposed the delimitation of a coastal reserve for the protection of fauna (i.e., a Fauna Reserve) in this area, to control the number and speed of dolphin-watching boats.

The proposal was supported by local residents, who feared boat traffic accidents, and also by boat owners, who feared increased competition from new boats entering the system. In addition, federal, state and municipal agencies acknowledged this proposal, since it was in accordance with "Projeto Orla"—a Federal Government Project run by a State agency since 2002, which aims to map coastal activities, deal with conflicts generated by tourism expansion, and increase capacity at the municipal level for environmental management by offering training courses to the municipal government staff. Finally, the local government used this proposal to increase Tibau do Sul's chances of being chosen by the Federal Government as one of the 65 main tourism hot spots in Brazil, which would garner increased financial support from federal agencies for local development.

On February 16th, 2006—only six months after academics presented a report on the impact of boat traffic on dolphin behavior—a Municipal Coastal Fauna Reserve (REFAUTS) was legally established, with the blessing and signature of government agencies at Federal, State, and Municipal levels, the Rio Grande do Norte Federal University and many local residents. This type of protected area is part of the Brazilian Sustainable-Use Conservation Units. A Consultative Management Council was elected with 16 seats, including delegates from local government (environment, health and education secretariat), federal and state governments, university, and civil society representatives (including two associations of boats owners, restaurant and hotel owners, fisheries communities, NGOs, beach bar owners, and community organizations). Mandatory rules to control boat traffic were designed and approved by the Consultative Council and later by the Municipal Council. Rule compliance has been monitored by scientists, but not enforced by the government. The Reserve Management Plan is still being crafted involving other stakeholders other than boat owners.

Although clear from the beginning, differences of interest became more acute during the two years since the Reserve creation, and after the initial objectives of each stakeholder had been achieved. Boat owners, while enjoying a market reserve due to limitations on the number of boats, resisted further reduction on their use of the bay. Many conflicts occurred between groups that respected the limitation of boat trips and those that repeatedly ignored it. At the same time, local politicians used promises for the enforcement or relaxation of norms to receive support from both sides on future elections. Federal and State agencies, although sending representatives to meetings, kept their distance for the most part, arguing that this was a local reserve and, therefore, should be managed and enforced by local people. Some of the delegates from civil society (e.g., fishers and beach bar owners) stopped attending meetings after the third one was held, arguing that they had nothing to do with the reserve. Despite all this, by the end of 2007, a relatively fair control of boat traffic was achieved—sufficient to at least minimize the impact of boat traffic on dolphin behavior (Tosi and Ferreira, online first).

One major dispute persists, however. This is a Municipal Fauna Reserve with a consultative management council; hence, all rules, decrees and laws shall be presented by the mayor to the Municipal Council to be approved by the latter. Despite holding 11 meetings during the first two years and the fact that the mayor had—since November 2006—a proposal of law to limit the number of boats in the Reserve and to delineate the routes they should follow, only in December 2007 did a law in this regard obtain approval by the Municipal Council. However, this law was not identical to the one jointly written and approved by the Reserve Management Council. It was revised and three new boats were allowed to enter the system—one of them the boat of the mayor's cousin, demonstrated the corruption of nepotism.

Despite initial difficulties of co-management, we understand that delay in regulating boat traffic was due to a normal maturation period, needed for all delegates to learn how to co-manage a common pool resource. Scientific data show that this reserve is effective for its conservation proposal: to prevent behavioral changes in dolphins due to tourism boat traffic. The relatively high frequency of the Reserve Management Council meetings (11 in 2 years) and the continuous alerts made by scientists of the predicted expulsion of dolphins from the area if the regulation of common-use failed to take place are important factors that fostered the co-management. Besides, it is explicitly stated by stakeholders that this reserve co-management is facilitated by the presence of a research team, which provides credibility of neutral, scientific support. Lack of support (financial, logistical and legal) from local, state and federal government agencies and the absence of a local police presence to punish violators (and free-riders) make rule enforcement difficult. Most importantly, however, the formation of a consultative management council hinders truly co-management. Management Council for Sustainable-Use Conservation Units should be deliberative and not only consultative, in order to minimize the ultimate control of decision-making power by the government. In addition, nepotism seems to be a cultural practice partially acknowledged by local communities in Northeastern Brazil. A direct dialogue between the Reserve Management Council and the Municipal Council should be pursued. Although the creation of any Management Council by definition necessarily facilitates debates and co-ordination of actions, management of protected areas by a consultative management council is not high-level co-management.

The Arraial do Cabo Marine Extractive Reserve, Rio de Janeiro[9]

The idea of creating the Marine Extractive Reserve in Arraial do Cabo (MER-AC) emerged in response to the industrial fishing fleets from large cities that have been invading the Arraial do Cabo sea since the 1980s, fishing in areas used by local artisanal fishers (Pimenta and Hargreaves 1999). Some of these areas were considered important recruitment and nursing sites. Establishing the MER-AC was expected to regulate access by outsiders, while asserting the use-rights of local fishers. The creation process was led by an agent of the local office of the Federal Environmental Agency (IBAMA)[10] with the support of academics. Despite two existing organizations representing fishers in Arraial do Cabo, another organization—the Arraial do Cabo MER organization (AREMAC)—was created to co-manage the MER-AC with IBAMA. From 1997 to 2002, all decisions regarding the MER-AC management were made at AREMAC assemblies, including the crafting of the MER-AC Utilization Plan, which regulated activities inside the MER-AC area. One problem in this scenario was that only traditional people, mainly fishers, had rights to vote at assemblies, while the MER-AC was established in an urban area with multiple uses such as tourism-related activities, industrial and port activities, research, etc. As a result, the Utilization Plan imposed restrictions on many of these activities, especially tourism. In sum, the MER-AC was created to solve a dispute between industrial large fishers and local artisanal fishers; but at the end, it increased certain disputes and even created new ones (Lobão 2000).

In 2002, new legislation (Federal Decree 4340/2002) instituted that any MER shall be managed by a Deliberative Council involving all major stakeholders. Some attempts have been made towards the formation of such a Council. Of particular importance was a workshop organized by IBAMA in June 2005, lasting five days and involving 163 attendees from various stakeholder groups such as fishers (different gear-groups), government (including municipal, state and federal levels), industry workers, port workers, academics, NGOs and other civil-society organizations. They debated the issues of conserving the MER-AC area and of constituting a Deliberative Management Council. Several seminars and talks were given by academics, government agents, and representatives of other stakeholders. On the one hand the workshop was informative for some people, promoted dialogue and made explicit the interests of various groups. On the other, (i) there was low participation of fishers and traditional people, because the call for the event was poorly disseminated; (ii) academics and government agents used jargons that fishers did not understand; and (iii) almost nothing substantial was decided—demonstrating that co-management is a long-term process.

Although it is expected to be more inclusive and participative, the Council has not been crafted due to numerous stakeholder conflicts and a lack of resources and preparedness on the part of many groups, including the government. For instance, the requested change in co-management arrangements has led to a political disengagement of local people (initially represented by AREMAC); on the other hand, the same people and other ones started to self-organize into smaller organizations—sometimes representing no more than 15 individuals—in order to have the right to a seat on the Deliberative Council. The number of fisher organizations has increased four-fold since the Reserve was created—not to mention the number of tourism organizations, which has also increased substantially.

Concerning the government's role in leading the formation of the deliberative council, it seems that serious attempts at formation occur as a result of pressure by higher-level government offices and/or depending on available funding to temporarily mobilize interested parties. Since 2007, however, a university group has been carrying out a project at the MER-AC which aims, among other things, to facilitate the formation of the deliberative council. The outcomes from this project have yet to be investigated.

We have observed so far that changes of key management figures have substantial implications for the management process. Since the MER-AC was created, the management system has had different managers with distinct profiles: IBAMA agent in charge of the MER management has changed four times (a fifth change took place since the end of our fieldwork) and AREMAC's president has changed three times. One problem with this lack of continuity is its effect on the co-management process; for instance, each time an IBAMA agent is changed it takes quite a while for him/her to learn about the local socio-ecological context and to promote continuity of the negotiations among stakeholders. The first and fourth IBAMA agents in charge of the Reserve management were the ones who remained the longest in the position; their profiles were perceived by interviewees in the following ways: the first manager was considered good or excellent by some people (mainly fishers) because he strongly enforced regulations; but he was considered unsatisfactory by others (mainly non-fishers) due to abuse of authority. On the other hand, those who approved the fourth manager (the one who organized the above-mentioned workshop) said he was more flexible, pursued dialogue with all stakeholder groups, had patience for negotiations, and did not inhibit fishers from entering the local office; while those who disapprove him (mainly fishers) complained about the lack of or ineffective rule enforcement and the damage it inflicted on their livelihood

The ultimate outcome of the MER-AC implementation regarding sustainability of fisheries is uncertain. On the one hand, the implementation of the Reserve Utilization Plan and its strong enforcement during the early years seem to have helped conserve the local habitat and increase fish catches. On the other, this relative increase in catch seems to be much more a resource re-allocation from industrial fisheries to small-scale artisanal fisheries (as the former were excluded from the system) than an increase in stocks, especially considering that most species caught in Arraial do Cabo are pelagic migratory fishes.

The Marine Extractive Reserve of Corumbau, Bahia[11]

The Marine Extractive Reserve of Corumbau—MERC—was created in September 2000, on the Southern coast of Bahia State. MERC was the first Marine Extractive Reserve (MER) to encompass coral reefs and reef fisheries (Moura et al. 2007). It is located on the Abrolhos Bank, along part of the coastline of Porto Seguro and Prado municipalities, covering 900 km^2 of marine areas (including beaches and mangroves) used by six communities (not including the villages themselves). The region is considered by the Ministry of the Environment as a priority area for Brazil's coastal and marine biodiversity conservation (Ministério do Meio Ambiente 2002). The main activities are artisanal fishing (by canoes and motorized boats) performed with several traditional (e.g., hook and line, gillnets) and "modern" fishing methods (e.g., diving, shrimp trawling) and tourism. MERC is a formal co-management arrangement that arose out of a history of conflict and community mobilization. In the early 1980's, motorized trawling was introduced in Southern Bahia, followed by a sharp decline in marine resources that was further perceived by the Ponta do Corumbau community. In the mid-1990s, the competition for marine resources (mainly shrimp) shot up; ice and diesel were exchanged for cheap fish and shrimp, leaving locals at a clear disadvantage compared to the better-equipped outsiders. In 1998, the first formal appeal for a MER creation was filed by villagers from Ponta do Corumbau, advised by a judge that frequented the area. Fishers from other two villages (Barra Velha and Caraíva), several of which related to Ponta do Corumbau's community, joined the movement later in that year. During 1998 and 1999 several meetings with resource harvesters (*extrativistas*), NGOs and academics were held and three other southern villages joined the process (Cumuruxatiba, Imbassuaba and Veleiro), despite some historical divergences. The process culminated with the closure of the Corumbau river mouth during a storm in order to exclude outsider from landing in Corumbau. In 2000, the Decree creating the MERC was finally signed.

MERC's Deliberative Council took place in 2002, and has met quarterly ever since. It is composed of 14 community representatives, constituting one more than half of the total seats, and by representatives of the Municipal, State and Federal governments, the legal system, the tourism sector and NGOs. The Council has been slow to receive governmental support, and the official recognition came only in 2006. The chair of the Council is a representative from ICMBio as in all other Extractive Reserves.

Although several fisheries and some tourism management measures have been decided in the scope of the Council, which even issued a Utilization Plan in 2002, one of the most important guiding principles towards the unit's Participatory Management Plan is still unresolved: who are the beneficiaries of the MERC? The SNUC law (Art. 18 § 1) states that the traditional populations of resource harvesters shall retain exclusive use-rights of natural resources, although local communities fail to immediately recognize themselves and others into this definition. Through participatory rural appraisal tools (Chambers 1999), debates about the relationship between identity and territory was deepened within each community. This resulted in six proposals for self-recognition, which were further discussed in a larger seminar, attended by representatives of each locality, when a consensus majority view was reached (Rodrigues et al., in press). Results of the seminar remain to be formally incorporated in MERC's much expected and needed Participatory Management Plan.

The main achievements of MERC's co-management arrangement include increased community empowerment, participatory monitoring of some fisheries resources (intermittent), and exchanges and integration among communities and with partners, accompanied by some positive biological indicators resulting from the unit's zoning scheme (Francini-Filho and Moura 2008). A noteworthy positive outcome is the strong presence of community representatives at Council meetings, whose scope is increasing: discussions are no longer restricted to fisheries and include livelihood and coastal issues. Partnership with NGOs and academics have advanced the co-management process and helped boost capacity building measures. Although MER provides some of the incentives needed to foster sustainable use of the resources system, there is still high heterogeneity of community participation in the Council, unsolved family and group conflicts within communities; unrealistic expectations about the implementation process and the rate of resource recovery (MERC is in its eighth year only); a lack of guidelines both for the implementation of the reserve and for its co-management (strong learning-by-doing process), an intermittent funding; a lack of explicit definitions of the rights and duties of co-management parties (including NGOs); and difficulties in marketing local products. Real state speculation and land tenure issues affecting most communities, as well as the depleted state of several fisheries resource, add to the challenge. Thus, exclusive use rights to the MER area have yet to translate into increased income for local residents, long-term conservation of the region's unique biodiversity, and the sustainability of its fisheries.

Crafting Co-Management Institutions in Brazil

Table 7-1 presents a summary of each case for cross-case comparison. Each case demonstrates a different stage of the co-management process.

Table 7-1. Synthesis of the five cases.

	CASE STUDIES				
	Amazon varzea, Santarém	**Upper-Medium São Francisco river**	**Tibau do Sul**	**Arraial do Cabo**	**Corumbau**
Geographical region	North	Southeast	Northeast	Southeast	Northeast
Water type	Freshwater	Freshwater	Marine	Marine	Marine
Target managment unit	Species (turtles, caimans)	Ecosystem (River Basin)	Species (dolphin)	Ecosystem (coastal area)	Ecosystem (coastal area)
Institutional arrangement	Participatory project towards wildlife species legal harvesting	Participatory project towards fishing accords	Fauna Reserve	Marine Extractive Reserve	Marine Extractive Reserve
Area	2600 km^2	5 municipalities, along about 240 km of the river	555 km^2 and 5.356 km^2 of buffer zone	56.769 km^2	900 km^2
Year started	2006	2004	2006	1997	2000
Phase	Assessment and planning	Implemented w/ consultative forum (GTPesca) [federal and state legislation reviewed and accorded]; no monitoring or management plan yet	Implemented w/ Consultative Council; critical rules approved and monitored; Management Plan is being crafted	Implemented, w/ Utilization Plan; no deliberative council or Management Plan yet	Implemented w/ deliberative council and Utilization Plan; Management Plan is being crafted

	CASE STUDIES				
	Amazon varzea, Santarém	**Upper-Medium São Francisco river**	**Tibau do Sul**	**Arraial do Cabo**	**Corumbau**
History	Initiated by university group and NGO partnership	Initiated by university group/NGO	Initiated by university group	Initiated by gov't agent with support from university group	Initiated by local communities received strong support from NGOs
Main stakeholders	Floodplain communities, NGO and university group	Fishing communities, fisher organizations, state and federal government, farmers, industry/hydro-electric companies	Sightseeing boat owners, municipal government	Fishers (several gear groups), federal and municipal government, tourism-related groups, port and industry	Fishers, NGOs, Tourism-related groups (tourism operators, hotels, real state speculators) federal government (ICMBio)

The Amazon varzea project and the Fauna Reserve at Tibau do Sul started only two years ago; the São Francisco project has been in place for about four years; while the two marine extractive reserves (MER Corumbau and MER Arraial do Cabo) have been in place for a much longer time—eight and eleven years, respectively.

Crafting institutions for commons management is a long-term process. In Brazil, it is still unclear what would be the timeframe for maturation of such complex process. At the Amazon varzea near Santarém, the participatory project is still mobilizing local communities to develop a management plan for aquatic resources and further submit it to government approval; that is, the initiative is in its early stage towards co-management and so far it has been an informal process. The process leading to the establishment of the Fauna Reserve at Tibau do Sul has taken about the same amount of time as the process studied in the Amazon varzea. At least three factors contributed to speed up the creation of a formal co-management agreement in Tibau do Sul: first, the proposed arrangement was a category of the National System of Nature Conservation Units framework with relatively well established guidelines for establishment and implementation; second, capacity for environmental management was previously established at the Municipal level; and third, the existence of Municipal government political will during the seminal steps of the process. In the Amazon varzea, on the other hand, the arrangement being pursued is not supported by any kind of institutional government framework and indeed it is in contradiction to the current legislation on wildlife use and management. Hence, the negotiation process leading to a co-management arrangement between resource users and government is expected to take much longer. The precedent in such type of arrangement in other areas in Brazil may eventually increase the likelihood of government approval for locally-devised management plans for wildlife species.

The São Francisco initiative may be categorized as in a more mature stage towards a co-management arrangement. It has been fostering community empowerment and has been able to mobilize stakeholders to come together into a Fisheries' Working Group, but no Fisheries Accord has yet been crafted. Despite the fact that Fishing Accords are "legitimized" co-management arrangements and there are many of these agreements in place in the Amazon area, the concept is quite new for resource users and government managers along the Upper-Middle São Francisco River, which implies that a lot of learning will be needed for all parties before an arrangement can be crafted.

Co-management processes within the marine extractive reserves of Corumbau (MERC) and Arraial do Cabo (MER-AC) are more advanced, although much still remains to be done. Again, the type of arrangement proposed (a Marine Extractive Reserve) helped to foster co-management[12]. In both cases, a Utilization Plan was crafted regulating access and use of the reserves area and resources. None of the two cases, however, have a Management Plan approved by a Council. In fact, at the MER-AC, the Deliberative Council has not yet been formed, nor has the Management Plan. At the MERC, the Council has been formed since 2002 and legitimized only four years later by the government; and its Management Plan is currently being crafted.

In studying co-management and assessing co-management initiatives, it is important to differentiate crafting management arrangements (e.g., conservation units) from crafting management measures (e.g., management plans). Often, after an arrangement is established, some immediate management measures are implemented in order to prevent further degradation of the resource pool and to reduce user conflicts (e.g., the Utilization Plan of the Extractive Reserves). These measures are hardly ever a consensus among all stakeholders and may indeed generate or aggravate conflicts over benefits; hence, their impacts should be monitored and they should be revised after a period of review. The MER-AC Utilization Plan is a good example of the need to revise management measures to accommodate all stakeholder interests—the plan was initially crafted considering mainly fishers' interests, but it is clear today that management must take into account other users' needs as well. Enforcement levels at the early stages, as well as biological and socio-economic monitoring of the outcomes are critical, but often neglected, components for such adaptation processes.

The adaptive co-management concept explicitly deals with this need to adapt management institutions to both socio-economic (e.g., benefit and sharing issues) and ecological feedbacks (e.g., stock assessments). This concept is embedded in a federal regulation (ICMBio Normative Instruction 1/2007, Art. 16), which asserts that any Participatory Management Plan approved by a Deliberative Council for Extractive Reserves and Sustainable Development Reserves shall be revised within a five-year period. Nevertheless, Extractive Reserves' experiences in Brazil show that a legal framework is a necessary but not sufficient condition to promote co-management processes, in particular, adaptive co-management. According the ICMBio (http://www.icmbio.gov.br/), there are currently about 50 Extractive Reserves in Brazil; but, only one (Chico Mendes Extractive Reserve) has already crafted its Management Plan, and around 20 have had their deliberative council legitimized by the government—a process lasting from two to seven years, depending on the reserve (Cordeiro and Curado 2007).

Advancements and Challenges towards Co-Management Processes in Brazil

In addition to the growing legal and institutional frameworks provided by the government to foster co-management processes, the cases described here highlight several other factors that help to foster such processes (Table 7-2). One major factor is the presence of supportive organizations that often facilitate dialogues between resource-users and government, bring essential funding and trigger resources and/or socio-economic monitoring (ideally within a participatory scheme). Supportive organizations are often academic groups (e.g., the Fauna Reserve, the MERC and the MER-AC in its first years) or NGOs (e.g., the São Francisco project, the MERC and the Amazon Varzea initiative). In fact, in many cases, such supportive organization acts as a trigger for the co-management process due to shared interests regarding conservation of biodiversity, preservation of traditional livelihoods, or local development. In other cases, they come onto the scene after the process has been initiated because they see an opportunity to carry out their particular agenda while contributing to co-management. Also, some NGOs have adjusted their agendas following the evidence that the inclusion of local residents in management is a more efficient way of achieving both conservation and livelihood objectives. The lack of such organizations for many years at the MER-AC may help explain why the Deliberative Council has not been formed yet. In sum, partnership between local communities, government agencies and supportive organizations appear crucial in fostering co-management in Brazil—a factor also ranked in first place in the study by Kalikoski et al. (in press) concerning community-based and co-management initiatives in Brazil. An exception, perhaps, is the "ProVárzea Project"—although a Federal Government-led initiative, it promotes co-management arrangements in the larger Amazonia Varzea realm, much like a supportive organization (Ruffino 2003).

It is interesting to note that many of these factors are in accordance with those identified by Seixas and Davy (2008), while the authors were studying factors that contribute to self-organization processes of successful initiatives managing commons.

Table 7-2. Factors that foster co-management of aquatic resources identified from five Brazilian cases.

	CASE STUDIES*				
Factors contributing to co-management	**STM**	**SFR**	**TS**	**AR**	**CO**
Partnership with supportive organizations (academics, NGOs)	√	√	√	√	√
Improved dialogue between resource users and government	√	√	√	√	√
Participatory planning	√	√	√	√	√
Overlapping interests among stakeholders	√	√	√	√	√
Economic Incentives; foreseen short-term benefits			√	√	√
Meetings on frequent basis for planning and conflict management		√	√		√
Building capacity for environmental management (at community and/or government level)	√	√	√		√
Building local capacity for resource monitoring	√				√
Community empowerment		√			√
Government decentralization		√	√		
Scientific information contributing to understand what is happening and what might happen in the resource system	√		√		√

*STM: Amazon Varzea in Santarém; SFR: São Francisco River; TS: Tibau do Sul; AR: Arraial do Cabo; CO: Corumbau

Table 7-3. Factors that hinder co-management of aquatic resource identified from five Brazilian cases.

	CASE STUDIES*				
Factors hampering co-management	**STM**	**SFR**	**TS**	**AR**	**CO**
Government agents' and resource-users' lack of previous experience in co-management – a lot of learning-by-doing	√	√	√	√	√
Government lack of support and infrastructure for rule enforcement	√	√	√	√	√
User-group conflicts, both within and among communities	√	√	√	√	√
Lack of information on resource stocks and on harvesting effort	√	√		√	√
Socio-economic disempowerment of local communities (at the initial stage)	√	√		√	√
Lack of guidelines both for implementing and co-managing a conservation unit (strong learning-by-doing process)			√	√	√
Cheating on locally devised-rules	√		√	√	
Government lack of ability/preparedness to work in a decentralized manner and deal with user conflicts		√		√	
Government bureaucracy hindering recognition/legitimization of agreements					√
Lack of legal basis for wildlife use	√				
Difficulty in defining who are the beneficiaries					√
Difficulty in estimating total number of users		√			
Use of jargons by academics and government staff during meetings involving resource user				√	
Unreal expectations about the implementation process and changes					√
Lack of explicit definition of rights and duties of co-management parties (including NGOs)					√

*STM: Amazon Varzea in Santarém; SFR: São Francisco River; TS: Tibau do Sul; AR: Arraial do Cabo; CO: Corumbau

The cases described here do not explicitly say much about funding or leadership, except that lack of funding (e.g., to promote meetings, enforce rules, and monitor resource stocks and use) may hinder the process and that depending on the profile of a manager, dialogue among parties may be hindered or fostered, as has happened in the MER-AC. However, all the cases are in place because there were funding opportunities and key people leading each process.Other factors exist that may hinder co-management (based on the experiences of the five cases presented in Table 7-3). One issue that cut across some of these hindering factors is the inexperience of not only local communities and government, but also of supportive organizations, in co-management processes. Such inexperience is demonstrated through the disempowerment of local communities in the initial stage of the process, as well as in the government's lack of ability/preparedness to work in a decentralized manner. On the supportive organization side, we noted that there is a strong learning-by-doing period for those organizations. In the cases of the Fauna Reserve, the MER-AC—during its establishment—and the MERC, some of the supportive organizations did not have previous experience with co-management. In the case of the Amazon Varzea in Santarém, the supportive organization had previous experience with fisheries co-management but not with wildlife co-management. In the São Francisco case, the supportive organizations had no previous experience in fisheries co-management; however, they went through a learning process as they established a partnership with another organization with a high track record in fisheries' co-management. In sum, it seems that all cases would benefit from capacity development at the level of community, government and supportive organization.

Conclusion

This chapter examined five initiatives for co-management of aquatic resources in Brazil. The existence of a legal framework provided by the government for establishing formal co-management arrangements have facilitated the creation of many initiatives (including, the Fauna Reserve in Tibau do Sul, the MER-AC and the MERC). Nevertheless, such framework has not guaranteed that on-going participatory planning, implementation, monitoring, evaluation and adaptation of co-management measures take place. We observed that co-management processes have advanced in large part due to efforts of supportive organizations such as academic groups and NGOs, and have become as important as legal instruments. However, support from such organizations is rarely maintained over the long-term, due to lack of funding. Also, despite the existence of legal framework for co-management, government bureaucracy and lack of preparedness and infrastructure often hinders co-management processes. On the other hand, we also observed a series of hindrances arising from the involvement of resource-users in co-management processes. Much has still to be done in order to build capacity and empower resource users to enter in co-management processes.

Capacity development at the level of community, government, and supportive organization could improve co-management processes in all cases. Capacity development may be broached through vertical transfer of information and knowledge (e.g., training courses, technical assistance, or mentorship) or horizontal exchange of experiences and knowledge (e.g., participatory research and monitoring, workshops, study and exchange tours, guided visits, or networks). Most efforts, however, depend on continued funding; hence, funding agencies should consider the long-term process of institution building and avoid rushing such process by providing funding for limited periods only. Indeed, the Federal Government should craft a specific policy on funding for implementing co-management processes, in addition to providing a framework to start such processes. That is, a policy to nurture co-management, and not just to deliver it. This policy should consider: (1) long time horizons; (2) rewarding initiatives where successful (i.c. accountable, transparent) collaboration with supportive organizations are in place; and (3) providing simple tracking indicators (i.e. reducing bureaucracy by avoiding demanding too much paperwork or reports with which communities fail to comply).

Finally, although most co-management processes begin by focusing on only one resource group—in our cases, aquatic resources—the evolution of each process has shown the need to consider other uses of the area, since most, if not all, of these cases could been seen as complex commons.

Endnotes:

[1]They include joint management, joint administration, participatory management, cooperative management and collaborative management.

[2]In some cases, the participation of civil society in co-management arrangements is not welcomed. For example, in the Quilombo do Frechal Extractive Reserve—an area that was formerly a slave community—there is a strong community council and they do not want to open the Deliberative Council to other stakeholders, since the community has its own cultural rules and agreements (related not only to resource use but also to social behavior) (Alexandre Z. Cordeiro, ICMBio, personal communication).

[3]"Participatory Management Plans", cf. ICMBio's IN 1/2007.

[4]By Juarez Pezzuti and David Gibbs McGrath.

[5]This project is part of the larger Varzea Project of a NGO–the Amazonian Environmental Research Institute (IPAM)–which promotes community-based management of lake fisheries, in place for over 12 years.

[6]By Ana Paula Glinfskoi Thé

[7]The project is led by the Canadian NGO World Fisheries Trust (WFT) and the Brazilian Federal University of São Carlos (UFSCar) and is funded by the Canadian International Development Agency (CIDA).

[8]By Renata G. Ferreira

[9]By Cristiana Simão Seixas and Paula Chamy

[10]In 2007 IBAMA was split in two: IBAMA, responsible for licensing and enforcement of environmental regulations and the Chico Mendes Institute for Biodiversity Conservation (ICMBio), responsible for Conservation Units.

[11]By Rodrigo Leão de Moura; Carolina V. Minte-Vera; Isabela B. Curado; Ronaldo B.Francini-Filho; Helio C. L. Rodrigues, Ronaldo Freitas Oliveira, Guilherme F. Dutra, Diego Corrêa Alves, and Francisco J. B. Souto.

[12]Before SNUC law passed in 2000, Extractive Reserves only had to have a Utilization Plan. Post-2000, the new legislation states that Utilization Plans shall be incorporated into Management Plans (see Cordeiro and Curado 2007).

CHAPTER EIGHT

Social-Ecological Systems Analysis in Coastal and Marine Areas: A Path toward Integration of Interdisciplinary Knowledge

BERNHARD GLAESER, KARL BRUCKMEIER, MARION GLASER AND GESCHE KRAUSE

We discuss the epistemological aspects of social-ecological systems (SES) analysis, particularly in relation to coastal and marine research and the integrated coastal zone management (ICZM), but also in relation to interdisciplinary approaches and frameworks such as sustainability science and sustainable resource management. From this discussion, we propose suggestions on how to develop SES as a framework for interdisciplinary social-ecological research.

We conclude that a knowledge-based strategy to systematically address complex problems can be constructed. A knowledge-based strategy can also study ecological and social processes at different system dimensions and scales, from local to global. SES analysis holds promise if it is developed along inclusive, interdisciplinary lines on an epistemologically balanced and theoretically and methodologically sound basis.

The use of resources in coastal and marine areas includes the social contexts of multiple use, different forms of ownership, and conflicts over the use of resources in water and land areas. Explicitly used knowledge systems include scientific, managerial and local knowledge from various stakeholders.

The development of interdisciplinary methods for coastal and marine research, stimulated through SES, can become an important area in SES analysis. In principle, a large number of methods and tools from the social and natural sciences may be used in interdisciplinary coastal and marine research. However, methods that will facilitate the integration and synthesis of knowledge from different disciplines are, at best, in infancy today. There are strong indications, however, that SES analysis has the potential to upgrade ICZM and incorporate it into an interdisciplinary social-ecological research framework.

ICZM: Historical Background and Current Trajectories

Since the early 1970s, coastal and marine research has quickly adopted the integrated coastal zone management (ICZM) approach. Early ICZM was driven by the practical requirements of resource use and management, and, therefore, it had limited scientific aspirations in conceptual, theoretical and methodological terms. To date, it has become the mainstream approach for coastal management and coastal and marine policies. Because ICZM comprehensively integrates natural and social science research, it also has the potential to become a "pilot concept" for social-ecological systems (SES) analysis.

A brief historical review of ICZM outcomes reveals a series of deficits that are related to the following: who should take the lead within an ICZM process; a lack of appreciation of context-based information; unclear perceptions of participation in the ICZM process and its local relevance; and an insufficient acknowledgement of the complexity of natural, social, economic and cultural factors. These deficits can be used to elaborate the SES framework for coastal and marine systems, and they have important methodological implications.

Although ICZM arrived in Europe only in the mid 1990s, European ICZM policy reflects a rather fast adoption of the approach. By learning from ICZM case studies (supported by 35 EU-ICZM demonstration programmes and their evaluation), the EU formulated and endorsed eight main principles for good ICZM:

1) Are there holistic thematic and geographic perspectives in the process?
2) Is a long-term perspective envisaged?
3) Is an adaptive management approach applied during a gradual process?
4) Is the process locally specific?
5) Does the ICZM approach respect and work with natural processes?
6) Is the process based on participatory planning and management?
7) Does the process support and involve all relevant administrative bodies?
8) Is there a balanced combination of instruments in planning and management?

However, when the EU Commission and Parliament recommended introducing national strategies for ICZM by 2006 (EU ICZM recommendation: 2002/413/EC), several issues surfaced. One open issue in ICZM practice is its interaction with research, specifically how much implementation depends on research and what kind of research is required to support and develop the approach. In addition, experience with ICZM suggests the need to develop new ideas, improved methods and scientific knowledge. Experience also suggests a need for management tools with sustainable coastal management. Because these objectives could not be achieved on its own, the ICZM core concept has been increasingly refined into an interdisciplinary research concept.

Interestingly, current debates in SES analysis resemble the European ICZM discussions during the early 1990s, when case studies in selected EU countries were undertaken (European Commission, 1999). These case studies, however, did not generate much discourse on the methodological and epistemological deficits of ICZM.

The special symposium on SES in Rio de Janeiro at the Society for Human Ecology (SHE) 2007 International Conference also suggested a lack of theoretical depth. We examined case studies of coastal regions around the globe, with the explicit objective of conducting comparative and integrated analyses of coastal transformation processes. The focal scale was the coastal social-ecological system (catchment to coast) at the local and/or regional level. The symposium showed that the SES conceptual framework is increasingly applied in interdisciplinary research on coastal and marine systems, but its theoretical status (particularly in epistemological and methodological terms) has yet to be clarified.

We will now discuss the epistemological aspects of SES analysis, particularly in relation to coastal and marine research and integrated coastal zone management (ICZM), but also to interdisciplinary approaches and frameworks such as sustainability science and sustainable resource management. From this discussion, we develop suggestions on how to develop SES as a framework for interdisciplinary social-ecological research and synthesis.

SES analysis emerged at a time when inter- and transdisciplinary concepts and approaches were quickly spreading in ecology. The broader cognitive stream of "new ecology" (Scoones, 1999) integrates concepts of social science. New ecology results from a criticism of prior ecological research, as paradigmatically formulated by human ecologists since the 1920s and by sustainability scientists since the turn of the millennium (Kates et al. 2000). It focuses on the dynamic interactions between nature and society. Recent decades have seen a substantial understanding of those interactions. This is because work in environmental science that considers the bi-directional relationship between human action and the environment. Additionally, there is a small but growing body of interdisciplinary work in social and development studies that seek to account for environmental influences. In order to shape a better understanding of the rapidly growing interdependence of nature and society, we need to move beyond the current literature.

The growing interdependence of nature and society is a historical *fait accompli,* and conclusions about interdisciplinary and applied ecological research are drawn on this basis. The interdependence of nature and society in past stages of human society is not further discussed. For the elaboration of SES analysis, it is more important to discuss the consequences of sustainability consequences related to current change processes. In this context and as a corollary to criticisms of linearity and environmental determinism (Kates et al. 2000: 2-3) a variety of systems concepts appear, including the following: linked social-ecological systems (Berkes and Folke 1998; Berkes et al. 2003), complex system dynamics (Kay et al. 1999), non-linear feedback and multiple scales of nature-society interaction (Gunderson and Holling 2002), and resilience in social and ecological systems (Scheffer et al. 2001).

The discussion about SES analysis is developed along two trajectories. First, SES analysis is a component of **interdisciplinary** and applied, participatory research. Second, SES analysis is part of an **epistemological** discourse about the limits of scientific knowledge.

We address some questions that arise from the above distinction in the following sections: "Use of SES in Coastal and Marine Research: The Rio Presentations" and "Developing SES: Future Options".

A discussion of the epistemological implications of the SES concept is needed to clarify whether SES analysis will improve our cognition of the real world for the benefit of resource management. The risk of a postmodernist farewell to realist and historical accounts of science is familiar: "Relativism now poses a serious threat to the whole possibility of intersubjective understanding and explanation, not just of society but of nature too. ...with that possibility goes the possibility of rational, democratic, emancipatory transformation of the world" (Lloyd 1989: 451). It is interesting to note that the two key methods used in SES analysis, modelling and scenario-writing, can be used with constructionist and realist epistemologies.

SES as a Conceptual System: Building Interdisciplinary Research Frameworks

The function of SES in coastal and marine research requires further methodological work, as well as a development of the conceptual and epistemological underpinnings. This can be achieved by connecting SES with research fields in which similar conceptual and theoretical advances are required. Some examples include global climate change (Parmesan

2006; Leduc 2007), natural resource management (Acheson 2006), the management of common pool resources including coastal ones (Janssen and Anderies 2007), and research on the quantitative dimensions of resource use in material and energy flow analyses or on the distribution conflicts that surround natural resource use (Fischer-Kowalski 1998; Fischer-Kowalski and Hüttler 1998; Martinez-Alier 1999; Fischer Kowalski and Haberl 2007).

When advances in interdisciplinary research occur along the route to integrating knowledge, it is imperative to synthesise and integrate research and knowledge from the natural and social sciences. One indicator of such integration might be the acceptance of more advanced social science knowledge and acceptance of concepts and theories in interdisciplinary research on sustainable human-nature relations surrounding (coastal) ecosystems. This objective has been formulated as bridging science and society, but it has been only partially achieved.

Noting Odum's call for a "new ecology" that emphasizes both holism and reductionism to address the difficult global situation, SES thinkers also need to incorporate human, purposive and fallible behaviour in their approach to systems analysis. According to Barton and Haslett, "...the combination of the open systems/ socio-ecological model and the dialectic process helps achieve this" (Barton and Haslett 2007: 153). It is not clear today, however, how systems theory or SES analysis will take up questions of human behaviour in different institutional contexts. These questions need more than conceptual modelling, and they require the sharing and combination of different types of knowledge. In recent debates about critical system heuristics (Richardson et al. 2005), such questions have been discussed, but only in terms of some of their normative implications. New transdisciplinary methods for knowledge sharing, such as participatory agent-individual-based modelling (abm/ibm), show promise, but their development for the coastal and marine realm is only just beginning.

In attempts to bridge the disciplinary gap in the analysis of human-nature relations, social science knowledge remains mainly guided by concepts and ideas from the natural sciences. Although systems theory is an interdisciplinary field that bridges various social and natural sciences, this field is also guided by the natural sciences. Resilience, for instance, is an originally ecological notion that appears to imply current "pristine" natural systems functions are to be conserved. Even when coupled with natural and social systems research, the predominance of natural science prevails (Kinzig et al. 2000: 7f), with the following areas recommended for interdisciplinary research: evolution and resilience of coupled social and ecological systems; ecosystem services; coping with uncertainty, complexity and exchange; environmental dimensions of human welfare, health and security; and communicating scientific information. All of these prospective research areas employ ecological perspectives and show no trace of awareness or adoption of theoretically oriented social science concepts. Even if social science theory could only be adopted for specific interdisciplinary social-ecological themes, which is by no means clear, the current practice of ignoring social science theory needs to be reviewed. The historical and political dimensions in managing natural resources need to be emphasised (Pauly 2001).

To support the process of building an interdisciplinary research framework over time, it seems useful to review relevant experience:

- It took a substantial amount of time before the rich field of conflict research was discovered to be important for interdisciplinary ecological research. The Swedish national research program SUCOZOMA made an effort to adopt conflict research (Bruckmeier 2005).
- It also took a substantial amount of time before pragmatic solutions were found to deal with the notoriously multi-semantic concept of power in social-environmental research, and before the mutual penetration of power and knowledge was clarified.
- It is important to prevent interdisciplinary ventures from developing a major (natural or social) science discipline surrounded by instrumentalised "supporting disciplines" whose agendas, methods and potentials are structurally–and often unintentionally–suppressed. Projects must be planned with a balanced distribution between the natural and social sciences in terms of senior positions, number of staff, budgets and planning responsibilities and rights. Ideally, programme administrations and funding should be oriented toward interdisciplinarity. Glaser et al. (2006) show that failure to do so can result in projects dominated by either natural or social science and, thus, a reduced potential for interdisciplinary synergies.
- When one does not discuss epistemological questions in research, it is not clear how the research design and methods may influence research results. An example discussed above was as follows: the predominance of one specific science, its concepts and perceptions, in interdisciplinary research (Kinzig et al. 2000: 7f).

The slowness and difficulty in developing conceptual frameworks for interdisciplinary social-ecological research shows certain parallels with the attempt to link the social and biophysical sciences. Incompatible epistemologies or philosophies of science and power differentials are, as identified by MacMynowski (2007), roadblocks to interdisciplinary research. Incompatible knowledge claims cannot simply be ironed out by setting up new conceptual frameworks or by bridging concepts. They also require strategies to develop interdisciplinary research agendas.

Box 8-1. Frameworks, theories and scientific discourses with affinities to SES analysis

1. Predominantly sociological or social science discourses
- "ecological modernization" (Spaargaren and Mol 1992)
- "reflexive modernization" (Beck et al. 1994)
- actor network theories (Latour 2005)
- political ecology (see Blaikie 1985; Springate-Baginski and Blaikie 2007)
- theories of unequal exchange and ecological distribution conflicts (Martinez-Alier 1999)
- ecofeminist theories pertaining to equity, equal access to resources and knowledge (Warren and Eskal 1997; Mellor 1997)

2. Interdisciplinary theories and discourses
- "coevolutionary development" (Norgaard 1984)
- common pool resource management (Ostrom et al. 1994)
- "societal metabolism" and human "colonization of nature" (Fischer-Kowalski and Haberl 1997)
- "metabolic rift" (Foster 1999; see Clausen and Clark 2005)
- theory of adaptive change (Holling and Gunderson 2002)
- sustainability science (Kates et al. 2000)
- earth system science (Schellnhuber et al. 2005)

Little trace of such strategies to build interdisciplinary and problem-oriented social-ecological research can be found today in coastal and marine research. These strategies are neither in the presentations given in the Rio symposium nor in the broader literature, with the exception of some methodological echoes of knowledge negotiation and integration problems such as the familiar issue of complexity (Presentation Mee 2007, Appendix A). Roux et al. (2006) use different wording to describe similar problems when they conclude that the co-production of knowledge in interdisciplinary research requires collaborative learning, knowledge interfacing, and knowledge sharing between experts and users as opposed to one-way knowledge transfers.

Negotiations over knowledge, concepts and worldviews are thus required. SES analysis stands at the very beginnings of such negotiations. This means that, to date, their necessity has been proclaimed (e.g., Berkes et al. 2003), blueprints for participatory SES analysis have been developed (Walker et al. 2002) and partially participatory SES analyses have been carried out (Krause et al. in preparation). Our suggestions to build SES research areas in a more structured way (by using ideas of Newell et al. 2005) are meant to facilitate further development of this emerging field of action-research oriented studies.

The most difficult and controversial issues in the SES debate surround theory:

> Social-ecological systems theory's point of departure is that in order to manage ecosystems sustainably, the combined functioning of the social-ecological system needs to be understood. This mind map … views human, societal and natural dynamics as part of one integrated system in which social-ecological interconnections are prominent and in which any delineation between social and natural systems is artificial and arbitrary …. The quest of social-ecological systems theory is "... to understand the source and the role of change in systems, particularly the kinds of changes that are transforming, in systems that are adaptive. Economic, ecological and social changes occurring at different speeds and spatial scales are the target of the analysis of adaptive change" …. The social dimension is thus an integral and inseparable, co-evolving part of the social-ecological system. (Glaser 2006: 133, including citation of Holling et al. 2002: 2)

The above quotation discusses a theoretical framework related to SES analysis and links this to epistemological questions and the normative background of these questions, a complex referred to as "mind maps" (Glaser 2006), pre-analytic visions, mental models or paradigmatic worldviews. Conceptual frameworks in ecology and human ecology seem to express an ever increasing complexity along the sequential stations of ecocentric and anthropocentric approaches (see also Purser et al. 1995: 1065ff), along with interdisciplinary and complex systems approaches (Glaser 2006: 135). This sequence of increasing complexity, however, should not be understood as a linear or directed process. The typology also does not clearly distinguish between predominantly axiological (value-based) approaches (ecocentric and anthropocentric) and predominantly problem-oriented approaches (interdisciplinary and complex systems).

As a minimum requirement for the further development of the SES-centred theoretical debate, we suggest that, before classifying and operationalising "theory" to guide research (see Ford 2000, from p. 45), it is important to distinguish the various possible meanings of the term:

- explanatory, interpretive, or synthesising;
- general or specific/limited with regard to the object of analysis;
- conceptual frameworks/models;
- analytical frameworks;
- deductively or inductively formulated; and
- one single theory or a combination of several theories.

Thus, it may be useful to begin with the classification of available social and natural science theories (Szostak 2003). In doing so, the somewhat non-standard debate about the epistemic challenges presented by biology to social theory and practice (Fuller 2002) can become less academic and more useful for research.

For the theoretical and epistemological development of the SES framework, its normative implications also need to be detailed. In research, more often than not, little attention is devoted to the normative background of new conceptual frameworks, theories and paradigms. Problem- or action-oriented research, rather than axiological or epistemological perspectives, is mostly emphasised (Presentation Santoro 2007, Appendix A). Beyond the systems-inspired SES-terminology, synthetic or interdisciplinary theoretical frameworks are rarely developed, which hampers further integration of knowledge. In the following two sections, we discuss the significance of SES for coastal and marine research and some procedures that may help integrate knowledge.

The Significance of SES for Coastal and Marine Research

The input paper for the SES Symposium in Rio demonstrates a conceptual framework (Presentation Glaeser and Glaser 2007, see Box 8-3 and Appendix A). The Symposium provided an opportunity to discuss strengths and weaknesses of the SES framework. However, continued reflection is important.

Box 8-2. Systematising Theory Types for SES Development

Different features of research are covered by the term "theory". In a preliminary classification, we distinguish between three types of theory and illustrate these with examples from human ecology and the social sciences. In a further step, not attempted here, biological and other natural science theories would need to be integrated on an equal footing.

(1) **Conceptual frameworks** (with little or no analytical or explanatory capacities) to systematically analyse a knowledge field or to systematise the available knowledge
- guiding concepts and heuristics that help to seek knowledge, guide description and interpretation, and find explanation (e.g., the human-ecology triangle)

- loose conceptual frameworks, typologies or conceptual models that guide the processes of analysis and synthesis (in human ecology, e.g., Parks "pyramid" [Parks 1936], Stewards "culture core" [Steward 1968], Duncans' POET-framework [Duncan 1953, 1973])
- systematic conceptual frameworks as steps toward building encompassing theories (in human ecology, e.g., Hawley 1968; Young 1974)

(2) **Substantive and explanatory disciplinary theories** (analytical, interpretive or explanatory)
- analytical frameworks/theories that allow for a systematic analysis of the whole system studied and its parts, such as sociological theories of social action and social systems (e.g., Parsons 1951) and Darwin's evolutionary theory in biology
- grounded theories: theories or explanations grounded in or derived from empirical knowledge or data by way of induction, often specific and thematically limited, not general theories of man or society (e.g., the social-psychological theories of human behavior by Glaser 2002, Strauss 1997)

(3) **Synthetic and interdisciplinary theories** can be developed as variants of (1) or (2), to allow for a systematic analysis of systems and their parts and for an integration/synthesis of the range of knowledge that can contribute to the understanding of systems such as society or nature such as in political economy and political ecology (with knowledge from different disciplines; logical and historical analysis of phenomena; and conceptualising development of social and ecosystems)

The ideal process of building theory involves "progressive synthesis" (Ford 2000), in which loose conceptual frameworks are advanced to explanatory theories that can be synthesised into interdisciplinary ones (Case 1 in this box; see also section "Conceptual Templates"). However, the latter are difficult to elaborate, and progress is slow. Recent post-modernist epistemological discourses have shown reduced aspirations for theory construction. Whether SES follows the ideal process is thus not self-evident.

Components of the SES Conceptual Frame: The Rio Input Paper

The guiding concepts in Box 8-3 include important components of the SES framework which, although of differing quality, are nonetheless part of one overarching concept.

- "Social-ecological system" is a theoretical conceptual framework. The identification of the properties of SES is complex and adaptive systems, and there is analytical bias created by the choice of ecosystems as the basic unit of departure for the definition of marine and coastal SES. This choice implies that the components of the social systems with which coastal and marine ecosystems interact remain incompletely covered.
- Resilience has quickly developed into an interdisciplinary concept that was transferred from ecological to social systems.
- Vulnerability seems to be a second order concept, not requiring theoretical elaboration. As an evaluative concept, however, it requires criteria and indicators.
- Adaptive capacity and transformability conceptualise social and ecological systems change. An in-depth theoretical discussion of the change concepts used by the social and natural sciences – such as social change, development, adaptation, accommodation, evolution and transformation – is required. Some of these debates took place in sociology and human ecology over a longer period of time. The more recent debate about path dependency and path transformation through development or change (Djelic and Quack 2007) can be viewed as part of that.
- Emergence is a complex term originating in holistic systems theory. Its operational use – what are the "emergent properties" in a functioning social-ecological system? – seems to be partially unresolved.

Box 8-3. The Conceptual Framework.

In order to further the integrated social-ecological analysis of coastal systems, the following concepts were employed, criticised and/or refined by all presenters in the Rio de Janeiro SES session:
A *social-ecological system* consists of a bio-geo-physical unit and its associated social actors and institutions. Social-ecological systems are complex, adaptive and delimited by spatial or functional boundaries surrounding particular ecosystems and their problem context.
Resilience is the capacity to handle whatever the future brings without the system being altered in undesirable ways. Resilience is necessary for a sustainable future and lies in self-reinforcing dynamics that prevent shifts into undesirable directions.

Vulnerability refers to the degree to which a system is unable to avoid undesirable consequences of change and/or transformations, such as resource degradation or climate change.
Adaptive capacity and *transformability* refers to the system's ability to change. The former, where the objective is to maintain existing systems, and the latter, where existing systems are found undesirable or characterised by emergent systemic failure, implies change.
Emergence is generated through the capacity of systems to self-organise without external direction. The open question of whether emergent phenomena can be derived at lower system levels needs to be distinguished from their exclusive occurrence at higher system levels.
Source: Glaeser and Glaser 2007 (Appendix A)

Use of SES in Coastal and Marine Research: the Rio Presentations

Participants of the symposium made only limited use of the invitation to critically reflect or refine the concepts, definitions and assumptions in Box 8-3. This may be an indicator of how difficult it is to trace SES concepts in a systematic, interdisciplinary fashion in both the social and ecological side of an integrated systems framework. The presentations in Rio revealed that SES analysis has not yet provided a coherent framework for international interdisciplinary and transdisciplinary coastal and marine research. This was the reason for establishing an opportunity in Rio de Janeiro that will help, in the long run, fill the gaps by connecting coastal researchers from all over the world. The terminology proposed by the session organisers should be taken as an indicator of the desire to create such a common framework rather than of its established existence. The Rio presentations made it clear that research practice was driven by issue-, discipline-, institution- or country-specific needs. Equally evident, however, was the demand for a common language to communicate across the borders of disciplines and specialised research areas. The papers presented in Rio allowed for a discussion of specific issues surrounding SES, which are as follows:
(a) innovative research methods for SES analysis;
(b) potentials and pitfalls in SES framework application; and
(c) unresolved issues in applying the framework.

(a) Innovative research methods for SES analysis. The development of interdisciplinary methods for coastal and marine research can be stimulated through SES analysis. In principle, a large number of methods and tools from the social and natural sciences may be used in interdisciplinary coastal

and marine research (see, for example the North Sea "Coastal Futures" project: http://coastal-futures.org/). In practice, the joint use of research tools or approaches by different disciplines for problems that require interdisciplinary research is still limited. Methods that can be applied in contexts with an "interdisciplinary inclination" exist. One example is network analysis (Watts 2004), which is discussed in Orbach's presentation along with the concept of "total ecology" (Presentation Orbach 2007, Appendix A). More complex methods that evoke high expectations in coastal research, such as predictive modelling in the ELME-project (Presentation Mee 2007, Appendix A), are at a less advanced state, with few new components and a need for further elaboration. The complexity of system processes generated not only motivations for modelling efforts but also served as barriers to the further development of these efforts. The limited progress of predictive models may not be related to the complexity of processes in social-ecological systems. The most important difficulty seems to be the insufficient recognition of the epistemic problems of prediction as well as oversimplified input descriptions of the social systems and processes involved. Very little knowledge from social-ecological research, and from the societal contexts in which lifestyles are embedded, is used in many of the social-ecological models constructed to date. We believe that it is a dangerously deceptive simplification to assume that, in environmental and resource management, all that counts is how much of what resource is consumed. We believe that a more refined state of description is necessary.

Other projects, such as the Chinese presentations with the guiding concepts of eco-province (Presentation Wang 2007 and Xue 2007, Appendix A) and adaptive management of coastal wetlands (Presentation Huang 2007, Appendix A), or the co-management approach proposed for the Indonesian Spermonde Islands (Presentation Yanurita et al. 2007, Appendix A), work with aims, concepts and assumptions that does not necessarily require an SES framework. It seems that, in SES analysis, the use of available methods is currently preferred to method development. The qualitative methods developed recently (see Bennett and Elman's 2006) do not seem to have had an impact on current research. Transdisciplinary agent- and individual-based modelling for SES analysis is in its initial stages (Castella et al. 2005). Our conclusion is that SES epistemology (especially the guiding concepts of complexity, uncertainty, and co-evolution) can stimulate methodological innovations. But we are at the beginning of this road; explicit SES research methods need to be developed (Presentation Bruckmeier 2007, Appendix A).

(b) Potentials and pitfalls in SES framework application. Similarities and differences in applying the SES framework do not only refer to the definition of terms. It is also important to have a common understanding of the entire framework: What are its core concepts and why? Are additional concepts required? Single concepts of the framework may be understood differently. For instance, in contrast to much ongoing SES work (e.g., Anderies et al. 2006, Resilience Alliance homepage-http://www.resalliance.org/1.php), one of the Brazilian papers (Presentation Blandtt 2007, Appendix A) argued that the resilience concept is "in a process of construction, requiring further academic efforts to reach a more complex theoretical definition with regard to social system analysis". Such divergent understandings underline the need to systematically trace the concepts in both components of linked ecological and social systems. This shows that the concept transfer needed between social and natural science to build an interdisciplinary framework is not yet completed. In another presentation (Presentation Das 2007, Appendix A), resilience was not treated as a theoretical topic but was operationalised to calculate cyclone damage functions. In this case, resilience was understood as a concept for practitioners to manage resource use.

Knowledge management was discussed in relation to fishery management, knowledge sharing, local knowledge and gendered knowledge (Presentation Maneschy 2007, Appendix A). Both themes of local and gendered knowledge are essential for SES research. The problems of knowledge sharing (instead of knowledge transfer) are not traced to the conflicts, in which claims about knowledge clash and result in power struggles. Only few presentations discussed the local knowledge theme, and none of these did so in depth, which may have been due to the fact that the theme was assigned to a different session. Still, there are difficulties in dealing with local, non-scientific knowledge in scientific research, excluding perhaps the simple case when local knowledge is collected in interview-based studies that adhere to scientific standards and routines. When it comes to the methodologically adequate treatment of local knowledge, however, the research is lacking despite the recent popularity of local knowledge in ecological research (e.g., Berkes and Folke 2002; Davis and Wagner 2003).

One of the more innovative undertakings that contributed to introducing the SES concept in coastal research is the MADAM[1] programme. Glaser (Presentation Glaser 2007, Appendix A) provided an overview of the development of the MADAM programme during its ten-year period of operation, in which an interdisciplinary research framework was developed around a spatially defined social-ecological system. In this case, the social-ecological system was a mangrove ecosystem on a coastal peninsula and its associated socio-economic impact area, i.e., the region populated by the direct human users of ecosystem resources (Krause et al. 2001; Glaser and Krause, in preparation; Krause and Glaser, in preparation). In their presentations, both Krause and Glaser (Presentation Krause 2007, Appendix A) showed, methodologically, the participatory development of desirable futures with local system users (Fontalvo et al. 2007). Integrative scenario development and problem-focused social-ecological analyses, such as on the sustainability of a major fishery and keystone ecosystem species (Glaser and Diele 2004), acted as important milestones for MADAM SES analysis.

As the most salient quality of SES analysis to date, the majority of presentations feature a new terminology to stimulate changes in thinking and perspectives about concepts and methods. The adoption of SES shows a new conceptual framework *in statu nascendi*, i.e., at its very beginning. Some questions seem to be answered in a pre-determined way. What can we know about complex systems? Echoing the postmodernist epistemic discourse about situated and incompatible knowledge, small narratives, subjectivity and multiple perspectives, this question is often responded to with reduced aspirations for research and the communication of its results. With the guiding concept of "adaptive management", a similar tendency is now in evidence in ecological research, a subject area that is otherwise less receptive to that what is seen as an epistemic, theoretical debate in the social sciences.

(c) *Unresolved issues in applying the framework.* The question of whether SES analysis means an epistemological paradigm shift in the study of the relationship between humans and nature is still under debate. SES analysis does not only imply changes in research and methodology because of complexity and uncertainty, but it also challenges pre-analytic visions, paradigms, worldviews or mind maps (Glaser 2006) as an explicit point of departure for integrated analyses.

At this stage, SES analysis appears dominated by the use of new concepts. These concepts (i.e., social-ecological resilience, transformability, emergence, adaptability) need to be interpreted and operationalised in particular SES contexts. New results conveyed through the use of new methods are not yet visible. Instead, the disciplinary traditions, the usual scientific points of departure, are more visible. A common point in the presentations, which echoes the principal message of the SES-framework, is that of the complexity of linked natural and human systems (Presentation Ramesh 2007, Appendix A). Is this yet another call for complexity analysis and management that generates little more than a lowering of expectations from research?

Finally, the question about innovation regarding SES analysis can be discussed in terms of unrealised possibilities and future options, and as a promise for research still to come. Promising issues that were not discussed in the symposium or in the more general SES discussions include the following: SES as a conceptual template for interdisciplinary and transdisciplinary research; SES as a specific theoretical approach; SES for knowledge synthesis; and SES as a conceptual framework for research and for resource management.

Developing SES: Future Options

The weak ties across research fields in which SES analysis is used demonstrate the future of the field. Deficits can be turned into new developments, which is a dialectical process. Thus, the Symposium organisers called for comparative and issue-related case studies to connect SES with research, to draw conclusions, to generalise, and to synthesise knowledge. In this sense,

(1) Knowledge management deficits will require us:

- to combine and synthesise different (scientific and non-scientific) types of knowledge;
- to learn how to live with change, risks, and uncertainty (not only how to investigate uncertainty scientifically); and
- to share knowledge instead of simply transferring it; to cooperate with stakeholders and their conflicting interests, at different levels and in complex situations characterised by significant knowledge deficits.

(2) Methodological deficits will require us:

- to use methods to integrate and synthesise empirical knowledge across disciplines and knowledge cultures;
- to design indicator systems to measure progress in sustainable development and resource management;
- to combine research methods derived from different disciplines; and
- to develop explicitly transdisciplinary methods that focus on social-ecological interactions and that empower the resource users in SES to work toward more sustainable directions.

(3) Conceptual and theoretical deficits require us:

- to integrate ICZM and other interdisciplinary approaches into the analysis of sustainable resource management with the overarching

SES framework, interpreting coastal systems as paradigmatic examples for SES;
- to adopt systems thinking and system concepts by creating a theory of human/nature-interaction; and
- to address the explanatory deficits of key concepts in SES and to look for corresponding theories that may be linked with SES.

(4) Epistemological deficits require rules and criteria
- for dealing with complexity and uncertainty in terms of concepts and methods (a main motive for SES);
- for generating knowledge and for integrating it in interdisciplinary studies that use SES concepts (transdisciplinarity, sustainability science, resilience science);
- for defining baselines and in explaining the desired state for whom and when; and
- for including norms and values into the analysis, i.e., for understanding the normative background of knowledge and decision making.

Elements of Interdisciplinary Knowledge Integration in SES Analysis

Promising issues that have not been discussed in the symposium or in the more general SES discussion elsewhere are as follows: SES as a conceptual template for interdisciplinary and transdisciplinary research, as a non-disciplinary theory approach; SES as a "bridge" for knowledge synthesis; a conceptual framework for research and for resource management. This final outlook can only hint what could be developed further.

Conceptual Templates

Interdisciplinary research does not work without a common language and a shared problem perception by the participating research communities. To develop the SES conceptual framework toward a language for integrative human-environment research, Newell et al. (2005) suggest a process model that includes the following steps in a first attempt to systematise knowledge integration:

Step 1. **Field of Enquiry.** Specifying the field of enquiry for interdisciplinary social-ecological research includes the description of "basic dynamical concepts, conceptual structures, and models of causality that various individuals and groups use to explain how the world works, how it responds to human pressures, and how to best manage human-environmental systems in a sustainable manner" (Newell et al. 2005: 301).

Step 2. **Discussion and Knowledge Sharing.** If the knowledge to be integrated "consists of disparate models of causality then the *integration* process cannot be simply a matter of building a 'shared language'" (MacMynowski 2007: 302). Further efforts to support mutual comprehension are required. It seems that the SES-debate has arrived at this point, and awareness is needed about how to avoid superficial approaches that facilitate communication but fail to support integration (Newell et al. 2005). In coherence with their ideas, this may include:

- Discussing thoroughly the conceptual framework shared by scientists representing different disciplines, which was the intention of the Rio de Janeiro symposium;
- Adopting strategies of knowledge sharing and integration as identified by MacMynowski, including conflict mitigation, negotiation, mutual tolerance; and
- Adopting "integrative methods that avoid arcane techniques and terms and that depend, as far as possible, on the kinds of everyday reasoning and language used by specialist and non-specialist alike" (MacMynowski 2007: 302).

Step 3. **Terminological Template.** The third step is to proceed with SES development into a truly integrative framework by creating a "conceptual template" (Newell et al. 2005) of agreed terminology. This template covers significant human-environment problems, and an associated list of potential nexus concepts for each of these (MacMynowski 2007: 302f). Newell's template can be used as a checklist to develop SES into an integrative concept.

Step 4. **Methods for Synthesis of Qualitative Research.** Step four is to build up a repertoire of methods for "interpretive synthesis" of qualitative research (Weed 2005). This step completes Newell's suggestions with the requirement of integration and synthesis methods for interdisciplinary research.

Step 5. **Progressive Synthesis.** A proposal for knowledge synthesis as a systematic component for ecological research is described by the method of "progressive synthesis" (Ford 2000: 467ff). This method can stimulate further SES development and direct it toward theory construction. The aim is to generate, from research in specific ecosystems, a theory to include causal explanations with four properties. The four properties are as follows: causal processes describe systems functions, they are consistent, they emerge through generalisation from similar events, and they create predictable responses. The criteria for coherence assessment include the following: testing propositions with data, proving consistency of concept definitions and explanations, limiting the number of propositions and applying explanations to broad knowledge areas.

Forms of Negotiating and Bridging Knowledge

The development of SES analysis in the knowledge bridging process has a key function in building a theoretically substantial knowledge framework for interdisciplinary human-nature research. There are several components from which to start:

(1) **"Bridge concepts"**. Sustainable development, to give an example, has served as a discursive platform for some time. It did not homogenise science and policy discourses, but allowed for a plurality of worldviews under a common umbrella. Whether or not it will continue to be used is a matter of convenience.

(2) **Joint learning and knowledge sharing**. These are ongoing processes between scientists, conservation and development practitioners, administrators and resource users. Knowledge integration by interaction has been called for in the transdisciplinarity and sustainability science discourses. Good examples probably exist in conservation and development practice, but they are underdocumented in the scientific literature.

(3) **Theoretical nodes**. Concepts such as co-evolution, adaptive management, nested systems, transformation (e.g., Holling et al. 2002; Norgaard 1984; Ostrom et al. 1994) are core components of an interdisciplinary SES analysis, serving as theoretical nodes in the discourse. They need to be clearly distinguished from discursive bridge platforms.

If we theoretically codify coastal and marine science as a field of research along the lines suggested above, we arrive at a "knowledge ecology" or an ecological mode of knowledge use. Complexity is then a question of knowledge management. Knowledge is more than its production, dissemination, application. Knowledge is developed in social practices, including scientific ones. The concept of "total ecology" (Presentation Orbach 2007, Appendix A) maps the linkages between biophysical, human, institutional, and ecological subsystems. Their interaction is an example of such an "ecological" use of knowledge.

Conclusions

The sources, concepts and methods for reconstructing the interactive behaviour of human and societal actors and natural elements in social-ecological systems require increased attention to knowledge use and to the normative reasoning behind it. Problems and deficits in interdisciplinary social-ecological research (see Section "Developing SES: Future Options") may be addressed with knowledge bridging methods and processes (use, integration, synthesis) that follow ecological–and social–complexities. A knowledge-based strategy to address complexity in a more systematic fashion can be constructed, along with the sequence of studying ecological and social processes at different system dimensions and scales. If developed along inclusive, interdisciplinary lines on an epistemologically balanced and theoretically and methodologically viable basis, SES analysis holds promise even if its conceptual framework is not fully suited for research applications. For our particular case of coastal and marine systems, the parallels and deficits of ICZM research mean that SES analysts can begin fast-track learning–rather than repeating mistakes already made–and thus may make important advances.

Resource use dimensions in coastal and marine areas include the social contexts of multiple use, different forms of ownership, and conflicts over water and land areas and the use of their resources. Conflicts arise with social complexity in the structuring of resource use via, for instance, access, conditions including property rights, or power relations. More systematic analysis of knowledge on the social structuring of resource use and management is required.

Explicitly used knowledge systems include scientific, managerial and local knowledge from various stakeholders. They pertain to resource management and research and may be formally defined or informally practised. Furthermore, there may be latent knowledge and rule systems for resource use and management (including normative "deep structures" of stakeholder knowledge) to be taken into account in SES analysis.

Stimulated through SES analysis, interdisciplinary methods for coastal and marine research can be further developed in several directions. In principle, the social and natural sciences provide many methods and tools. However, the integration of knowledge from different disciplines and from sources beyond the disciplines that are often closer to the SES in question is in its infancy today. We have shown here that SES analysis has the potential to upgrade ICZM and incorporate it into an interdisciplinary social-ecological research framework. However, this work has only just begun.

Endnote:

[1]Mangrove Dynamics and Management Programme (1995-2005) a Brazilian-Germany interdisciplinary research cooperation with its study region on the north coast of Brazil. For details see Saint-Paul and Schneider, forthcoming).

Section III

Integrating Human Ecology

CHAPTER NINE

Human Ecology and Health

FERNANDO DIAS DE AVILA-PIRES

Human and natural sciences have much to offer to, and to profit from, general ecological theory. Human ecology is transdisciplinary in essence, but so far, no proper methodology is available for the complex analyses it requires. The prognostic is that no single method will be able to satisfy the needs or to attend to, the requirements, particularities, and differences in outlook of the distinct fields of knowledge involved in human-ecological research. Human ecologists must learn the basic principles, specific methods and special techniques of both natural and human sciences, striving to acquire a global understanding, rather than a compartmentalized knowledge of complex problems. The future of the profession lies in the preparation of technical and political advisors in matters related to human impacts upon the environment. Those professionals must be able to provide global overviews, identify important factors, and suggest appropriate courses of action for eliminating or minimizing ecological problems.

Human ecology differs in many aspects from plant-and animal ecology. Complex and diversified social organization and culture in man are powerful ecological factors: they pervade all human actions and decisions. Primitive, rural, urban, and marginal populations show different degrees and types of relationships with the environment, and their analysis must bear this important aspect in mind. They represent distinct ecological taxa or categories. Ecological relationships of man in distinct ecological environments and conditions are here analyzed.

As to Man's impact upon nature and natural resources, a novel concept of responsibility must be reached, where subpopulations, at whatever level, must be responsible for the economical use and preservation of resources at their sources, wherever they come from: in the immediate vicinity, as in primitive geosystems, or in distant systems, when imported by technologically advanced urban communities.

Introduction

Human ecology is the most humanistic among the sciences. In its broadest definition it embraces the totality of human knowledge, as it aims to unravel the relationships of man with the biosphere. Similar claims might be made by geography–the study of our spaceship with its passengers, and by anthropology–the study of man. A more manageable and circumscribed definition limits its objective to **the study of man's relationships its physical, social and cultural adaptations to and its impact upon, the environment.** Throughout this text I will use **nature** for natural productions, and **resources** or **natural resources** to designate exploitable products. **Environment** is employed in its wide sense of the sum of all factors, biotic, abiotic, and social, which affect an organism, and is affected by it.

The scope of human ecology as here defined is vast enough and truly systemic and transdisciplinary in essence, as we shall see.

The human species has achieved a progressive independence from environmental constrains through cultural evolution. Limiting factors, as the seasonally of primary production and the availability of food, water, temperature, light, predators and parasites, all have been made permanently accessible, suitably modulated, or successfully controlled.

Leaving its original foyer in the African continent, *Homo sapiens* travelled across land masses and oceans to colonize the Earth. Man-the-hunter became farmer, built urban settlements, and succeeded in establishing new types of relationships within the biosphere (IUCN 1964).

Taxonomic and Ecological Referentials

The classification of organisms in Linnaeus' hierarchical system in the 18th Century–the Century of the Systems–was intended to be an inventory of the types or kinds of organisms created by God: a sort of passenger manifest of Noah's Arch, plus the fishes and vermin not allowed on board.

The allocation of organisms to slots or pigeonhole in a hierarchical system is intended to show kinship or phylogenetic relationships, i.e., common origin or descent. It strives to reveal phylogenetic relationships, but not ecological affinity. To achieve this goal we must identify the **niche** or the role of the organism in its community. The basic unity in ecology is the **biotic community** plus its **physical environment,** not the **species**. So, in the study of human ecology, the basic units would be the biotic communities where man is present.

The concept of species is based on the interbreeding potential of bi-sexual organisms in a natural population. All ontogenetic stages of organisms is taken into consideration. In practice, certain stages of development furnish taxonomists with better diagnostic characters better than others, but no one–not even strict typologists as Linnaeus–would allocate different stages of development or sexes of an organism, in distinct species.

During the historical period of Iberian epic oceanic explorations that lasted from the 15th to the 18th century, new land masses were discovered, teeming with unknown plants, animals, and strange men. The New World had no place in the Scriptures, and there was no explanation for the presence of man in isolated continents, cut off from Mount Ararat, where Noah's Arch was supposed to have come aground. Answers to this riddle had to be reached in time, by reasoning not by consultation to biblical writings, which gave rise in the 17th century to curious theories suggesting a reconfiguration of the continents and made reference to the lost tribes of Israel (Abreu 1930; Vásquez 1948). In the 15th century, Pope Paulus III decreed, in the encyclical *Veritas Ipsa*, that natives from the New World were human beings, and belonged to the species that, two hundred years later would be officially named *Homo sapiens* by Linnaeus (with six *varietas*). Differences in skin colour, the shape of the head and the contour of facial features were then explained in terms of lamarckian climatic influence.

Much of the discussions concerning the relationships of human ecology with general ecology arise either from an attempt to smooth the differences, or from a failure to distinguish ecological from taxonomic relationships. Rappaport (1990) tried his best to apply the ecosystem concept to anthropology, at the same time suggesting the adoption of a "regional system concept", in human ecology. Regional systems would be composed of "interactions among ecologically similar populations, that is, distinct populations of the same species occupying similar or equivalent ecological niches". In his argumentation there are two important partial misconceptions. One arises from his definition, when he considers ecosystems as "essentially, systems of matter and energy exchanges among unlike species". The second concerns local populations, and transhumance. In ecology, individuals of different sex and age may integrate distinct biomes. Also the geographical and ecological distribution of a species is, in most instances, discontinuous, fragmented in a checkerboard of local, mendelian or interbreeding populations, known as demes (Gilmour and Gregor 1939; Huxley 1942). Individuals move from one sub population to another as the young abandon their parents, and adult transhumants search for other territories to defend, and new home ranges to satisfy their trophic and reproductive needs maintaining a genetic flow.

The issues involving taxonomic concepts and the ecologic referential does not stop here. In many groups of animals, male*s* and females of the same species differ significantly in morphology, habits, and habitat. They may live in distinct habitats and fill different ecological niches. The case of organisms that go through radical metamorphosis during their ontogenetic development is enlightening. Amphibian tadpoles are mostly aquatic, gill-breathing animals; mosquito larvae are also fresh-water and detritivorous. As De Beer (1954) pointed out, "It is a corollary of such adaptations that the better the larva is suited for its mode of life, the greater will be the difference between it and its adult." Coupled with the ability to reproduce in the larval stage, this is indeed a pathway for divergence and evolution by a process called paedomorphosis, as shown in prose by Hardy (1954), De Beer (1954), Gould (1977), McNamara (1997), Amundsen (2005), and in verse by Garstang (1951). Adult frogs breathe through lungs and live on dry land. Adult mosquitoes breathe through tracheae; the females are blood-sucking while the males feed on plant juices. Many marine invertebrates have planktonic and free-swimming larvae, while adults are sessile and benthonic. In these instances, young and adults of the same species live in different biomes and are part of very distinct biotic communities (Berrill 1955; Jägersten 1972). Selective pressures tend to force them to adapt more and more to their respective habitats, and diverge in shape and habits.

All human beings belong to the same taxonomic species, but distinct human populations impact differently upon the environment and should be considered distinct **ecological** taxa (Avila-Pires et al. 2000).

By the end of the 19th century the old theological notion of the balance of nature was replaced by the idea of the dynamic relationships of organisms with environmental factors. Creationism postulated that species were shaped to fit natural slots in a planned system. Adaptations showed the divine plan, and variations were considered to be providentially provided for minute adjustments. All nature lived in the permanent state of a climax. The idea of an evolutionary process guided by the hazards of natural selection of arbitrary and random variations, while homeostasis emerged from chaos, changed our whole view of nature and led eventually to the new science of ecology.

As Charles Darwin implied (1859), and David Wallace (1984) aptly recognized, "...humanity is not **destined** for anything. Evolution has always been open to new possibilities, which is why it has been so chaotic and devious".

Ecosystem theory is based upon the idea of **nutrient recycling and energy transfer in natural communities**.

Human ecology finds justification as a discipline of its own by the fact that relationships of man with the physical and biotic environment are mediated by culture. Culture is not necessarily adaptive, it may even hinder survival, but it permeates all relationships through individual and social behaviour. As it was perfectly summarized by Rappaport (1990), here "We are concerned with a species that lives in terms of meanings in a universe devoid of meaning but subject to physical law".

While accepting that there is a recognizable field called **human ecology**, there is no such a thing as a **human ecosystem**. The foundation of community hierarchy is the trophic pyramid not the phylogenetic relationships of organisms. Closely related species actually compete, because they use the same resources, while distinct species of different clades may substitute one another and occupy the same niche, as ecological equivalents or vicariants, although they share no common ancestral.

It is usual to define figuratively in textbooks the ecological niche of an organism as the role or profession of an individual. Along its evolutionary history, man acquired the power to be a dominant in its communities, and recently, a powerful agent of change of the biosphere.

Several technological revolutions were needed before the first human settlements were established. The most important overall acquisition along its evolutionary path was the capacity to express, transmit and store acquired experience and culture. Several anthropologists contend that population growth in human populations act as a drive towards technological change. If this is true, this is a very distinctive trait of human culture, history, and evolutionary process, not shared by other animals.

Articulated language was the first key to progress, leading to cumulative knowledge and diffusion of learned experience. This unique characteristic was responsible for a novel path in human ecology and evolution, distinct from genetic variation, mutation, recombination association, and selection. It led to the inheritance of accumulated acquired knowledge, experience, and wisdom. Progressive increase in the rate of diffusion accelerated the tempo of the eminently human cultural evolution. After language, abstract thinking played an important role in human evolution.

Human Cultural Evolution

Along man's history, the overall emerging pattern was one of progressive independence vis-à-vis the constraints of ecological factors.

Hawley (1982) agree that the limitations imposed by the physical environment depend on technological progress. In fact, invention, technology, and science made all the difference. Culture mediates the interrelationships of human populations with their environment and intergroup relations, and is influenced physiologically, psychologically, and ecologically by it. We may say that culture is the distinctive ecological niche of man. But, if cultural evolution led the path to ecological independence, it brought about a new type of limitation in the shape of taboos and prejudices which prevent the use of available potential trophic products. These products do not constitute resources when they are proscribed, by social or individual belief, custom, religion, or prejudice.

During the 1970's McKeown (1988) presented a carefully documented critique of modern medicine, trying to show how little medical procedures and discoveries had been responsible for the increase in life expectancy and control of certain diseases. He suggested that along the path of human evolution the patterns of diseases changed. Prior to communal organization, humanity had suffered mainly from accidents and injuries; in contrast, the domestication of animals and the first human groups were affected mostly by infections of zoonotic origin; finally, with the advent of affluence, hygiene, and better nutrition, degenerative diseases prevailed. McKeown's major error in his otherwise well documented work was to confuse ecology and evolution. Along the history of humanity we find contemporaneous communities representing all the stages that we can discern, in its rise from the hunting-gathering to urbanization and affluence. Furthermore, the pattern of disease responds to ecology and not only to evolution. Individuals and communities will show a prevalence of those diseases that occur under the **ecological** conditions they live in. Accidents threaten whoever engages in perilous procedures, either when hunting a tiger or when racing a Formula 1 sports car, and zoonotic infections are one of the dangers of associating with animals, today as it did at the dawn of communal living.

Primitive Populations

Lacking advanced technology and technical resources to change the environment in a permanent or extensive way, and using limited amounts of energy, our ancestors and the contemporaneous aborigines, explorers, hunters, and campers living off the land, with restricted resources and little outside contact, may be roughly considered ecologically primitive.

In the history of man, the first technical revolution came with the arrival of what archeologists call *Man-the-toolmaker,* and one of the best introductions to the techniques employed by our ancestors to turn rock and bone into artifacts is a book first published in 1934 and thoroughly revised in 1953 by Leakey (1960). Pleistocene and early recent man was part of local ecosystems, their populations depending solely on local resources.

Hunter-gatherers lived–and indigenous peoples subsist today–in small groups, and subjected to the vagaries of climate and season, like other higher mammals. With a low life-expectancy-at-birth, nomadic by necessity, they were prey to predators, prone to accidents and violent death, susceptible to acute prionic, viral, bacterial, parasitic infections, many of those acquired from animals used as food or as sources of sub products as hides, skins, bones, as well as from pets or *xerimbabos* kept around or inside their dwellings. Victims of periodic episodes of famine and cycles of starvation and abundance, primitive people left little mark upon the face of the Earth (Truswell 1977; Karlen 1995). Tools provided them with the means to hunt, gather roots, and defend themselves, but not with the power to change the landscape.

On the positive side, nomadism prevents dangerous environmental contamination, by not allowing geohelminths to complete their life cycles. Populations tended to be too small to maintain endemic diseases that cause lasting immunity, as *herd-diseases*, and *density-dependent* infections.

Pleistocene and early recent man were part of local ecosystems, their populations self-sustained and dependent solely on local resources, and could hardly be considered as ecological dominants.

Rêgo (1999) describes well this condition as he proposed a new way to ensure sustainable development in the Brazilian Amazons. He showed how traditional populations utilize their resources as they diversify cultivation, harvesting, and animal raising thus ensuring harmonic relationships with the environment. He explains that due to the relative isolation of human populations in forested areas leads to cultural differentiation. Their habits are regulated by the natural cycles.

Cultural anthropologists and naturalists of the 19th century thought they could equate contemporaneous technologically primitive peoples (aboriginal, hunter-gatherers) with early primitive man, and learn about cultural traits, mores, mentality, and social structure through the study of modern tribes. It has been recognized that present day indigenous peoples are no living fossils.

José Rodrigues Coura's (2007) researches on Chagas disease in the Amazon Region show that the enzootic cycle of *Trypanosoma cruzi* in nature was probably maintained during millions of years by vector transmission, and/or by oral route through the ingestion of infected triatomines by edentates (Xenarthra), rodents, marsupials, primates and also by carnivorous. Carlos Chagas in 1912 was the first to describe both armadillos, *Dasypus novencinctus* and the triatomine *P. geniculatus* infected with *T. cruzi* living in the same burrow. In 1924 he found this parasite infecting a squirrel monkey, *Saimiri sciureus* (Chagas 1924). The adaptation of triatomines to human dwellings in endemic areas may have begun with deforestation during the agriculture cycle or for cattle raising in the last 200-300 years, when Chagas diseases became an endemic zoonosis.

There is clear evidence that the Amazon region presents a great variety of vectors and wild reservoirs of *Trypanosoma cruzi*. In that region Chagas disease is now considered a zoonosis – a disease transmitted from animals to infected humans who colonize the sylvatic ecotopes and wild triatomines invade human dwellings.

Rural Populations

Agriculture and domestication of animals brought about the second major revolution in the history of mankind, and changed forever the relationship of man with the environment. A sustained source of animal protein and a steady supply of grains improved human health conditions. Archaeologists agree that in the last ten thousand years a primitive form of agriculture appeared in several parts of the world. There are clear evidences of at least three main centres of active development of early agriculture: the alluvial plains between the Tigris and Euphrates, or Mesopotamia, extending to today's Syria, Lebanon and Israel; Northwestern China; and Mesoamerica. The earliest evidences of a primitive agriculture are in The **Fertile Crescent**, in Mesopotamia. China sprang some seven thousand years ago, and Mesoamerica, two thousand years later. The rapid increment of human populations coincided with this revolution and with the retreat of the ice, at the end of the last glacial age. The consequent elevation of the sea level submerged the coastal plains. The milder climate allowed the expansion of forests. Most "theories" that try to explain the birth of agriculture involve circular reasoning or tautological explanations. Cultivation implies radical changes in living habits, and community organization, and those habits derived from the adoption of a new life style.

From the existing millions of different animal and plant species in the biosphere, a handful of terrestrial forms became domestic, through a long empirical process of artificial selection. (Caras 1997). Early selection procedures would lead eventually to the gene manipulation and genetic engineering of nowadays.

Hunters usually caught old, weak or diseased animals and an impaired hunter could not provide for his family as long as he was in poor health. As a consequence, forced nomadism became unnecessary.

The early settlements allowed for new types of relationships amongst its inhabitants as the establishment of clans, social hierarchy, stratification, and division of labor and the accumulation of lasting wealth in the form of land, crops, and animal herds. Artisans, artists, witch-doctors, and warriors had a chance to become full time professionals. Availability of large amounts of staple food was balanced against the accumulation of waste, nutritional excesses, and change in overall epidemiological patterns (Nnochiri 1968; Cockburn 1977; Tyrrell 1977). But water storage and irrigation systems became a source of parasitic infections; soil-borne helminthes were able to complete their biological cycles in sedentary human hosts. Agricultural soil would become impoverished by repeated cultivation and a system of slash-and-burn or itinerant agriculture was developed. Rotation of cultures was devised, introducing periodic changes in the ecosystem. But energy and fuel consumption remained low.

The invention of methods for preserving food allowed the regulation of stocks, the end of famines, population growth, and one important step further from the mere local economy of subsistence, threatened by natural seasonality and periodic vagaries of climatic conditions and natural disasters. It also permitted extended travel and exploration of new lands and seas. It led to surplus production, accumulation of wealth and trade, to cooperation, interchange and diffusion of ideas, inventions, and discoveries. In time, it would affect distant systems, by way of trade and the import of foreign products.

Hawley (1982) recognizes that technologically primitive populations are restrained by the conditions imposed by the **local** habitat. The development of means of transportation and communications widens their cultural horizons, beyond the local constrains.

New techniques, population growth and diffusion of knowledge were the outcome of the agricultural revolution.

The only viable explanation for such a change to have occurred is the change in ecological conditions after the end of the ice age: the expansion of fast growing, high yielding grasses, which offered easily collected seeds, and which were to be artificially selected to become the cereals of to-day.

In a paper on the rural ecology and development in Java, Soemarwoto (1974) analyzed some aspects related to the flow of energy, the cycling of minerals, and biodiversity. Some of his comments are pertinent to the question of delimitation of ecological systems:

> The more isolated a village is located, the less are the imports and exports of energy of a village. In Java no village is completely isolated from cities. But bad conditions of roads in some areas significantly reduce the imports and exports. […] A situation was created in which one ecosystem, the city, lived at the expense of another ecosystem, the village.

The author shows how the interruption of the former cycle: man-fish-man, where human excreta were canalized into the ponds resulted in the loss of protein, sewage disposal, and eutrophication. In addition, "The disturbance of the mineral cycle by the use of fertilizers also makes the rural people more dependent on energy subsides, which have to be imported from outside the ecosystem".

As to the economy of production, and its influence upon population size, Willis (1980) registers that, in Tanzania,

> Fipa villages are densely concentrated, and normally separated from other villages by distances of from two to five miles. Because of their development of a highly productive system of compost mounding, Fipa are able to maintain spatially stable villages over long periods. The average village population of 250 is also about two-and-a-half times that reported for the frequently shifting villages of the nearby Bemba, who practice a form of slash-and-burn agriculture.

A network of trade links rural settlements. Local scarcity of food items are met by import from neighboring villages. Specialization arises, some tribes producing pottery, canoes, or other handicraft, for trade. Boundaries of trade networks are more easily delimited, in the rural level of human ecology.

The analysis of the role of simple trade networks linking rural settlements, and its implications upon the more theoretical aspects of the ecosystem concept was aptly presented by Ellen (1970) for pre-Columbian people in North American and by Fish and Fish (1970) for the Moluccas' archipelago.

In terms of human and animal health seasonal, annual, and perennial crops exert a great impact upon the ecology of zoonoses. Parasites and infectious agents that have no animal reservoirs need large population densities to establish themselves and not to be extinguished due to the lack of susceptible individuals. Sedentarism brought about the increase of human local populations. Abundance of crops raised the trophic potential for the ruderal fauna, and the herds of domestic animals helped to increase the populations of those parasites common to non-human and human hosts. Elimination of large predators results in the increase of unwanted commensals.

In rural communities, due to intimate contact with both domestic and wild animals, arise helminthiases, fungal, bacterial and viral infections due to soil and water contamination, amebiasis, toxoplasmosis, hidatidosis, filariasis, schistosomiasis, dracontiasis, paragonimiasis, leptospirosis, bubonic plague, malaria, Chagas disease and accidents caused by poisonous animals (McKeown 1988; Karlen 1995).

Several rodents are the hosts of robovirus, a term applied to rodent-borne viral diseases as viral hemorrhagic fevers and hantavirus infections caused by Arenavirus and Hantavirus, respectively. Sabiá virus is a unique Arenavirus associated with viral hemorrhagic fevers in Brazil. This fever was first identified in 1990 in a fatal case in São Paulo State and in two laboratory workers that were infected while handling the virus and survived. Until now, the primary rodent reservoir of the Sabiá virus is unknown.

Hantavirus is a genus in the family Bunyaviridae, in which several members have gained more and more attention after 1993 in the last decades, mainly in the American hemisphere, when the first cases of Hantavirus Cardio-Pulmonary Syndrome (HCPS) were described.

In Brazil, after the first identification of HCPS more than 800 cases have been reported until April, 2007. Juquitiba virus carried by the black-footed pigmy rice rat *Oligoryzomys nigripes;* Araraquara virus, carried by the hairy-tailed mouse *Necromys lasiurus;* and Castelo dos Sonhos virus, whose reservoir is unknown, are genotypes associated with human disease.

Urban Populations

The development of agriculture, the accumulation of wealth through trade, and the permanence of dense settlements would lead eventually to a new level of ecological integration, the urban, highly technological civilization.

Some authors came to recognize the importance of this distinction, as Moran (1990), who reminded anthropologists of the importance in recognizing the differences in scale between the densities and technologies of contemporary societies and those which have been the subject of most of their studies, if naive recommendations were to be avoided. These differences are valid today, and they concern the very nature of the relationships with, and impact upon, the environment.

A new concept emerged for this novel level of integration, the urban technological community, namely the concept of the **geosystem**.

Urban settlements are not self-sustained ecological units, and the expression **urban ecosystem** may only be used figuratively. There is no trophic structure with a nutrient recycling system operating, in cities. Nutrients and energy are imported from agricultural areas, power stations, oil fields, coal mines, or atomic plants. We may speak of an **urban ecology**, but not of an **urban ecosystem**.

There are ecosystems inside urban limits: natural green spaces, parks, lakes and ruderal areas, but a city itself does not qualify as a recycling ecological unit. Urban man is an ecological novelty. Box and Harrisson (1993) recently discussed the roles and policies related to what they have denominated *natural green spaces*. The ecological role, the size, location and the risks involved in creating or maintaining natural green spaces inside urban limits is a topic of discussion nowadays.

Analyses of this kind do not take into account the particular aspects related to local social attitudes, ecological requirements and health implications involved in the presence of natural habitats inside the city limits. Temperate parks are very distinct from equatorial forests, swamps, mangroves, and other types of biomes. Guttenberg (1975) also remarked that "The value of the urban environment, however, is a matter of continual controversy".

For the purposes of this analysis, we shall concern ourselves with the resurgence of urbanization in Europe and generalization of trade which took place during the renaissance of the 12th century, and which gave rise to our contemporaneous urban civilization. The urbs became a new "environment", quite distinct from the old burgs, which were merely administrative centers in a fundamentally agricultural society.

Among the changes brought about by the urban revolution were the rise of a mercantile economy based upon a monetary system of trade; a new class of citizens with social mobility; industrial development with the accumulation of surplus products intended for trade; and eventually the admission of the profit as an acceptable practice, not a sin against religion. The expansion of urbanization in Europe, after the Middle Ages brought the end of feudalism and an increase of intensive agriculture to support the growing demand for food by a new class of citizens with the means to pay for it.

Cities are part of economic units, regional, national, and/or international. Urban civilization demands increasing amounts of energy for transportation, storage, distribution, packing, and preparing, food; for disposal of domestic and industrial wastes, for purifying the water it pollutes, and to run the course of daily modern life. Cheap sources of energy as animal traction, wind, water, were substituted by steam engines, fossil fuels, electricity, and nuclear power.

The impact of urban settlements is not felt only locally as it interferes with distant ecosystems. Technological man changes coastlines, river courses and their regimes. It modifies the patterns of geographic distribution and the population dynamics of plants, animals and microorganisms. It changes the landscape by draining lands, flooding valleys, mining, irrigating dry areas. It pollutes air, water and soil with organic and inorganic substances. Domesticated and highly selected strains of plants and animals become high energy consumers: they need fertilizers, pesticides as a protection against competitors and parasites, and are expensive to maintain, harvest, store, preserve, and transport.

Cultural evolution and advanced technology freed man from some ecological limitations, but brought about other constrains-and responsibilities. It is our contention that urban populations must be held responsible for the amount of change and impact they inflict upon natural ecosystems, wherever these impacts are felt.

Several factors remained unchanged. Like all the other vertebrates, man depends upon primary producers, the green plants, and symbiont microorganisms to synthesize nutrients and vitamins. The laws of thermodynamics still govern the transfer of energy, and the longer the ecological chain, the greater the loss.

As it is the rule with organic evolution, there is no upward trend or goal in human evolution. There is random variation, adaptation diversification.

The capacity of soil, air and water to absorb, degrade, and recycle wastes is limited. A sensitive question nowadays is the overload of wastes that pollute the environment. Global change is the responsibility of urban man, as it is the conservation of resources wherever they come from. Urban man must be aware of where the resources he uses originate, and what he can do to contribute to lessening his impact upon the environment. This includes the question of loss of biodiversity through land reclamation for industrial agriculture and forestry, for human occupation, degradation, and the exploitation of natural products.

In cities, ecological factors must be analyzed in meso- and micro-scales. Mesoclimate, microclimate, geographic micro distribution, provide and explain differences in the epidemiological patterns of incidence and prevalence of diseases, and in the assessment of risks.There is an increasing rate of death by mugging, terrorism, and drugs. An epidemiological study of hypertension in a suburban area in Rio de Janeiro (Governor's Island) showed that while in 1980 cardiovascular diseases (CVA) were responsible for 40.7% of all deaths, in 1991 external causes as car accidents, violence and homicides, became the leading cause of death in males below 50 years of age. (Klein et al. 1995). Also there is stress, malnutrition, and obesity.

Crowding is responsible for the accumulation of waste, pollution, close interpersonal contact, easy transmission of infection agents from host to host, and for the increase in the numbers of commensal animal vectors and reservoirs of diseases, as rats, mice, flies, noxious insects, pigeons, stray dogs and cats. Cosmopolitan people also mean a risk of cosmopolitan diseases. City parks offer conditions for bacteria, protozoa, geohelminths, to develop and complete their biological cycles. Urbanization of rural diseases may change their character and ways of transmission. Blood parasites may be transmitted through blood transfusion, instead of insect vectors, as it is the case with malaria and Chagas disease. Mott et al. (1990) showed that the rates of infected blood with Chagas disease parasites in blood banks is 6%-20% in Argentina; 15%-25% in Brazil; and as high as 63% in Bolivia.

Contact with wild animals, reservoirs or vectors of pathogenic microorganisms is reduced or eliminated. Zoonoses of *domestic* species and of *cosmopolitan commensals* become important, especially for those groups of individuals more at risk, as children, aged and the immunologically compromised. Cage birds, sparrows, pigeons are present in large numbers in towns and cities. Public markets, as we find everywhere may be the source of some zoonoses. In the Orient, the promiscuity between pigs and fowl gave rise recently to SARS. Epidemic diseases that are density-dependent must be carefully prevented.

Endogenous, hereditary, and degenerative diseases become prevalent. Also important are infectious diseases as rabies, transmitted by dogs, cats and bats, psittacosis, toxoplasmosis, scabies, lice, larva migrans on beaches and children parks, leishmaniasis and calazar, dengue fever and the risk of urban yellow fever, both transmitted by the urban mosquito *Aedes aegypti*. In 1993 the Director General of the World Health Organization opened the 46th World Health Assembly recognizing that

> Dengue and DHF are the most serious and rapidly rising arbovirus infections on the world. ... Most cases occur in densely populated urban areas. ... At present dengue and DHF threaten one fifth of the world's population, or approximately one billion people residing in urban areas ... (WHO 1993).

Subsequent developments proved him right, and rather modest in his estimates.

Among cultural factors we list cultural shock, as new migrants come to town, breaking family patterns and ties. Habits brought from rural areas to the city may be responsible for health problems linked to uncontrolled pet animals, waste disposal, discarded containers, open water reservoirs, which offer a breeding ground for vectors.

We must bear in mind that modern means of transportation allows for the rapid circulation of germs from wild or unsanitary areas to large cities, even before symptoms become evident, threatening to bring back old scourges, as the recent re-emergence of several diseases attest.

Favourable aspects include a lower growth rate, access to better education, medical assistance, including better diagnostic facilities and hospital treatment.

The worldwide integration of humanity and the compression of both the temporal and spatial dimensions known as globalization affect multiple aspects of human lives including infectious disease transmission. Chagas disease or American Trypanosomiasis, although discovered a century ago, still stands as one of the main threats to public health in Latin America and several aspects of its epidemiology has been deeply influenced by either ancestral or current globalization phenomenon.

Presently, it is estimated that 12 to 14 millions people in 18 endemic countries are infected with *Trypanosoma cruzi,* the causative agent of this disease. Over the years, several distinct epidemiological changes related to Chagas disease transmission and distribution have been witnessed; most of them influenced mainly by rapid and dramatic environmental changes, ecologic pattern of the triatomines insect vectors, and the migration of human populations. Social and political aspects are also determinant factors in the incidence of the disease, and all factors mentioned are linked throughout Latin America. Here, we discuss features related to globalization, highlighting current achievements and future challenges in terms of epidemiology, control, and treatment of Chagas disease predominantly in Brazil but also present in other countries.

Marginal Populations

Low income, marginal pockets, favelas, barreadas, slums, whether in the periphery, in derelict surroundings may be responsible for the existence of conditions distinct from those prevalent in upper class districts or city and town suburbs. It must be noted that the term **suburb/suburbia** has opposite meanings in third world countries and in developed ones. In the first, it describes the site where marginal populations concentrate. In the second, where housing development offers better living conditions.

Urban marginal populations of squatters living in slums are "urban" in the geographical sense, but not in terms of environmental impact.

Compared to rural communities, they are in disadvantage, when we consider food and water supply and sanitation. Soil and water contamination due to deficient or inexistent drains and trash collection is the rule. Vectors and reservoirs of pathogenic parasites and microorganisms are rampant. Run-off from the slopes end in the streets and avenues, especially during heavy rains, carrying litter, faeces, and night soil.

Marginal populations are deprived of natural resources to supplement their diet.

There are few animals from the native fauna. Predators were long eliminated, and ruderal, commensal, cosmopolitan species abound without proper natural or efficient planned control. Cats, dogs, and rats roam free and in large numbers.

Prevalent are the helminthiases, toxoplasmosis, filariasis, leptospirosis, malaria, leishmaniasis and all common diseases present in large cities and towns.

Coming into the city proper, groups of children, many with a home and family, but spending days and nights in parks or sheltered on sidewalks are ignored by health authorities. They may recur to public health facilities, often using false IDs and declaring a non-existent address. If school children and workers are required by Brazilian law to show a vaccination card, there is a large sector of the population who is ignored by epidemiological surveys, and is an unknown quantity in health statistics.

Conclusion

As early as 1939, Clements and Shelford remarked that the very essence of ecology is its synthetic nature, and noted that the trend towards specialization in training and in methods of research was having an adverse effect …"reflected in a hostile or indifferent attitude to an approach vital to the ecological study of man." They further pointed out that ... "students of ecology will continue to be trained primarily as botanists, zoologists, sociologists, or economists for some time to come-probably as long as university departments are organized on the present basis."

Their comment applies to all fields of transdisciplinary knowledge. In most cases, professionals in one area claim the field for themselves, usually adopting a reductionism approach. The emergence of the theory of systems provided many a good example (Bertalanffy 1968).

Since the scientific revolution, the trend towards specialization has been constant and necessary. Definition of disciplines and demarcation of territorial rights began during the 17th Century, following the profissionalization of scientists and after the advent of the learned academies and their journals in Italy, France, England, and Germany. It found its way into the universities, through the establishment of chairs, departments and specialized curricula being welcomed by the professional guilds. The somewhat artificial borders of the distinct fields of knowledge were to be challenged after the rise of new intermediate areas, as it was the case with the birth of physic-chemistry, in the 19th Century. Furthermore, the extension of research methods and theories from one specific field to another has a great epistemological potential. Pasteur's application of his knowledge of crystallography to solve the mystery of fermentations, and from hence to unravel the origin of infectious diseases, constitutes a good example.

Applying scientific laws and principles to solve pragmatic problems and to develop new technologies used to take a long period of time, sufficient in itself to allow for the evaluation of its effects, and preparation for the changes it would bring to society in general, usually in the next generation. This timing has changed and resulting impacts are immediate. New ideas spread rapidly, thanks to the revolution in the communications media. The financial rewards behind them brought about a new industrial revolution, the results of which are being felt worldwide. Technological and industrial research occupies nowadays an important role, and is responsible for novel aspects of social development, and political directives. As a result, man's impact upon his social and natural environments acquired a distinct dimension. Also his responsibilities.

Traditional university undergraduate curricula are insufficient to provide the necessary abilities demanded by the science of human ecology. The fallacy of the two cultures-scientific and humanistic-so aptly criticized by Snow (1959), is at the root of our present problems, and Aldous Huxley (1962) dismissed our clumsy attempts at solving them:

> Your cure for too much scientific specialization is a few more courses in the humanities....But don't let us be fooled by the name. ...They're simply another form of specialization on the symbolic level.

Problems are seldom simple, and those resulting from the interaction of man with environmental factors, never are. In 1977, Maldague addressed all these questions in detail, to reach the conclusion that what we need is to prepare experts with as open mind. Actually what we need is less people saying what we need, and more people explaining how to accomplish our quest for transdisciplinarity.

No single, magic, all-purpose method or research instrument will permit the manifold analysis needed for the understanding of the type of problems that confront and befuddle human ecologists. These problems result from a combination of factors which can be analyzed with precision, plus actions that depend upon the whims of individuals and the ways of societies. As a result, the training of a human ecologist must start with professionals of different backgrounds, and supplement their specialized knowledge with a selection of basic principles, methods and techniques of analysis used by the so-called exact, natural, and social sciences.

In his capacity as an analyst and advisor, a human ecologist must be able to evaluate all aspects of a problem with the same degree of detail. He should know how to draw a composite picture of natural settings, climatic patterns, biota structure, human elements, social structure and institutions, cultural patterns, historical antecedents, economic imperatives, ethical and political developments. He must rely on specialists to further his preliminary analysis and to supplement his data, but he must be able to communicate with them, to ask the right questions, and understand their answers. He should be careful to steer away from speculative ideas such as those of ecological determinism and ecological catastrofism. In short, he should be able to acquire a holistic vision of the question.

When addressing a particular problem, he may begin his analysis from the standpoint of his special professional knowledge. But he should be able to overstep the limitations of his field of expertise and to delineate as complete a picture as possible of the situation in all its complexity. From his initial analysis, he must be able to determine how detailed he must make his inquiries in each field. He should the be trained to identify the relative role and weight of the each contributing factor; and to recognize the weakest link in the chain of events, where intervention will bring the desired results. The final picture he draws must show the dynamics of synergetic relationships, not a mosaic of independent elements and factors. As an ecologist, his main concern should be with the **interrelationships** of the distinct factors, more than with the listing and description of each of them.

Transdisciplinarity means the construct of an overview, systemic in essence, explanatory in its objective, which goes beyond the mere juxtaposition of partial analysis. They are important in specialized fields of knowledge, and are needed to clarify specific issues. General surveys, on the other side, lack detail and precision to allow for policy decisions. Up to the present, though, there is no single methodology for doing a transidisciplinary study, and I doubt if there will ever be one. Strict rationalism applies to the analysis of physical problems but not to the understanding of personal choices or social mores and decisions. Myths, traditions and individual preferences are as important to the health and survival of human populations as their physiological trophic needs and requirements-if not more so. Subjective factors are important in human behaviour, but are difficult to quantify. They fit badly, if at all, in mathematical models.

The best we could hope is to devise a set of different, but complementary tools-like a Swiss-army knife-and train a new generation of open-minded professionals in the competent use of each, to fit the distinct needs of a transdisciplinary quest.

The original concept of the ecosystem, as expanded and popularized by Odum (1953), was extremely useful as an archetypal model and as a didactic tool. It was utopic and theoretical. From general ecological theory, the idea of perennial, dynamic, adjustment, and not of equilibrium, is applicable to human ecology. The idea of a balanced nature that pervaded protoecological thought for several centuries was incorporated in the explanation and justification of social and political structures up to the 19th century, to be questioned by the liberal revolutionaries of last century. The concept of a natural equilibrium was generally accepted, possibly because it led to a comfortable feeling of order in nature. Its dismissal disrupted this feeling, as the idea of an evolutionary process based on chance mutations with no final goal or progress felt uncomfortable for the human mind and unacceptable for Christians. But Moran (1990), quoting Margalef (1981), admitted that, “Increasingly, it has become more productive to conceive of ecosystems as thermodynamically open systems that are out of equilibrium”.

Man's place in nature has been a theme for arguments and philosophical discussions from antiquity to our days. Man's dual role, as an organism-a primate-and one in possession of special attributes acquired through the process of cultural evolution, is a source of conflicting views, accentuated when the discussion involve ecological matters, as they pertain to the relations with the environment or "nature".

While in a study in basic ecology the emphasis or focus is upon the interrelationships of the manifold factors involved, in human ecology the central object is man, and we try and study man's relationships with, and impact upon, the environment. Rappaport (1990) recognized that

> Earlier analyses in ecological anthropology, reflecting a general deficiency in cultural and social anthropology, did not pay sufficient attention to the purposes motivating individual actions ... nor did they pay sufficient attention to behavioral variations among individuals and the groups to which they belonged.

Another important notion, and a trend in fashion, is that of the limits to growth, in terms of population and production, translated in the rather vague notion of sustainable development, a sort of "economic climax" defined by the older idea of rational exploitation of natural resources.

Finally, I agree with Rapapport (1990) in that “There are grounds for distinguishing ecosystems and regional systems”. In human ecology, the concept of **geosystem** or **regional system** is of fundamental importance. This distinction, coupled with the notion that primitive, rural, urban and marginal subpopulations show distinct ecological properties and types of relationships with the environment, must for the basis for sound ecological theory.

Acknowledgements

I am indebted to José Rodrigues Coura, Elba Regina Sampaio de Lemos, Jane Margaret Costa von Sydow, Adauto José Gonçalves de Araujo, Luiz Fernando Rocha Ferreira da Silva, and Karl Reinhard who provided pertinent data for this chapter.

CHAPTER TEN

Human–Marine Nature Interactions: What Kind of Valuing? Towards an Integrative Modelling Approach

PIERRE FAILLER, SERGE COLLET AND ALIOU SALL

World fisheries are currently largely depleted across the globe. Despite the intent of governments and international organisations to effectively manage fisheries—and thereby reduce adverse impacts on marine ecosystems, it appears that oceans are evermore under threat. One option to reverse this accelerating depletion of marine resources is to seek to authorise only those fishing practices and policies that have a positive societal benefit. However, in order to do this, one needs to measure the societal costs and revenues of each fishing practice and each fishery policy, compare them, and then use the findings to inform public policy. The novelty of this approach, presently being developed in an international research project (ECOST, 2006/2010), is the way in which ecological, economic and social costs are combined to give a true value of the real cost of fishing activity—–a cost that is not captured by current market mechanisms. The modelling work in process that supports this approach seeks to integrate different values (market values and non market values) in a holistic manner. Nevertheless, non market values are very difficult to incorporate into the model and there is a risk that attempts to do so (as in the case of option values) may simply undermine the strength and meaning of factors possessing non-market values. Therefore, a parallel aspect of the research demands a deep study of the characteristics of the human valuing processes in order to grasp the epistemès embedding them, the values that society gives to the sea and the entities therein. In the last analysis, these values, in a sense, function as underpinnings to our modelling activity—and bring a more profound meaning to the measurement of the societal costs of fishing practices and fishery policies. This chapter lays down the foundations of this modelling approach on the interactions between ecology, economy and society and examines briefly its significance for policy implementation.

Introduction

In spite of the application of the principles of sustainable development, the promotion of responsible fishing, the adoption of the precautionary approach and the more recent Johannesburg Plan of Implementation, marine ecosystems continue to show signs of decline. Two main reasons can be called to explain this situation. The first is linked to the fact that, in developing nations, fishing policies still operate according to a developmental paradigm while in developed countries, efforts are directed at the maintenance of fleets and employment without investment in ecosystem rebuilding. The second reason lies in the logic of fisheries management applied which consists of maximising the private benefits of fishing companies without considering the side effects on marine ecosystems and society. Thus, the question of reconciling the development (or maintenance) of a fisheries sector with the well-being of the underlying ecosystems becomes central. Therefore, the first question to be addressed is the following: is it possible to develop tools that aid decision-making within the fisheries sector in a way that will allow ecosystems to be restored? Secondly, if so, how should this be executed so as to maximize the chances of success?

The structure of the paper that follows addresses these two questions. First, we detail how the market mechanism presently fails to incorporate the ecosystem effectively within the development of fishing policies. This leads us to suggest that current notions of value need to be extended so as to embrace the concept of societal cost. A review of valuing processes from the anthropological-philosophical perspective to the economical point of view is presented. We then apply these ideas by coupling the concept of societal cost to the ecosystem model Ecopath. This forms the basis for the ECOST model approach which seeks to model aspects of the link between ecology, economy and sociology. A brief conclusion completes the chapter.

Traditional Management Approaches, Market Failure and External Costs

The techniques of bio-economic modelling of fisheries, developed largely from the 1950s onwards, draw primarily upon standard economic theory. Yet while the use of such models permitted significant theoretical advances in the practice of fishery management, their actual application failed to fulfil the hopes of economists (Meuriot 1987). Maximization of individual profit and the fishery rent under technical and resource scarcity constraints, and the adjustment of supply and demand through the mechanism of prices, had seemed to offer insight into effective fisheries management. The development of the concepts of resource and market equilibrium (“Maximum Sustainable Yield" and "Maximum Economic Yield") were applied to the management of commercial species and, in the majority of cases, helped to explain stock decline. Notwithstanding the availability of increasingly strong and sophisticated computing capacities, the models developed remained attached to the analysis of mono-specific stocks. It was about herring of the North Sea, the cod of Newfoundland, the sardine of Morocco, the tuna of the Gulf of Guinea, etc. By ignoring interactions between the species, and of the species with the marine environment, the bio-economic models fostered:

- the reduction of the reality to a simple problem of revenue maximization (fishing management becomes a technical exercise—requiring the adjustment of fishing effort to the availability of the resource);
- the illusion of the simplicity of the functioning of the marine environment (the system is presumed to be in/moving towards equilibrium—and so fails to take into account resource variability over time and space and the diversity of living resources within an eco-system);
- the partitioning of reality (insofar as the biologist deals only with biological aspects such as stock recruitment, the economist deals with the economic agenda, while the manager assumes responsibility for the introduction of given policies);
- the mistaking of the object of fishing management (insofar as the focus became managing the species of commercial value—without considering the place of these species within an ecosystem).

The reductionism of contemporary management models therefore leads to an impasse. This impasse can be seen in the following figure (Figure 10-1), which presents, in a simplified manner, how fisheries are generally perceived by traditional managers, economists and biologists. Relations between the entities are unilateral. The government impacts on the fleet and their activities through management measures such as prohibiting certain types of gear, introducing quotas and setting total allowable catch quotas (TACs). It also acts in the market by means of regulations relating to food safety, food quality and/or price control. The ecosystem is reduced to a residue, to be impacted upon, as fish stocks fluctuate in line with the policies introduced.

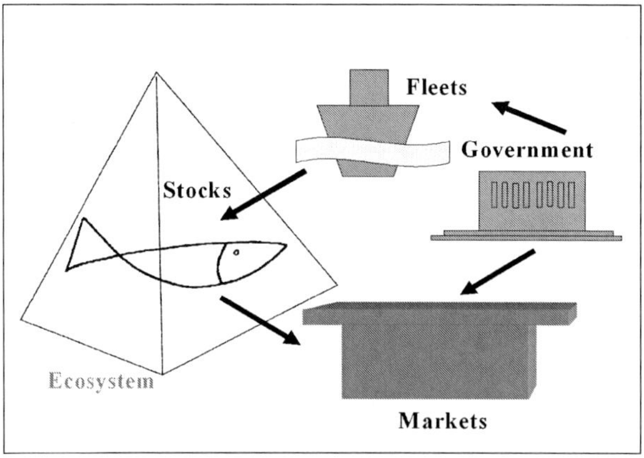

Figure 10-1. Reality of fishing as perceived by the traditional managers, economists and biologists (from Failler and des Clers 2002).

Such a perception of reality ensured that only costs of production, processing and marketing are considered (referred to henceforth as "production" costs). These costs are of a private nature and are the only ones incurred by producers—even if the productive activity generates a series of indirect effects, both on the marine environment and on civil society in general. Economists generally refer to these indirect effects as "externalities". Externalities occur when the action of one agent impacts on the wellbeing of other agents (and such agents are not compensated for this impact via the market). Externalities can be positive (effect of a vaccination campaign for example), or negative (pollution for example) (Boncoeur and Olivier 1996). In the case of fisheries there are a number of such externalities which, while not integrated into the production sphere as private costs, are crucial for the well-being of the fishery.

In the economic and social sphere:

- The costs of fishing management (primarily research, design and application of management measures, control and monitoring)—which Arnason (1999) found reached 30% of the value of landings in the USA, and 25% of the value in Newfoundland.
- The public subsidy of fishing activity (this can take the form of financial support for the construction and modernisation of boats, exemption from taxes and customs duties, state payments for access rights to the Exclusive Economic Zone (EEZ) of third countries, etc.) (OECD 2000, 2003);

- The opportunity costs related to the extraction of marine resources through fishing, rather than their exploitation in other ways (eco-tourism ventures, for example);

In the natural sphere:

- By-catch (Wiium [1999], for example, suggests that 14 kg and 7 kg of fish respectively are discarded as by-catch for each kilo of shrimps or octopi landed, by-catch being equivalent to as much as 26% of world landings).
- The destruction of natural services (oceanic function as a carbon and pollution "sink" the stability of the marine environment, etc) which ensure marine patrimony.
- The destruction of particular properties and functions of the ecosystem (for example, Pauly et al. [1999] and Cury et al. [2001] note how the over-fishing of certain levels of the trophic chain generates disturbances which are reflected through the whole ecosystem)
- Irreversibility of the damage caused (the resilience of the environment and ecosystems is very variable, and more sensitive systems may be irrevocably damaged—to the detriment of both current and future generations).

The market price is supposed, in standard economic theory, to act as indicator of scarcity (value) and as a behavioural signal. In the fisheries case, price can account for the increasing scarcity of the resource—but it will not be capable of securing the ecosystem against over-fishing of species. It does not therefore function as an effective mechanism regarding the incidence or impact of fishing—both at the societal and natural level—since it ignores or belittles the external costs of fishing activity. The limits of traditional fishing management approaches are therefore reached.

Values and Valuing Processes: the Difficult Connection between the Social Science Perspective and the Economy

Values and valuing are indeed the most studied of social phenomena, even as early as the fifth century BC by Greek philosophers such as Aristotle and Plato. The social dimension of valuation is at the very heart of the thinking of Durkheim and Mauss in France, of Weber in Germany and of Parsons in the USA (Parsons 1937; Parsons and Shills 1951). From the Cornell group of cross-cultural studies of values to the recent European and international inquiry (Boudon 1999), the tracking of values in all their aspects and processes of change seems to be or to constitute the very essence of the work of the broad church of social sciences.

For the last decade, there has been no politician or decision-maker, strongly influenced by what is considered a very conservative rhetoric in Europe, who does not speak of generally reified values. In 2003 for example, the people of the world were asked to hear and to see the "American values in action" in a declaration by Powells at the very beginning of the Iraq war. However, beyond this dominant political rhetoric, what are values? Wrongly or rightly, these values and disvalues as subjective structures stem from social practices. As mental constructs (goal-oriented subjective constructions), they are fundamentally linked with desires, needs, interests and habitus (Collet 1979: 18–250). If values include or contain cognitive elements, they involve strong affective components too. The more deeply rooted a value is, the more central a place it takes within a system, or configuration, and the more intensely it is lived, arousing feelings. Values can mobilise vehement energies. From a social-science perspective, the first definition elaborated was that of the cultural anthropologist C. Kluckhon in 1951:

> A value is a conception, explicit or implicit, distinctive of an individual or characteristic of a group, of the desirable which influences the selection from available modes, means and ends of actors (1951: 395).

Thus, value includes three structural dimensions: affective (desirable), cognitive (conception) and conative (selection). In such a theoretical framework, values would be patterns of regulation accepted as desirable by subjects or persons in a given culture or sub-social setting ,the family for example, which would serve as guiding precepts in their lives, and even more so within the complex process of reflexive identity construction. Values develop within the process of social learning in everyday routines and practices. Criticising the "juridicism" inherent in "culturalism", Pierre Bourdieu, the great French sociologist who died at the end of January 2002, wrote in the Logic of Practice that humans never imitate values but acts and body gestures instead. The body is the living unconscious depository of culture. Spinoza (1994), in his Ethics, speaks of affects imitation as the main dynamism in the diffusion of passion and social emotions. Thus values imbedded in acts and practices are incorporated in all spheres of social practice and social systems, making social order both possible and resistant to change.

As social subjective structures, they encode the variety of the modes of perceiving and acting. They differ from goals in that they provide a general rational for goals, which are more specific, and motivate the pursuit of these goals through particular means or methods. They provide a means of self-regulation of impulses. In summation, everything that social actors appreciate (appreciation schemes), wish to obtain, appraise, set up or propose as an ideal can be defined as a value, and each value has an object that is qualified by a judgment. Values become norms when they commend and/or regulate social conducts or prescribe a course of action, and values provide the grounds for accepting or rejecting particular norms. Core values correspond to what is called the "ethos" of a culture; they are the dominant values of these people, which, once reified, can clash with the values of other people.

Undoubtedly, values are enshrined in desire, the cathectic dimension of human life. As Spinoza stated in the Ethics (Ethics III, Affects definition I), well before Sigmund Freud, "desire is the very essence of man". It is not by chance that the great philosopher insisted on cooperation processes and argued against all forms of religious fanaticism and ignorance, for the necessity of tolerance and love of the understanding of nature, of which human being is a small part (Spinoza 1994, Political Treaties II: 8). The existence of plural, diverse and opposite values, which can clash, thus calls for tolerance and deliberation in the majority of social contexts. Social peace, consensus and functioning of institutions are built upon core values, which, as systems are constantly challenged by the evolving social context. Of course values are not static and can sometimes undergo dramatic changes: for example in the Western world the growing ecological awareness since the seventies. By contrast with expanding radical individualism (Dumont 1983), such a value as human solidarity has been declared to be romantic, altered or dying. The terrible catastrophe of the Tsunami of 26 December 2004 has proved, in the concrete behaviour of people all over the world, e.g. the true locus in which to "observe" values (Firth 1964), that solidarity is a human ever-existing value, which, in certain circumstances of disaster, can suddenly be awakened or warmed up.

To conclude this brief scanning of values from a social-science perspective, it is worth taking into consideration that values are not the sole subjective structures steering social practices and more broadly human behaviour. Desire, needs, habitus, interests are other individual structures determining social actors and practices, which can come into conflict. Such an approach makes clear that the elaboration of a theoretical interdisciplinary unifying framework for a cross-disciplinary notion of value is by no means simple. Figure 10-2 demonstrates these complexities and discordances in the full valuing of the marine-human heritage and relationships between its component dimensions.

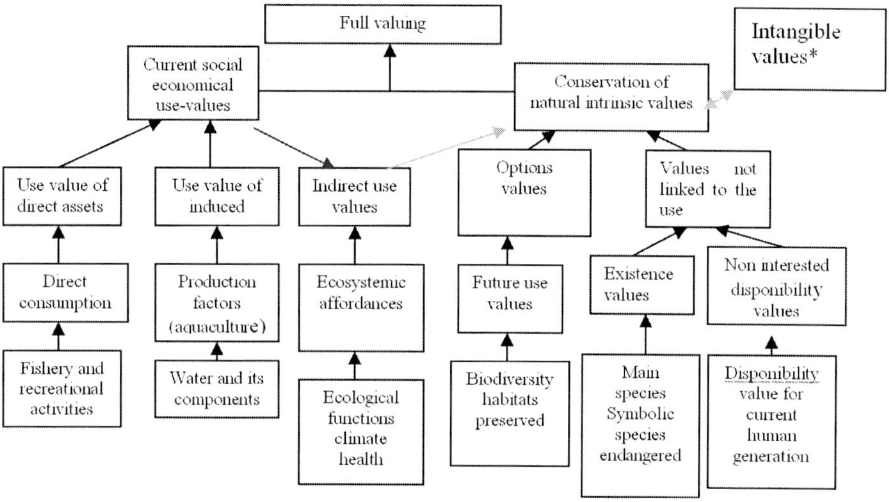

Figure 10-2. Full valuing of the marine-human heritage (re-elaborated from the Figure 10-3).

Note: Intangible marine values are:

- re-creative values: intrinsic qualities of the sea that interact with humans to restore or create new stimulations of the mind, body and soul.
- spiritual values: qualities of the sea places that inspire people to relate with reverence to the "wholeness-holiness" (Rappaport 1999) of marine nature. They correspond to an intimate connection with sea places.
- identity values: the sense of marine place, of belonging, which is fostered through experiential situations. Marine entities, especially favourite sea places link people to these places through experiential settings, myths, legends or histories.
- aesthetic values: appreciation of the beauty, harmony, strength and ascribing deep meanings to the prismatic faces of the marine nature.
- artistic values: all the qualities of marine scapes and fishing gestures which induce feelings inspire imagination in creative expressions.

- educational values: the qualities of marine nature which foster expanding enlightenment, understanding and respect between humans, and between humans and sea entities.
- therapeutic and appeasement values: the complex relationships between maritime places and people which enhance or restore physical and psychological well being (Collet 2007b; Sall 2007).

What has been explored up to now is, broadly speaking, the social elements of the human valuing process and its centrality in human life. However if we pay some attention, it allows the building of bridges with the reflections of philosophers about objectivity, subjectivity of values and their role in human affairs.

From a social representation perspective, values belong to the Lebenswelt (the world of living). They are objective and subjective universal constructs, which, as we have seen, cannot be reduced to the rationality of instrumental goal-oriented action in the economic sphere e.g. Homo economicus who looks to maximise his self-interest or utilities. The constitutive process of values does not follow from rational behaviour. Moreover, values are not aggregations of individual desires reduced to preferences; public goods, for example, are much more than the aggregation of individual instrumental values achieved through the exploitation of non-human entities constituted as resources. Indeed, resources, or utilities considered as such in social networks, embed a valuation process.

In ancient Greece of the ninth century BC, fish was not at all considered a resource to commodify. In this ancient Greek culture, the marine environment was the wildest realm of nature; the sea world was pre-eminently the realm of anti-nature. The valuable portion of nature was the agricultural-pastoral community of the Oikos, which was dangerous to leave and for which warriors were willing to die. In Homer's poetry, and more particularly in the Odyssey, to die at sea "devoured by fishes" or sea monsters like Skylla was the most tragic of all human fates (this is still the case today in many cultures). Fish was the wildest, most inhuman food, fit only for seagulls. Marine life enjoyed a completely different status and was held sacred in the worldviews of Mesopotamia, Bronze-Age Egypt or the Syrian and Palestinian coast in the Iron Ages I and II; deified and mummified fish came to symbolise wisdom and knowledge given to humans (Oanes) if not peace and justice (Nanse) (Collet 1995).

Thus the archaeo-history of the symbolic forms of appropriations of sea entities reveals how past cultures, over the long span of time and in the enlarged Mediterranean space, have constructed a relationship with the sea, which includes not only a valuing process but also an ethical dimension. A brief look at insular cultures that are engaged in, and forced into, a difficult process of transition opens the broad field of the intangible: the cultural and spiritual dimensions associated with the marine world. Within the Torres Strait or the Solomon Islands (Melanesia),

> the seascape is a living history with associated myths, stories and legends that provide moral and cultural guidance. It is the store house of social identity for the islanders (Cordell 1984: 307).

Such a marine environment is fundamental to the process of the constitution of identities and personhood. In popular folktale, such as that of Aukun of Mabuiag Island, the heroine is a fisher woman who travels along the sea bottom and encounters and names a myriad of reefs around the island sea territory. At every site she distributes baskets of various fish species. Used as mapping references, these story places are considered and highly valued as cultural sites. The marine universe operates as a "true cultural property" (Cordell 2001), fostering caring feelings towards these marine places, in which humans are a part of nature-pars-naturae (Spinoza 1994).

In Senegal (West Africa), in the context of integration of the inshore fisheries to the global market (Sall 2007), the author analyses how the syncretic Muslim animist's beliefs and values continue to structure the way fishing communities are looking at the sea and its patrimony. Guet-Ndarian fishermen consult the marabout who has the power to bless the coming season and to ensure sustainable access to fish. The launch of a canoe and opening of the fishing season are all marked by propitiatory rituals and offerings such as the immolation of an animal to the sea goddess Daour or Mame Coumba Bang. Much more, the ascendancy of the holy sea power, Rappaport's wholeness-holiness (1999: 459) over human being can be gauged to the fact, that well beyond the sphere of traditional fishing communities, the re-establishment of a broken alliance between a human being and a spirit arousing the feeling of the alteration of the psychic state—is realised through ritual bathing in the sea.

In contrast to the rich fields of these symbolic relationships to the sea and its entities, the dominant hyper-modern vision is structured by the dualism, subject/object (the powerful technologies labouring the natural world); nature is forced "to deliver its mysteries", as Bacon or Descartes would have commentated. In the current Baconian Cartesian naturalist view, nature "delivers its services". Nevertheless, in that world, still ordered by the paradigm of male technical and political domination, new conservationist attitudes have emerged. Marine entities can be imbued with strong meanings. Today, many kinds of whale and other charismatic species are no longer tapped as resources and exchangeable utilities, although there are exceptions such as Japan, Norway and the bloody ritual in Faroe (van Ginkel 2007). If we consider that, in the early sixties, whale meat represented 16% of the world fish catch (3.8 million tons) and that in 1997 whale-watching tourism represented, on a world scale, an activity amounting to a "monetised price" of three billion dollars, we can observe that in the space of just over forty years, a radical change of modes of conceiving and perceiving these types of marine entities has occurred. Humanistic, naturalistic, symbolic valuations have in fact superseded utilitarian ones. Northern societies have replaced instrumental values by preservation values. Near extinct, whale species became values for themselves. The living Koine, as companionship between human and other species, has been enlarged, and consequently the modern Western chasm between man and nature has been partly filled.

Therefore, today some marine species enjoy, in some yet precarious ways, their existence for their own purpose. This preservation value has been humanly conferred at the price of social struggles, pressures on institutions and international agreements; but it is worth noting here that it is no longer homocentric. The demanding reflection of J.B. Callicott on the intrinsic value of non-human species is there at stake. How to define the intrinsic value of a whale, or still better a child for parent's love?

> ... for example, a new born infant is valuable for its own sake, for itself but not in itself. In and of itself an infant child is as value—neutral as a stone or hydrogen atom [....] An intrinsically valuable thing on this reading is valuable for its own sake, for itself, but it is not valuable in itself, i.e., completely independently of any consciousness. (Callicott 1986: 162).

This type of value is intrinsic but in a "truncated sense". Scientifically or affectively, the subject of valuation is always, and cannot be something other than a product of the human mind. Preservation values correspond similarly to this approach to the intrinsic value.

Two other philosophers have differently developed a theorisation of intrinsic values. One is Holmes Rolston (1988), an environmental philosopher; the other is H. Jonas (1984[1979]). In order to allocate an intrinsic value to species and ecosystems, Rolston emphasises that every life centre teleogically organised, e.g. pursuing a goal "or telos", in a Darwinian acceptation, consists in reproducing its kind or its species. The "good" that every organism tends to express and to reproduce is nothing else than its kind or species. If we are right, we see the "good" as grounded in the reproduction of the being as an existential process labouring in the prodigious richness of existing entities. The reproduction of the being is thus the value in itself, this trend or this striving towards the necessity of being, something which refer to Bergson's Creative Evolution or Spinoza's conatus (Ethics III, prop. 6). H. Jonas, in Chapters III and IV of his book, revisits the very thorny issue of the relations between values and ends or purposes, with a reference to Spinoza: Spinoza's harsh criticism of the Aristotle's final causes, which he qualifies as "asylum ignorantiae" (Ethic I, Appendix). The end of Jonas' Chapter II is an implicit dialogue with the Spinozian concept of effective causality or immanent causality. For Jonas, the world harbours values, unlike Kant, for whom the natural world is free of values. Purposes are immanent to the modes of being and they are not necessarily subjective, i.e. mental. Jonas speaks of a "kind of subjectivity without subject" (1984: 77). He does not define values as "values for whom", nor their utilitarian aspect, but for themselves: "Values are the good belonging to the order of immanent purpose", and the philosopher adds "… that in bringing forth life, nature, evinces at least one determinate purpose: life itself" (ibid.: 74). For those who have condemned Jonas for an absolutist way of thinking and a kind of religious finalism, the philosopher takes care to add that "… entirely unconscious and involuntary, as the purpose of digestion and its apparatus in the totality of the living body", it makes sense to speak of a "labouring in nature and to say that nature along its tortuous path is labouring towards something, or this something multifariously struggles in it to come to light" (ibid.:75).

In our opinion, we are very close to the Spinozian concept of natura naturans, or nature naturing understood as the productive activity of things, the immanent self poiesis, that one of "god or nature", the infinite productivity (i.e. open to the whole nature). All nature, therefore marine nature, harbours values. A necessity of being finds its "tortuous pathway through the long and hard process of evolution" (Jonas 1984: 76). The Jonassian statement is not a metaphysical one which would proceed from an essentialism of life. Indeed it is this necessity of being that can be heard in the cry of the totally vulnerable human newborn, which would be at risk without hurried and complete care. However, humanity in many parts of the world is deprived of exercising this responsibility in decent conditions. In all human societies, this is an absolutely asymmetrical relationship in which the survival of the totally vulnerable being depends upon the gift of very long-term care and love. The continuation of life is dependant on this primordial gift, which is the archetype of responsibility, the primordial responsibility. Gift is the key issue; there is no social—natural life without gift, without "affordance". (Ingold 2000: 166-167). Giving for protecting is the strategic operator for conserving, for restoring social-natural life (Collet 2002: 546-549)

In contrast to Hume's natural fallacy, from the existence of the vulnerable being, a duty arises. Necessity (a natural one) is transmuted into duty: the duty of care. Jonas unveils paradigmatically the intrinsic value and good of being, and further, how values are enshrined in nature. Such a philosophical concept and approach would not be far from those of the biologist. We cannot be certain if Wilson has read Jonas and Spinoza, however, he has read A. Damasio (2003), who, in a Spinozian perspective, brilliantly demonstrates how our feelings are enshrined in neurobiological regulations of our nature, thus putting a definitive end to Cartesian dualism. The conclusion of Wilson's book Consilience deals with the absolute necessity to develop ethics based on intrinsic value, thus pointing to an "existential conservatism". Again, in The Future of Life, the Harvard biologist comes back to this issue in Chapters VI and VII, proposing concrete solutions, in particular extended protected areas, in order to halt the involutive destruction of life. Therefore we have a core area of convergence in social sciences, philosophy and biology, and at the same time a divergence with environmental economics. The thinking of Aldo Leopold (1949) on the inadequacy of economics to take into account the healthy functioning of land ecosystems from a "... hopelessly and lopsided perspective of economic self-interest" (neoclassic economics paradigm) finds an echo in the thinking of Wilson, who writes in Consilience:

> The time has come for economists and business leaders who so haughtily pride themselves as masters of the real world, to acknowledge the existence of the real real world (1999: 326).

Economists will reply to Leopold and Wilson by saying that in a market economy, the majority of goods and services are assigned by price. Price acts as a signal to producers and consumers—and allows them to adjust their production and consumption decisions accordingly. It is an indicator of relative scarcity under conditions of current (and anticipated) supply and demand. However, the resources and services offered by oceans, and nature in general (fauna and wild flora, water, air, ecosystems, etc), are outside the market. They don't have a price. And, without a price indicating the importance of the sacrifices made in order to obtain or conserve them, economic agents have the tendency to presume that their price is zero. There are, as a consequence, innumerable instances where natural assets have been sacrificed because their intrinsic values have been ignored in economic calculations. Yet if economists attempt to put a price on natural assets, anthropologists and philosophers urge them to exercise caution in seeking to quantify valuing epistemologies regarding nature. If Sen (1987) has weakly restored an ethical dimension to economics, Collet (2001, 2002,

2007a), R. Larrère (1994) and C. Larrère (1997) have pointed out the necessity to go a step further and analyse our western anthropocentric relationship with nature—first by discarding dualistic approaches to humankind and nature (the intrinsic value of nature approach), and second by redefining the interactive relationship between humans and nature using the notion of ecocentrism.

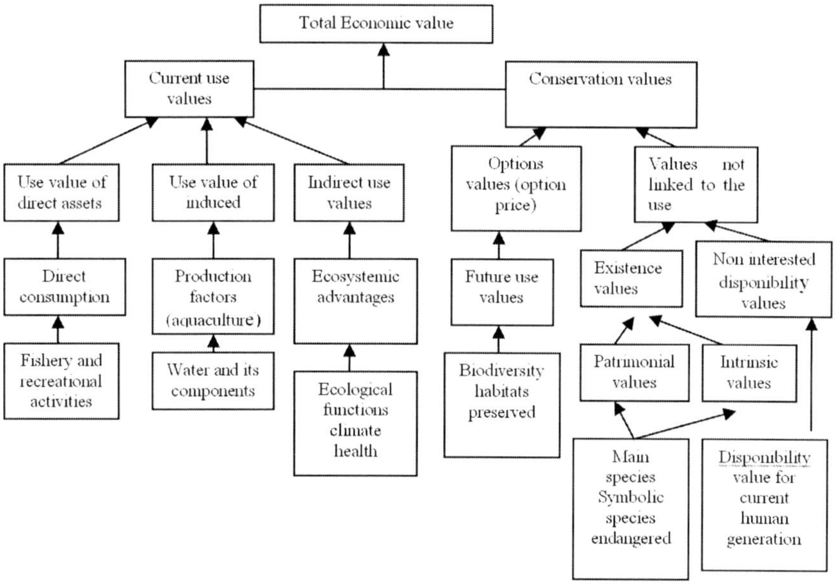

Figure 10-3. Economic values attributed to the marine patrimony (adapted from P. Point [1998]).

From an economist's perspective, an asymmetry exists and develops between the use of marine resources as consumption products and as generators of environmental services due to the shortcomings of the market pricing system. Such a situation presents a very real challenge to the public authorities as inertia would almost certainly lead to resource overexploitation—and consequently it is imperative to take affirmative action so as to compensate for the shortcomings of the market pricing system. In some instances, the creation of a market of negotiable quotas—starting from a given global quantitative constraint—can help reveal a market price which better reflects the marginal opportunity cost of the fishing activity. Yet the opportunity cost disclosed does not necessarily reflect the full value of the ecosystem, it merely provides additional information on one or more possible productive uses of the ecosystem. There is thus a gap between what the opportunity cost (marginal) of fishing suggests—and which managers use to optimise the allocation of resources—and external costs, as noted above. To remedy this we need to include prices reflecting external societal and natural effects. This requires knowledge of all the various use values which can be encountered in the marine environment. The diagram above tries to give an indication of these from the neoclassic economic perspective.

According to this dominant vision which remains somewhat anthropocentric if not anthropomorphic for anthropologists and philosophers, the marine environment delivers several broad services; the extraction of commercial and subsistence resources, the provision of factors of production, and ecosystem services. If the first two functionalities are measurable using indicators of price, the third is not so easily reducible to a monetary variable. Yet these services are of primary importance for the functioning of the marine environment and, in particular, the production of marine resources. Fortunately, it seems that there is a possibility of integrating those external costs (which contribute to the degradation of the ecosystem and its services) into an evaluation of the "true" cost of fishing policies through recourse to the notion of social—or societal—cost. The concept of social cost, first enunciated by Pigou (1920) and subsequently developed by Coase (1960), was formulated to allow for the "internalisation" of externalities. They suggested that compensation was in order in those instances when the activities of economic agents caused a nuisance (externality) to the well-being of others. Although the concept of social cost as formulated by Coase and his successors only measured nuisance in human terms, we contend that the approach is equally applicable when considering disturbances caused to ecological systems.

Modelling the Complexity of Ecological, Economical and Sociological Interactions: an Approach in Construction

Taking into account the external effects associated with a fishing activity requires a change in our understanding of the operational dynamics of fisheries. The following diagram illustrates the complexity of incorporating such effects within the marine environment. Now a social request is added to the production demand of the fishery to reflect not just intergenerational concerns, but also current non-market values of the ecosystem (this includes the range of use values given in the right hand boxes of Figure 10-3). The ecosystem is thus a new actor in the fishery landscape. Formerly reduced to the various stocks exploited by the fishery, the ecosystem is now placed centrally within the landscape and viewed as having functions and properties which it is advisable to preserve (see Figure 10-4 below).

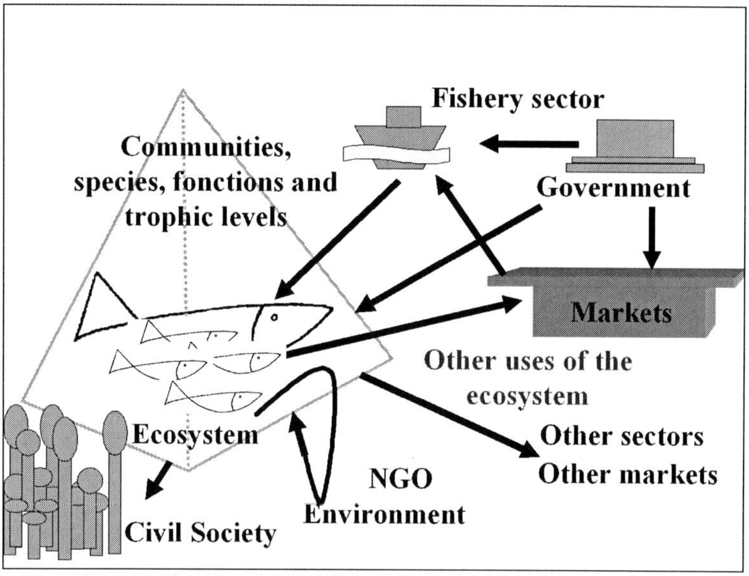

Figure 10-4. Complexification of the reality of fishing as perceived by the traditional managers, economists and biologists (from Failler and des Clers, 2002).

Thus, the analysis in terms of rights to pursue a certain strategy is simply one option within a portfolio of competing rights which, in the fisheries context, might include:

- the right to catch fish, and to reject/discard those which are not wanted (the by-catch);
- the right to continue fishing, despite evidence suggesting that stocks are being degraded/are in ill-health;
- the right to one to contribute to the loss of functionality of the ecosystem by destroying some of its properties, functions or elements.

The question then is the following: how is it possible to limit—even suppress—the rights of the ship-owners to degrade stocks and the ecosystem? The contemplation of a framework which integrates societal costs with a dynamic ecological model which reflects the relationships between the various trophic levels while contextualizing the economic drivers of fishing activity may then offer some prospects for capturing the non-market effects associated with the activity and, in this way, come to offer a better view of a complex reality. The Ecopath model, as delineated by Christensen et al. (2007), is a functional representation of the ecosystem which can be used as a starting point for developing a framework to evaluate the social cost of fishing activities and, more particularly, to inform development projects and management plans. The basic idea is to apply the iterative process which characterises Ecopath as a method of measuring—in an incremental way—what occurs in the ecosystem when a fisherman uses a particular fishing gear (such as a trawl). The following figure shows the ecological costs that result from the first iteration.

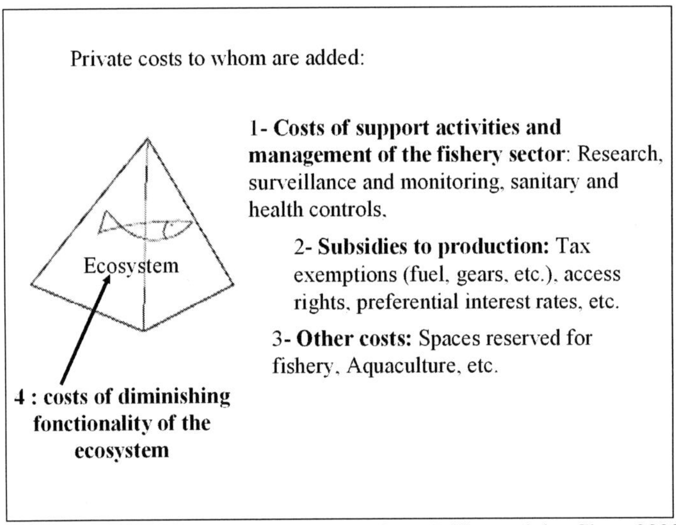

Figure 10-5. Social costs related to a first tide (from Failler and des Clers, 2002).

The costs related to the activities of management, the public subsidy of fishing activity etc. are costs which can be expressed in monetary form. They are thus easily added to the private costs. On the other hand, the reduction in the functionality of the ecosystem due to the use of trawl nets is more difficult to express in monetary terms. While it may be possible to allot a monetary value, this value is still likely to only capture those ecosystem goods and services which have a commercial value, and other non-market outcomes such as ecosystem degradation will continue to remain unpriced (and so undervalued). It is of course possible to compare what the ecosystem in good health (situation before fishing) produces and relate this to the ex-post scenario of fishing and over-fishing to obtain the difference in expected incomes. However, a reduction in trophic levels (or a shortening of the food chain) can lead to an increase in biological productivity (Palomares and Pauly 2004), or even to the subsequent extraction of species such as shellfish or cephalopods with greater commercial value. Even here though, in simply conceiving of the ecosystem as a natural entity intended to produce fish for human consumption, we are guilty of disregarding its remit in the regulation of climate, the absorption of CO2, etc. Thus, ecological costs related to the degradation of the ecosystem through over-fishing must ultimately be quantified if the resulting model is to be a robust one.

The renewal/continuation of fishing activities will be reflected (as shown by figure 10-6) by both an increase in private costs due to the increased scarcity of the targeted species (while this is compensated to some degree by rising sales prices, declining marginal elasticities of the demand will not fully offset the revenue loss), but also by ecosystem change. The immediate consequence of such a situation, which is becoming increasingly commonplace across the world today, is that the cessation of fishing for a particular species—or at a particular level of the trophic chain—sees fishing effort redirected towards other species lower down the trophic chain. In other words, fishers target increasingly smaller and smaller species.

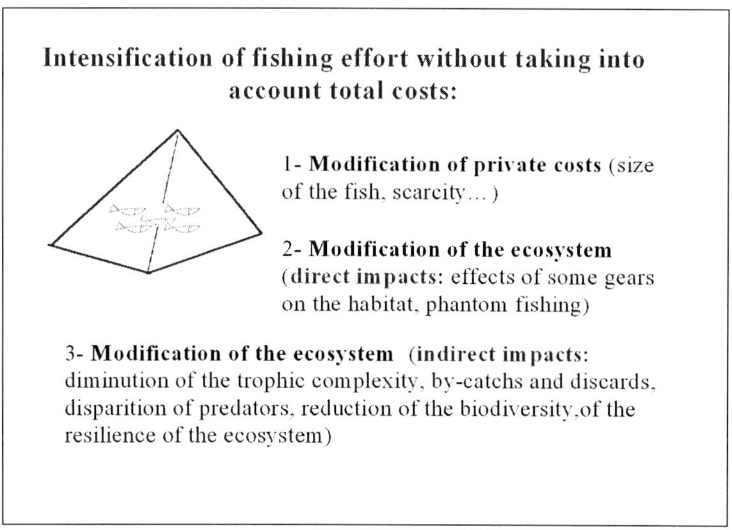

Figure 10-6. Social costs related to a second tide and followings in situation of non sustainable fishing (from Failler and des Clers 2002).

The choice of the model is fundamental to the management of natural resources. Until now, the main models used have been neo-classical models (initially developed by Gordon-Schaefer, later extended by Clark, Munro, Bjørndal, see Anderson (1986) for a general review). They are usually used to measure the economic effects of fishing activities through reference to the stock level and fishing capacities. The advantage of such models resides in their simplicity and transparency, it being possible to reproduce them once a certain amount of basic data has been collected, and they have been widely applied in Europe and throughout the world. While they have, in effect, "proved" themselves through such widespread use, such use has also highlighted their limitations—particularly as regards their (in)ability to give a complete picture of the full impact of fishing activities. Accounts of fishing's effect on the natural environment are either omitted completely, or considered solely in terms of the target species under consideration. The effects on the underlying ecosystem are never taken into account, consideration of their economic and social impact is limited to the creation of jobs, value added, and effect on public revenues—while the costs borne by civil society in management and/or restoring damaged ecosystems are invariably ignored (Failler and des Clers 2002). The primary purpose of such models therefore is to consider the private profit involved, without taking into account the costs borne by society and the ecosystems themselves.

Since the beginning of the 1990s, new types of ecosystem models have emerged which provide a more comprehensive understanding of the structure, function, and regulation of major ecosystems (see Mann 1988; Pahl-Wostl 1993; and Gaedke 1995, for example). Mass-balance biomass models (such as Ecopath and the inverse methods model) are being used globally to systematically describe ecosystems and to explore their properties (see Vézina and Platt 1988; Christensen and Pauly 1992, 1993; and Pauly and Christensen 1996). Mass-balance descriptions of trophic interactions between all the functional groups of the ecosystem recognise the complexity of the habitat and provide valuable information on the health of the marine environment.

Integrated economic-biological-ecological models have also been developed since the beginning of the 1990s. Integration is possible because of the predominant use of mathematics in biology, economics and ecology. Mathematical models are used to analyse and study relationships in each of the disciplines, and this commonality provides a conducive environment for integration. Linkages between the different disciplines generally precede in one of two ways. One method is to extend the originally formulated model towards the other spheres and integrate the "foreign components" into a modified model which retains a degree of disciplinary bias. Ecologists, for instance, have developed what we might term "ecology-cum-economic" models that rely on the use of system dynamics to investigate how the ecological system behaves under a specified set of policy instruments. Economists, on the other hand, have produced "economic-cum-ecology" models to help determine optimal policy responses within a given system. However, while they have been diligent in modelling robust and dynamic economic systems, they have been as guilty as the ecologists in downplaying the dynamics of the "other" system. A second method is to construct a new model through interdisciplinary work, each discipline bringing its own tools and ideas into a common core framework. The consilience principle can also act as a bridge, helping to link facts and fact-based theories across disciplines so as to create a common groundwork (Wilson 1999). Nevertheless, and despite the impressive recent advances in computer technology, truly integrative models have only been developed within the climate change community. Natural and marine resources management models are largely still designed according to neo-classical norms.

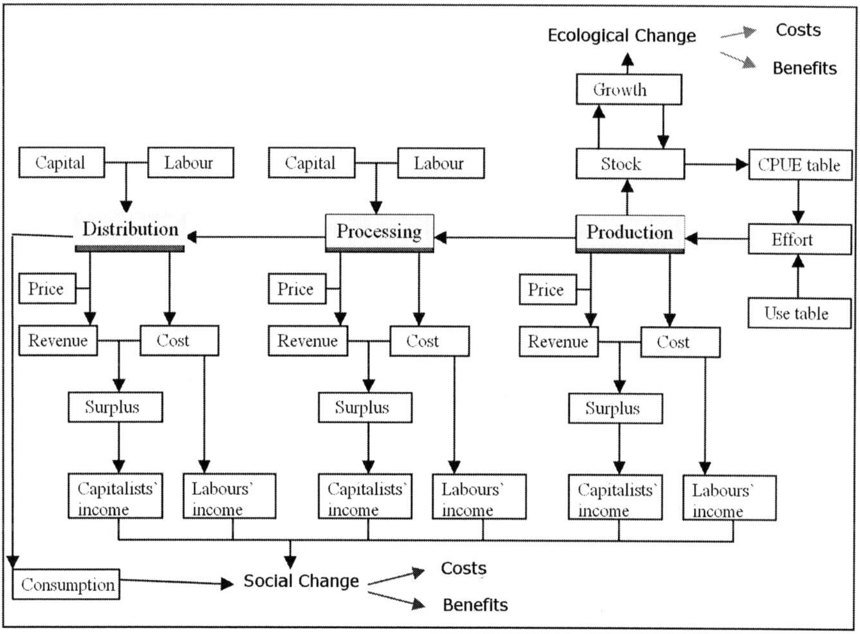

Figure 10-7. Relationships among social, economic and ecological systems.

The ECOST project attempts to move the discussion further forward, developing a methodology that can assist policy or decision makers in assessing fishery practice and thus inform fisheries management. Not only does it embrace the integrated assessment models developed over the last twenty years, it seeks to extend such models into the societal or social domain. It does this by constructing a new model through interdisciplinary endeavour, using the consilience principle as an analytic cement to fuse social, economic and ecological systems into an integrated assessment model that can address the interrelationships between the three systems while maintaining detailed disciplinary descriptions of each of the systems. Figure 10-7 above shows that the economic system consists of production, processing, and distribution. Production has an impact on the biomass of the species and thus affects the ecological system. While economic activity creates products, it also generates and distributes income among factors of production (incomes which are subsequently distributed among people).

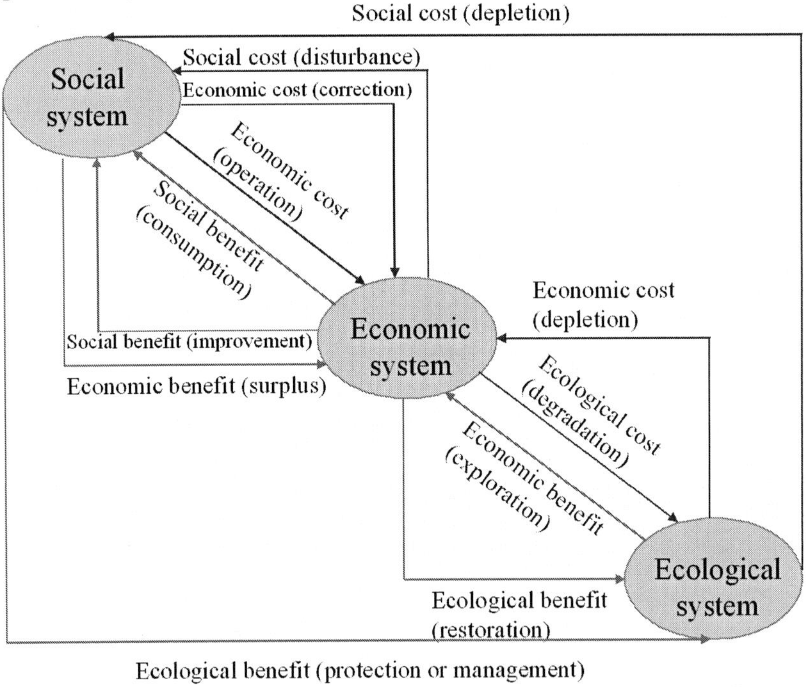

Figure 10-8. Relationships among social, economic and ecological systems.

The distribution of personal income can have profound implications for the social system. Equally, fishing effort varies across fishing metiers and the selection (non-selection) of these will have differential impacts on the ecological system, employment and functional income in the sector, besides further affecting other parts of the economic and social systems.

The economic activity of fishing occurs because the social system demands the consumption of fish. The social system benefits from consumption of fish (social benefit of consumption) while incurring a number of operational costs—production, processing, distribution and service costs (economic cost of operation). The functional income distribution resulting from the fishing activity may disturb the social system, thereby imposing costs (social cost of disturbance). If the economic system intends to correct the disturbance, the correction introduces a cost (economic cost of correction), although socially the correction is beneficial (social benefit of improvement).

Equally, economic activity may lead to a degradation of the ecological system (ecological cost of degradation), as resource stocks are identified (economic benefit of exploration). If action is taken to reverse the degradation, costs will be incurred (economic cost of restoration), although the ecological system will benefit from ecosystem restoration (ecological benefit of restoration). In the absence of policies to reverse degradation, degradation (or depletion) may trigger greater fishing effort that reduces the social value of the marine environment, imposing further costs (economic and social cost of depletion). The ecological system can also benefit through resource protection or management (ecological benefit of protection or management). Figure 10-8 above summarises all these interactions.

While economic costs can be measured in monetary terms, ecological cost is assessed by comparing current catches with historic levels of stock biomass, and social costs are incorporated through reference to measures such as poverty indices, consumption ratios, nutritional intake and gender (in)equity. In the ECOST model we seek to convert social and ecological costs into monetary terms—despite recognising the existence of non-monetary values and intangible values such as the ones described earlier in the chapter. Values which are unable to be incorporated directly into the ECOST model will be used to "screen" the results generated in order to ensure that the outcomes "make sense" from both society and fishing communities point of views.

Conclusion

Ecological assets are common pool resources (public goods), although the consequent externalities associated with their use are not usually well accounted for in market mechanisms. In response, introducing the concept of societal cost into the policy debate demands the capture (and hence allows the internalisation) of these externalities. However, the notion of societal cost is not widely addressed in the scientific literature. While Nadav (2000) and Garcia-Alte et al. (2002) have employed the concept to discuss sociological change at the societal level, we contend that it should take a broader meaning. Not only should it embrace; (i) “total economic value” — the measurement of the monetary and non-monetary values of the services rendered by the environment (actual use value), of services that it may potentially be able to render in the future (optional value) and the value of the existence of the former (existence value) (Pearce 1996); (ii) the “marginal cost of replacement”—the cost of replacing the services currently provided by a piece of nature (Arrow et al. 2000); but also (iii) the functional value of a species in an ecosystem—and the function of one particular ecosystem within the wider ecosystem (Christensen and Pauly 1993); and the human values regarding the seas that do not fit into the economic evaluation such as cultural and intrinsic values. Thus, the societal cost can be interpreted as a shared concept uniting ecology, economics and sociology. It can be beneficially employed in helping to analyse and then address some of the issues that marine ecosystems and fishing communities are commonly facing all around the world. As Kurien (2003:10) pointed out, the dimension of the fishery problem can no longer be confined to a local level:

> the macro trend of globalisation and the counteracting micro trend of localisation in many ‘tropical-majority’ world countries give rise to the need for new approaches to governance at both levels.

ECOST—the project and its end-product (the ECOST model)—is intended to help address this. Central to the project is the logic of the Johannesburg Plan of Implementation (JPoI): to restore as much as possible marine ecosystems by 2015, a logic that is also in accord with the philosophy of the Code of Conduct for Responsible Fisheries (Doulman 2007). To meet this challenge, the project mobilises, on top of the sixteen renowned scientific organisations entrusted with carrying out the applied research, seven regional and international UN development and management organisations. It also maintains close links to key NGOs in order to facilitate interaction with social actors and the dissemination of findings and policy prescriptions to the various levels of decision-making (local, national, regional and international). In this way, the concept of societal costs—which this paper presents preliminary sketches thereof—trekked through fishing activities and fishery policies, will contribute to the more effective management of the oceans.

Acknowledgment

This paper has been carried out with financial support from the Commission of the European Communities, specific RTD programme "International Research in Co-operation" (INCO-DEV), "Ecosystems, Societies, Consilience, Precautionary principle: development of an assessment method of the societal cost for best fishing practices and efficient public policies" (ECOST). It does not necessarily reflect its views and in no way anticipates the Commission's future policy in this area.

CHAPTER ELEVEN

New Directions in Human Ecology Education

ROBERT DYBALL, RICHARD J. BORDEN AND WOLFGANG SERBSER

This chapter is an outcome of the "New Directions in Human Ecology Education" workshop at the XVth International Conference of the Society for Human Ecology in Rio de Janeiro, 2007. Like similar gatherings at past human ecology meetings, our plan was to review selected institutional backgrounds and to explore current models and practices. Three of the presentations focused on ways established organizations are bringing new ideas into fresh combinations and incorporating these changes. In an open dialog, workshop participants discussed additional opportunities for collegial innovation and enhancement of human ecology education.

Examples of curricular innovations were presented from two of the oldest academic programs in human ecology, the Australian National University and the College of the Atlantic in Maine, USA. A third case presentation—by the equally old German Society for Human Ecology—described how local pressures have led to a proposal for a new free-standing college of human ecology in Europe. These three cases are presented below.

First though, the chapter reviews the common roots of human ecology, noting how key metaphors have been appropriated over time from different source domains in order to give meaning to the novel concepts that human ecology develops and applies. Then, using Tengström's 1985 typography, different phases of human ecology are discussed in terms of the extent of their disciplinary integration and presumptions about the nature of the human condition and the methods by which it can by understood. The chapter then discusses the challenge of modern interdisciplinary human ecology in the context of education and cross-institutional learning partnerships more generally, and discusses new directions and avenues for learning and collaboration. Following this, the chapter turns to the three case studies in order to illustrate how the individual historical backgrounds and local contexts in which human ecology has arisen results in diverse approaches to, and practice in, the subject. The chapter concludes with a discussion of how common frameworks and communication pathways are developing that might further enhance collaborative approaches in human ecology that are enriched by the diverse backgrounds of the participants.

Historical Roots of Human Ecology

As Kormandy notes, a degree of arbitrariness surrounds the selection of a starting point when tracing the historical roots of any intellectual development (Kormandy 1978: 1292). The origins of Human Ecology can be traced back into ecology itself, a concept which in turn has its roots in the metaphor of "the economy of nature" (Wooster 1977). At its foundation lies a concept from the social sciences, "economy", projected into the environmental domain. At the time that Linnaeus invoked it in his 1749 treatise, the "economy" metaphor carried with it the comforting implication of being "managed for someone's (typically human's) greater utility" (Bowler 1992) and so fitted well with prevailing beliefs that nature was God's creation, balanced, designed and ordered for mankind's use. In 1866, when Haeckel coined the term "ecology", he also invoked the "economy" metaphor. However, over the intervening century "economy", as both practice and a phenomenon of study in the social sciences, had become dominated by capitalist means of production, distribution and consumption. So, in defining his ecology as "the body of knowledge concerning the economy of nature", the "economy" metaphor Haeckel uses is infused with a rather different suite of inferences than it had in Linnaeus's time (Grusin 1993). As a disciple of Darwin, whose evolutionary ideas were also influenced by metaphors from capitalist economic processes, Haeckel's definition includes "the study of all those complex interrelations referred to by Darwin as the conditions of the struggle for existence" (quoted in Wooster 1977). Consequently, the concept of ecology in its origins was strongly influenced by the then current social science ideas of competition for finite resources, accumulation of wealth and notions of progress and efficiency.

Once naturalized in the biological sciences, "ecology as economy" concepts were mapped back into the social sciences—geography and anthropology most notably—to account for cultural differences amongst the peoples of the world. However, to view human behaviour as merely responses to ecological conditions—"environmental determinism" in its various forms—grossly failed to give an adequate account of the creativity, diversity and inventiveness of human culture. The inevitable reaction within the social sciences, in the forms of "possibilism" and other kinds of strong cultural relativism, tended to exclude or downplay any role at all for ecosystems in constraining human behaviour (Schutkowski 2006). These extremes of emphasis on either "environment and biology" or "culture and society" have provided an unhelpful and divisive backdrop to the development of human ecology. To this day terminology, methods or emphasis that seems to preference one over the other is sure to provoke argument amongst human ecologists.

Mono-disciplinary Human Ecology

This ebb and flow of concepts moving between the environmental and the social sciences characterizes much of the history of human ecology (Gaziano 1996). The term "human ecology" is first encountered in introductory sociology text (Park and Burgess 1921) where ecological concepts appropriated from the environmental sciences are found useful in developing understanding within the discipline of sociology. Other social sciences soon followed, with the term claimed by first one then another; including geography (e.g.: Barrows 1923), sociology (eg, McKenzie 1926), ethnology (e.g. Forde 1934) and anthropology (e.g.: Bews 1935), amongst others. Each of these disciplines employed terms such as "competition", "succession", "web of life" and "mutual interdependence" in reference to patterns or processes in human contexts, but beyond this there was little or no common meaning, methodologies, or collaboration between these versions of human ecology. This is what Tengström terms the subject's "monodisciplinary" phase (Tengström 1985).

Multi-disciplinary Human Ecology

Towards the middle of the twentieth century, new developments in the environmental domain of ecology began to cross over and manifest in approaches to human ecology. Notably the term "ecosystem", coined by Tansley in 1935, introduced a holistic, systems based understanding of inter-relationships and process. The focus became "not only the organism-complex, but also the whole complex of physical factors forming the environment" (quoted in Evans 1956). From this perspective, an organism is seen as embedded in a network of relationships that both constrain its numbers, distribution and behaviour and which the organism itself affects as it lives its life.

It is this systems-influenced understanding of human ecology that came to dominate the subject, employing concepts derived from Tansley and subsequent writers such as Bertalanffy (1950a, b), and Odum, who emphasised fundamental constraints of ecological energetics (1953 and refined in subsequent editions). Never without its critics, who saw the spectre of environmental determinism, these approaches stressed the fact that, like any other species, humans are embedded in ecosystems and are ultimately dependent on these ecological processes for their health and well-being. This caution against human hubris resonated well in many quarters, as concern about human impacts on the environment, and the future well being of the planet, grew across the late 1960s and 1970s. From this period onwards human ecology typically concerned itself with the "Sustainability" of a given practice or process, again adopting a term that economics had used in relation to scarcity and growth (Pezzey and Toman 2005).

The influence of these ecological approaches was also particularly widespread in many of the applied professions. In regional planning, for example, Ian McHarg pioneered an ecological model in his widely regarded book "Design with Nature" (McHarg 1969). His approach was further expanded as the basis for the human ecology planning program at the University of Pennsylvania (McHarg 1981), which became the training ground for a generation of professional planners. Similar metamorphoses took place in architecture, agriculture, impact assessment, public health and epidemiology (Cf. Sargent 1965, 1974) A common feature of these approaches was their applied dimension, as well as a growing awareness of "the costs of not knowing" and the broader ecological impacts that can result from "solutions" that are defined against very narrow criteria. Consequently, they tended to view problems and their possible solutions from a number of perspectives, and by expanding the range of disciplinary content, each of these professions substantially enlarged the scope of their inquiry and recommendations for application. For the most part, however, these were summary collations of expert knowledge, imported directly from the represented disciplines with little or no blending of that knowledge and little understanding of the inter-relations between the parts of the whole. The cumulative results of these various endeavors have been characterized by Tengstrom (1985) as "multi-disciplinary human ecology"—its second stage of development.

Interdisciplinary Human Ecology

The final decade or so of the 20th century was characterized by a growing recognition of the need for an interdisciplinary approach to human ecology—its third phase. The special feature of this interdisciplinary reorientation came from a realization of the necessity to move beyond extant (disciplinary) contributions. It was increasingly clear that a bona fide human ecological understanding of human-environmental relations required more integrative methods. The "problems" with which human ecology was concerned were typically "ill-defined, unstructured, involve high decision stakes and societal interest, high uncertainty, and [often are] dynamic and embedded in a dynamic setting" (Mobbs and Crabb 2002). In such cases, the use of discipline-based contributions, by themselves, was insufficient. Interdisciplinarity called for a qualitatively different and originatively undertaking—based on disciplinary input—yet also capable of combining, transforming and unifying it.

Throughout this "third phase" of interdisciplinary human ecology there have been some who have been calling for an intellectual framework that will enable a coherent "blending" of knowledge, and the associated emergence of new understanding that distinguishes the particular approach to complex human-ecological interactions conducted in the name of "Human Ecology" (Hawley 1986, Young 1989. For a discussion on "blending knowledges" see Newell et al. 2005). To date, no such unified approach has yet been adopted, and there remains a wide range of perspectives that call themselves "human ecology", each informed by their own local and institutional contexts and history. This paper will now illustrate this diversity through a comparison of the development path of Human Ecology in Australia, North America and Europe, and a description of current and future directions of three programs on these continents.

Human Ecology Education

Interdisciplinary approaches to understanding presented a particular challenge to education—especially higher education. Whereas the practical professions were clearly enriched by multidisciplinary participation, their project-based time frames seldom kept participants engaged in extended dialogs. Education was different. As university professors and administrators began to embrace the ideals of human ecology education, they rapidly discovered that design of interdisciplinary curricula demanded more integrative models. Students needed not only a broad exposure to multiple fields of knowledge. They also required entirely new pedagogical models. Their courses, curriculum and teachers had to prepare them with the additional skills of integrated learning. This recognition was further highlighted by two stubborn problems. On the one hand, worldwide concern was unfolding about the causes and possible solutions to environmental problems. From the outset, environmental problems were acknowledged to be complex—usually involving an intricate interplay of ecological and human dimensions. On the other hand, educators were confronted by the long-standing separation of knowledge into academic departments. The segregation of disciplines was further reinforced by the very design of higher education. This was a problem both at the level of institutional architecture which physically separated units, as well as in the historic identification of academics with their professional fields—nationally and internationally. In short, the fundamental nature of higher education and university-based research were at odds with the aims of human ecology curriculum development. Higher education and the academic world offered a universe of disciplinary perspectives. But there were few places prepared to support its interdisciplinary integration.

Nonetheless, human ecology's intuitive appeal found an initial foothold in many forms, despite the impediments. The early leaders of human ecology education discovered institutional colleagues in highly diverse educational settings. Working together, they fashioned new courses and preliminary curricula that attracted a new generation of students. In the 1960s and 1970s—long before the World Wide Web and email—most of these initiatives took place without any knowledge of one another. One of the curious features of this period was the widespread—and generally independent—choice of "human ecology" as the founding term for these emerging programs.

Most academic fields can trace their origins to key individuals or institutions. Growth and development of the discipline then follows from these sources to new institutions, academic journals and professional organizations. The history of human ecology is different. It seems somehow to have hatched, more or less spontaneously, in separate and distinctive ways at about the same time worldwide, although many of these early programs were stimulated by common concerns and reference similar gestational ideas and thinkers. However, the inter-institutional and international connections among these diverse beginnings would be late in coming.

Networking Human Ecology Worldwide

One of the first professional organizations for human ecologists was the Commonwealth Human Ecology Council (CHEC). Registered in Britain in 1969 as a charitable trust, CHEC was established as a central coordinating body for human ecology initiatives and activities worldwide. Its first conference—focused on human ecology education—was held in Malta in 1970. A few years later, in 1973, a survey conducted by P. F. Rogers (1974) summarized over 100 courses of study in 92 institutions in 16 countries. Since then, CHEC's activities have been amplified through additional international meetings, numerous books and other publications, and in development, education and applied projects in Great Britain, Australia, Canada and India, as well as Africa and the Caribbean.

Another attempt to organize human ecologists on an international basis took place through the International Organization for Human Ecology (IOHE) which held its first meeting in Vienna in 1975. A number of national and regional societies resulted as spin-offs of IOHE activities, including: The German Society for Human Ecology, the Nordic Society for Human Ecology (NSHE) and the Austrian Society for Human Ecology. The IOHE and NSHE are no longer active, whereas others have emerged, including the Italian Society for Human Ecology and the multi-national European Association of Human Ecology (EAHE). The EAHE pioneered an "International Certificate in Human Ecology" with partner institutions in France, Portugal, Switzerland, Italy and Belgium. The flagship program on this consortium is based at the Vrije Universiteit Brussel (VUB) in Belgium, where it has offered graduate level education in human ecology since 1972. Other programs established at this time (e.g. at Gothenburg and Lund Universities in Sweden) also continue to offer advanced studies. A number of others have been developed elsewhere. In sum, academic degree programs in human ecology now appear to be in place in most European countries.

Irrespective of a long and varied history during human ecology's mono-disciplinary phase (Cf. Shepard 1967; Young 1974, 1983; Borden 1989, 1991), the United States was a late comer in the trend toward inter-institutional organization. The United States-based Society for Human Ecology (SHE) did not become active until the mid 1980s. Since then, however, it has been increasingly influential at both national and international levels. The primary aim of SHE has been to provide a collaborative forum for human ecology scholars, educators and practitioners. The backbone of these activities is a consistent schedule of international conferences, beginning in 1985 at the University of Maryland and running through to SHE XVI hosted by the Huxley College of the Environment in Bellingham, Washington in 2008. SHE's meetings have been held across the United States, in Canada and Mexico, as well as in Europe and South America. The organization's initial goal was to support collaboration among United States institutions of higher education. By the mid-1970s dozens of academic programs of human ecology had sprung up across the country. Some were newly re-organized colleges within large land-grant universities[1] (such as Cornell, Rutgers and Michigan State Universities, among many others). Others grew out of scientific ecology programs (e.g.: University of California—Davis) and within a range of other smaller universities and colleges. College of the Atlantic (COA) in Bar Harbor, Maine—founded in 1969 as the first institution of higher education fully dedicated to the interdisciplinary study of human ecology—would also play a major role. In addition to support for human ecologists through conferences and professional networking, SHE has also been a stimulus for scholarly publication—with edited volumes from many of its conferences, as well as through a high quality peer-reviewed journal "Human Ecology Review".

Even from the beginning SHE has played a dual role: one which follows its founding mission within the United States; and a second function, as a semi-formal nexus for broad-based international coordination. This began as early as SHE's first conferences, which attracted leaders in human ecology education from around the globe. One of the first international partnerships growing out of these events was the UNESCO sponsored seminar on the integration of environmental education into general university teaching held in Brussels in 1989 (Susanne et al. 1989). Subsequent exchanges have continued a rich cross-fertilization among educators and inter-institutional collaborations. Some of these collaborations—for instance between EAHE, the Italian Society for Human Ecology, the Japanese Society for Human Ecology and the International Association of Ecology—created additional forums for exploring educational issues within INTECOL's International Congresses of Ecology. These events, and the resulting books (Suzuki et al. 1991; Hens et al. 1998), provide in-depth summaries of educational models for human ecology education around the world. (For a more detailed historical account of the Society for Human Ecology and its collaborative activities, see Borden 2008.)

Beyond Humans and Nature—Entering the Virtual World

Looking back across the brief history of interdisciplinary human ecology, we are reminded of how much the world of communication has changed. In the 1970s and 1980s the interchange of ideas was slow—limited by the rhythms of periodic face-to-face conferences, printed communications, and what is now called "snail mail". By the 1990s things were rapidly changing. The introduction of e-mail was a tremendous facilitator of interpersonal communication. Not long thereafter, the World Wide Web brought all of us even closer together. The search for academic literature and institutional information was at our finger tips. We are in the midst of discovering the possibilities and promise of these innovations. Their potential throughout society is only beginning to register; and within higher education what we teach and how we teach is in a state of vibrant flux.

For human ecologists, these are especially dramatic changes. Long standing difficulties of finding and communicating with colleagues globally are evaporating. Powerful search engines greatly simplify discoveries of complex, multi-layered topics. What only a few years ago could take countless hours to find is now available almost instantly. A recent internet search of the term "human ecology education" returned, in a matter of seconds, nearly 2,000,000 hits! More compound searches—across any sub-themes of interest—are as easily rewarded. For an interdisciplinary, international field like human ecology, these are indeed powerful tools.

Conversely, there is also a two-way flow. Not only can we find things more readily. It is far easier to be found—to extend our interests, share views and coordinate communication. The impact of these opportunities has already altered the community of human ecologists. Recent SHE conferences have been entirely planned and coordinated via email and web-based programming. All surface and air mail has been eliminated. The program on which this chapter is based—even though its final presentation was done in person before an audience in beautiful Rio de Janeiro—was formulated, re-fashioned and rehearsed within an interactive "virtual" learning environment.

These are spectacular changes in how we do our work. The possibilities for collaboration and mutual exchanges are enormous. They are transforming how we—as educators and as human ecologists—can build the future of our profession. They also raise compelling questions about how we can (and must) modify our teaching methods. Our students are no longer living in the world in which their teachers were educated. As educators, therefore, we not only must catch up and keep up with our students; we must also invent the means to guide them in preparing for a wholly new kind of world.

The standard definition of human ecology proffers itself as the study of the inter-relations among humans and their environment—the complex world of which humans are a part. We are only beginning to discover the powerful tools this "virtual" world has given for that understanding. But there is more, perhaps: this new and rapidly expanding domain of informational connectivity may be itself a new world. However we choose to view it, there is little doubt that it is world-changing. We should not be blind to its limitations, or too ready to celebrate new technology. However, there is exciting potential that the internet and related methods of communication has for building learning partnerships between the traditional bastions of human ecology in North America and Europe with new centres in Asia, Africa, Latin America and Oceania. The forms of these learning collaborations are currently evolving and developments will no doubt be reported to future SHE conferences and to other fora and through other publications.

The discussion now moves to the background, current activities and new and future directions of three old institutions for human ecology, in order to provide models of the kinds of approaches to human ecology education that could be adopted or adapted in a future of enhanced international collaboration. The discussion starts with the Human Ecology program at the Australian National University.

Models for Institutional Change—Bringing New Ideas into Practice

The Australian National University

The ANU is Australia's only National University, funded by the federal government. It is located in the nation's capital city, Canberra, on 145 hectares of buildings amongst gardens, sporting fields and bushland. It is relatively small, with around 13,500 students and some 3,600 staff. It is a research intensive university and prides itself on melding its teaching activities with that research.

Origins of Human Ecology at the ANU

Founded in 1946 in the spirit of post-World War Two nation-building, the ANU was set up as a small community of well funded elite researchers, with a broad and largely independent mandate to work in the "national interest". In the 1960s the ANU was merged, rather unhappily at the time for both parties, with a pre-existing teaching college—the University College of Canberra. Although the combined entity was called "The Australian National University" for many years it continued to function as two separate institutions, with research activities concentrated in the elite

Institutes of Advanced Studies (IAS) and undergraduate teaching conducted by academics in the School of General Studies.

This research-teaching division within the ANU was influential in the formation of a program of study in Human Ecology by Stephen Boyden in the 1970s. Boyden was recruited into the IAS at the John Curtin School of Medical Research, in 1960 to work on cell biology and immunology (Fenner and Curtis 2001: 73). However, his interests were on much broader understandings of human health and well-being. These interests developed, in part, through conversations with Rene Dubos whilst they were both researchers at the Rockefeller Institute for Medical Research in New York. Dubos and Boyden were concerned with a holistic approach to understanding the "human condition". Their common background as biologists led them to start from fundamental evolutionary principles, but they realized that their thinking had to embrace the cultural domain of human knowledge, values and beliefs (Boyden 2003). Dubos went on to make a number of foundational contributions to "the sustainability debate", including involvement with the United Nations Stockholm conference on humans and the environment, at which the slogan "think globally, act locally" was launched.

The intellectual freedom that was encouraged within the IAS, allowed Boyden to continue to pursue these lines of thought, helped in no small way by a sympathetic Head of Department, Frank Fenner (who had himself worked with Dubos) and the encouragement of the then Vice Chancellor, Leonard Huxley. This latitude extended to allowing Boyden to spend a year researching the cultural dimensions of health at the Research School of Social Science, whilst still employed as a Professorial Fellow in Experimental Pathology. It is questionable whether such arrangements would be tolerated, let alone encouraged, in today's climate of institutional "efficiency".

Following this, Boyden formed an Urban Biology Group, later known as the Human Ecology Group, based within a new Department of Human Biology (Fenner and Curtis 2001). In 1970 he organized an international conference on "Education and the Environmental Crisis" as a response to emerging concern about the state of the global environment (Brown 1978). The conference resolved that tertiary education had a duty to research and learning into environmental problems, and it urged participating universities to seek funds "to establish courses designed to give a comprehensive understanding of the human situation" (quoted in Brown 1978: 171). As a consequence, Boyden proposed that the ANU should confer a new degree, the Bachelor of Human Ecology, as a program of study dedicated to interdisciplinary studies of human-environmental relationships.

However, within the institutional structure of the ANU, undergraduate teaching was the responsibility of the School of General Studies, and new programs and degrees required the approval of the disciplinary-based faculties. Whilst the interdisciplinary ideas behind the proposed human ecology program might have been tolerated within the IAS, they were far less welcome to the faculties. The Faculty of Arts objected that there was "little chance of the proposed integrating units being accepted, because not enough attention would be paid to the cultural aspects of the study of man". The Faculty of Science response was that there would be "insufficient attention to scientific method" and would not approve such a program unless there was "more scientific rigour". Both found agreement that "if a course was Science it could not be Arts and vice versa" (quotes are all from Brown 1978).

It took the enthusiasm of a few key academics on the ground floor, who were prepared to put in their time and effort, and the support of a few key senior officers, to overcome faculty resistance to this new approach. The presence of a small number of powerful figures to "pull" change, in combination with separate "push group" of junior players willing and able to deliver outcomes, seems not uncommon in overcoming structural inertia to institutional change (Barnett and Brown 1983).

Eventually, in 1973, a program in Human Ecology was first offered within both Arts and Science as:

> A sequence of interdisciplinary units on the interplay between human societies and the environment designed to encourage students to work towards an integration of their previous and existing training in various other disciplines. (ANU 1973)

The courses in the program were from the very beginning based on an integrative conceptual framework which encouraged thinking and communication about the dynamic interrelationships in human ecosystems. This framework, which was an essential feature of the program, recognised the crucial role of human culture as a powerful force in the human ecosystems. Consequently, because of the nature of the phenomena that they sort to understand, the courses were necessarily transdisciplinary. The courses were very much concerned with the underlying principles governing the cause-and-effect interrelationships in the system, and this approach was later covered in a number of books, notably *Western Civilisation in Biological Perspective* (Boyden, 1987). Furthermore, the group that developed the program was at the same time engaged in designing and running the first study of the ecology of an urban ecosystem, as recorded in *The Ecology of a City and its People: the Case of Hong Kong*

(Boyden et al. 1981). This research had an enormous impact on the content and intellectual nature of the undergraduate program, and established the tight relationship between research and teaching as a feature of the program from the beginning.[2]

Today

The Human Ecology program has gone on to offer all levels of study from undergraduate majors to PhD. In this time the cultural context in which the program has operated has, of course, changed often. Teaching and learning about "stable societies" (Goldsmith 1970) and the "structure of permanence" (Clarke 1977) has changed over time to the concern with "sustainability", which now in turn is changing further to notions "eco-literacy", or "biosensitive society" (Boyden 2004) and guiding the transitions to such societies. This dynamic has also changed some of the primary objectives of the program and the means by which its learning outcomes have been delivered. However, broadly speaking the program and its various courses has been characterized by an emphasis on the following key notions:

- Human Ecology is a legitimate and valid knowledge domain in its own right.
- Important though they are, the world is not divided into "disciplines"—human thinking is. Integrating these disciplinary (and other) knowledges is a distinct methodological challenge.
- Systems-thinking provides a powerful way to "get at" an understanding of the complex, multi-scaled, interactions that characterize human-ecological situations.
- Students learning should be made personally meaningful through field trips and other experiential activities, including the students' everyday lives.
- "Normal" values and everyday behaviour account for much of a society's ecological signature, so encouraging critical engagement with the "taken for granted" is crucial to imagining alternatives
- Overall, the courses are not so much concerned with cataloguing the problems of today as envisioning ethical, sustainable and worthwhile futures

Boyden once remarked that "for some reason, human ecology has often elicited quite extraordinary opposition from elsewhere in academia", a phenomena he notes is "not restricted to the Australian National University" (Boyden 1995). True as this claim was at the time of the program's founding, and for much of its history, there are now some encouraging signs that this hostility is abating. There seems to be a growing recognition of the need for dedicated interdisciplinary centres for "sustainability science", or similarly named programs, to work alongside the more traditional disciplines in tackling the complex environment-society problems of humankind today. New directions at the ANU have seen the recent establishment of the Fenner School of Environment and Society, named after Frank Fenner, and now home to the Human Ecology program. The mission of the Fenner School is "to build and sustain a world-class, nationally-distinctive School at the ANU for transdisciplinary[3] research and education on complex environment-society systems" (Steffen 2008). The Fenner School houses a wide range of both social and environmental scientists, as well as Human Ecologists, who collaboratively research and teach into various aspects of "sustainability science", very broadly understood. The school also functions as a "hub" or nexus point for collaborative research across the campus, and with partner institutions globally.

These teaching and research activities are matched by a commitment from the ANU to "practise what it preaches", or "walk the talk". Amongst other commitments to sustainability practice the ANU is a signatory to the Talloires Declaration and plays a lead role in Australasian Campuses Towards Sustainability (ACTS), a regional umbrella body for promoting principles of ecologically sustainable development in the tertiary education sector. Major building and works initiatives have been initiated to deliver campus sustainability measures. These developments are joined by environmental management programs aimed at reducing the environmental impact of the ANU campus operations, including targets areas in greenhouse gas emission reduction, water conservation, urban biodiversity management, sustainable transport, environmental risk and so forth. Wherever possible, and in keeping with the education principles discussed above, student research into these campus sustainability projects is encouraged. Consequently, the operational arm of the university—which is all too often only seen as a "support service" for academics—is harnessed as a teaching and learning resource. Students from a broad range of disciplines have been involved in projects over the years, including students of human ecology as well as economics, accounting, visual arts and music, creative writing, ecology, resource management, marketing, law and engineering. In many cases student projects can be made suitable for academic assessment at various levels, from 1st year essays, later year coursework projects, independent research projects, through to Honours and Masters theses. Above all, though, students derive immense satisfaction from seeing their project findings put into "real practice" and this strongly reinforces their learning outcomes (Dyball and Carpenter 2006; Carpenter and Dyball 2006; McMillin and Dyball 2008).

The Future

It would seem that the time is now very conducive to the approaches to learning and research that characterize Human Ecology. Ideas and proposals that would have been dismissed as "irrational" or "unfeasible" at the time the program was founded, and through much of its life, are now seemingly much more acceptable, both within academia and in community, businesses, farming and government circles. For many issues it is a time to deliver on the arguments and advocacy that have been made over the previous decades; to "bring home the bacon", as would be said in Australia. This is all welcome indeed, but it is important that in delivering on the ground prepared for us by yesterday's thinkers we continue to prepare future grounds for today's students to move into when their time comes.

The particular circumstances that surround human ecology at the ANU are distinct from those that framed human ecology at the College of the Atlantic, yet a number of parallel influences and kindred motivations can be detected, as this next account discusses.

College of the Atlantic—Mission and Background

College of the Atlantic was founded in 1969 as the first institution of higher education in the United States fully dedicated to the study of human ecology. The college is located on Mount Desert Island, Maine—250 miles north-east of Boston. Its ocean-side campus is in close proximity to the scenic Acadia National Park and the popular resort town of Bar Harbor. From the outset, COA has sought to develop human ecology as both an academic and applied field. It also has been an active participant in many of the international human ecology networking activities described above.

Founded on the premise that a new educational philosophy was needed to prepare students to address the world's social and environmental challenges, the college is patterned around several distinctive principles. One of them is its unique, interdisciplinary faculty structure. The college has no departments or academic ranks in the traditional sense. Instead, faculty work together on design and implementation of the curriculum as a whole. Another distinctive feature is the college's commitment to self-directed studies. While all students obtain either a Bachelor of Arts (B.A.) or Masters of Philosophy (M. Phil.) in Human Ecology, no two students take the same course of study.

The bridge between the resources of a non-departmentalized, interdisciplinary faculty and students' individually designed degree plans is provided by the college's problem-focused mission. This takes place at several levels. First, each student works closely with academic advisors to define their own focal themes and educational path. They individually explore and engage in effective ways of interrelating disciplines and of relating thought and practice. This process begins with a first-year interdisciplinary, foundational curriculum which expands in the second and third years through self-directed studies. As it gains focus, it culminates in a term-long internship experience, a self-reflective philosophical "human ecology essay" and a senior thesis. Second, students are given opportunities to work with other students and faculty in a wide variety of issue-centred teams of research, outreach and applied projects. Finally, as an institution, the college holds a long-standing commitment to innovation and leadership in education and applied human ecology—regionally, nationally and internationally. Beyond these interdisciplinary and integrative academic features, COA is also distinctive in terms of student diversity. Through a partnership with the Davis International Scholars Program, which supports highly qualified students from around the globe, the COA community contains the greatest percent of international students of any four-year liberal arts college in the United States.

These elements have combined into a remarkable record over the past four decades. In the first year of the college, for instance, COA students were instrumental in getting a groundbreaking beverage container recycling bill through the Maine state legislature—the first in the country. Many other achievements have followed: construction of the first solar home in the region to qualify for a conventional mortgage; leadership in collaborative planning and ecological policy development from local to international levels; and notable accomplishments in conservation and applied ecology—ranging from endangered plant protection and reintroduction of multiple avian species, to creation of the "Allied Whale" research centre for international marine mammal research and policy studies (Rabineau and Borden 1991; Hall 1994; Clark 1997; Koffman and Borden 1998).

Sustainability—Leadership and Practice

As has been discussed above, "sustainability" is a relatively recent term in the environmental lexicon. Nonetheless, its underlying ideals have been a hallmark of the college from the beginning. They have been a continuous thread throughout the curriculum, as well as a significant dimension of outreach initiatives. Over the years, the college community has maintained an unbroken dialog on "ecologically responsible" institutional standards and practices. The range is broad: chlorine and dioxin free office supplies, energy-efficient lighting and heating, organic and regional food sources (increasingly from the college's farm and local growers' cooperatives), recycling, composting, sustainably-managed forest products for building materials, and so on.

In the 1990s these themes became increasingly popular throughout higher education elsewhere—not only as an object of study in environmental studies programs, but also as an arena for institutional leadership at the highest levels. A diverse constellation of inter-institutional initiatives began to form, e.g., the Talliores Declaration of university presidents committed to sustainability practices, the National Wildlife Federation's "Campus Ecology Program", Second Nature's models of best practices, the Earth Charter's 16 principles of sustainability (begun in Rio in 1992 and ratified in 2000 in the Hague) and, of course, the 1997 UNESCO Thessaloniki Declaration's decade of education for sustainability.

For College of the Atlantic this was an exciting opening and a significant challenge. The college began to find a new audience for its educational mission. It was a time to speak up, to stand out, and to walk the talk. What follows is a brief outline of the college's institutional response.

- 1996 - membership in the Campus Environmental Initiative for the design and sharing of college policies about purchasing practices, waste management, energy efficiency, etc.
- 2002—signatory member of the Talliores Declaration
- 2003—first college to join Maine's STEP-UP program to reduce institutional environmental footprints, contract for renewable electricity services, monitor greenhouse gas emissions, etc.
- 2005—first campus to hold a "zero waste graduation"
  —hiring of a permanent, full-time director of sustainability to ensure a continuing commitments to a sustainable future
  —ratification of the Earth Charter by the college's All College Meeting
- 2006—Board of trustees pledge to establish a Net Zero carbon emissions policy—the first college or university to do so.

—Hosting of the XIV International Conference of the Society for Human Ecology with funds for carbon emission offsets for all participants' travel invested in a green energy savings account for renewable energy sources

- 2007—Net Zero pledge achieved—all greenhouse gas emissions reduced, avoided or offset and third party verifiable—first college or university to achieve carbon neutrality
  —Founding member—American College and University Presidents Climate Commitment (ACUPCC)
  —deemed the "greenest college in the world" by Grist Magazine
- 2008—opening of state-of-the-art residence complex of six low environmental impact, super insulated, energy efficient buildings.

Taken together, these and other activities have brought COA's educational mission full circle. As the college's curricular goals and institutional operations merge toward a seamless integration, they have entered the forefront and broader aims of business and society. What once were considered alternative educational ideas are now in the main stream of higher education.

Leadership in sustainability practices has brought considerable recognition, including the Sustainable Endowments Institute's first "Sustainability Innovator Award". Attention from the electronic and print media has followed as well (Butler 2007; Lewin 2007; Monastersky 2007). The media spotlight is a mixed blessing. Attending to the press can be time consuming and distracting in educational settings. Some visitors seem unable to grasp the bigger picture of human ecology or focus on somewhat unimportant or minor features. But there is another side to the coin. There is satisfaction in seeing the work of building a college validated through the eyes of others. It is gratifying when the Chronicle of Higher Education proclaims "when it comes to sustainability, the College of the Atlantic seems to be walking while the others are merely talking" (Monastersky 2007). It is even better when the New York Times goes further: "While sustainability and the physical environment are certainly major components, the concept of human ecology goes much further, encompassing almost anything...." (Lewin 2007).

The German Society for Human Ecology

The German Society for Human Ecology was founded in the mid 1970s mainly by public health and environment oriented movement of scientists. The society developed into an interdisciplinary group spanning a broad field of disciplines, with participation from Austria, Germany and Switzerland with some branches in Scandinavia and Belgium. The membership was networked across central Europe by annual meetings, the journal GAIA and regular yearbooks. By the end of the 1980s there were chairs in Human Ecology established at some universities, including Zurich and Berlin, but no independent program of study had been established. However, the opportunity for further raising the profile of Human Ecology and the possibility of establishing an independent program was dashed by the restructuring of German universities that followed the so called Bologna process[4]. A project or problem oriented program of study, as promoted by human ecology, now seems to be impossible inside the system of public universities in Germany. Yet there is evidence of a continuing and strong interest by students for a program in Human Ecology. Evidence of this interest is the success of the Austrian program Master in Social and Human Ecology in Vienna and Klagenfurt in 2007. Consequently, there has been a proposal to form and to fund a dedicated College of Human Ecology, not only for Germany but for all of Europe.

The possibility for such a college has become favourable due to recent developments in how institutions for education can establish themselves. In Germany since the 90s private colleges and schools have created a new, exciting and expanding market in response to perceptions that State universities have become overcrowded and ineffective. There are now 74 state-approved—and about a dozen unapproved—private universities and colleges in Germany. Degree programmes in economics, law and social sciences dominate the programs that these private universities offer, but other subject areas, such as art and theology, also exist. Less common are programs in subject areas such as engineering and the natural sciences. Demand for these programs of study is strong and on the increase, with interest from international students, including Austria and Switzerland, augmenting that of native German students (see Brauns 2003). However, as yet there is no private collage offering an interdisciplinary program of study in which students can flexibly put together their own individual curriculum tailored to their own interests and needs. As discussed, such an interdisciplinary, flexible and student oriented approach is beyond the mainstream State institutions.

A College of Human Ecology for Europe

These are the circumstances which, in about 2007, there was a proposal for the foundation of a College of Human Ecology in German-speaking countries—following the model of the College of the Atlantic. The reactions were manifold, but overall the response has developed into a growing campaign of support, both in Germany and beyond. After an initially cautious, reticent or sometimes critical response the proposal has enjoyed an increasing number of much more positive attitudes, and there is now a number of activities underway to promote the venture (Serbser and Mrzljak 2006). Key issues currently being negotiated include the formulation of a curriculum; the location and thus the catchment from which students will be drawn, including the size of the local and international pool and its demographic; the funding conditions and possibilities within Germany and Switzerland, particularly regarding tuition fees, against the costs associated with running a private college, and especially the conditions of providing grants, scholarships and other possibilities of financial sourcing for candidates. In the German context the approach requires a major paradigm change, with the focus switching from an institution being first and foremost concerned for itself, to an institutional approach that is primarily concerned with its students.

To further promote the scheme a study-group within the German Society for Human Ecology was formed. The first meeting of the group outlined the basic principles for the mission and vision of the proposed College, modelled on those of the College of the Atlantic, as refined over 30 years of operating and contributing successfully to the world of education and beyond. A first 5 year start-up funding model for a college, initially based on tuition fees, was generated, with indicative figures that seem very plausible. In essence, a German College of Human Ecology can be expected to operate efficiently and economically with as few as 25 new students enrolments per year. Moreover, the college can also be expected to establish itself as a centre for scientific research in the field of human ecology. However, there is a need to generate founding sources beyond tuition fees in order to make the College accessible to students on low incomes. Again, the model to look at to make this happen is the College of Atlantic, where some level of scholarships are provided to 70 percent of the students enrolled.

The outcome of the meeting was an official statement, the "Higher Education Initiative in Human Ecology", which was designed to promote consideration of the following aspects of the proposed studies in the field of human ecology:

- Teaching methods in social and natural sciences including artistic techniques and creative media;
- important contents of scientific and artistic ways of investigating nature, environment and society;

- the enhancement of expertise in system-thinking and systemic action;
- the development of individual competences within the course of lifelong learning, public campaigning, understanding oriented communication, creative practice of collective problem solving and responsible critical thinking connected to an awareness about the limits of knowledge.

The College Initiative at Work

In September 2007 a working symposium with David Hales, president of the COA, was organized in Berlin. An important question that was discussed at the symposium was, whether a college of human ecology could offer any additional services within the already diversified landscape of study choices in Germany and Switzerland. Providing a project oriented curriculum, the college would offer an almost revolutionary concept regarding the German education system: From day one, students themselves are responsible for the course of study they choose. They are responsible for the study strategies they develop to guide and monitor their very personal route of success, and the college providing just the structural patterns of learning. The tenor of the discussion was that students who reach the Bachelor level in human ecology will understand creativity as an integral part of science and therefore will approach a professional career in a more undeterred way (see Serbser and Mrzljak 2007).

The initiative was eventually presented at SHE in Rio de Janeiro, where important further academic support for the concept was forthcoming. The proposal will continue to be developed and refined, including undertaking a feasibility study and financial budgeting, as further steps towards founding a College of Human Ecology for Europe.

Conclusion

The purpose of education is to prepare young people to anticipate and respond to the needs of society. But the growing complexity of human society and the interlacing relations of humans within the living world are challenging educators, researchers and practitioners everywhere to reorganize their thinking. At all levels, from local to global, people are recognizing a need for entirely new conceptual frameworks and forums for their development. It is a curious fact that the term “human ecology” has been used throughout the world as a name for this confluence of scientific ecology, interdisciplinary education and environmental concerns. No single statement can sum up the many meanings of human ecology. Yet for nearly a century it has been applied to a diverse family of speculative and scientific lines of thought. In a general sense, it points to certain problem areas manifesting the influence of ecological thinking and to various intellectual movements cutting across traditional academic disciplines.

In this chapter we have offered an overview of this history. We have further highlighted its origins, its development and a recent initiative of two established human ecology programs—along with some first steps of an entirely new undertaking. Taken together, a number of conclusions are suggested. First, looking back across the early history of human ecology (e.g., at its mono-disciplinary and multi-disciplinary phases), it is surprising how many separate starting points there have been. As this history was reconstructed by scholars of the field, a robust foundation has been discovered. What were once isolated bodies of theory and research are now gathered together and readily shared among contemporary human ecologists.

Secondly, as human ecology has moved towards a bona fide interdisciplinary field, the integration of its origins has substantially enriched its credibility and status. Along with these changes, the nature of inquiry and teaching has likewise transformed. In the 1980s, when individuals from separate institutions and distant continents began to meet, the central questions revolved around broad philosophical and pedagogical issues, such as "what do you mean by human ecology?" In recent years patterns of thinking about human ecology have become more nuanced. For most people, the living world of ecology—and humans' place in it—is now generally assumed. The search for a unifying theory has been replaced by a shared perspective. This then forms the basis of a multiplicity of foci and richly contextual ways of thinking. As fresh ideas and approaches emerge within this framework, the questions now are increasingly ones of "how to do human ecology?" These features are apparent in recent conferences and publications, which are structured around a growing family of human ecological problems and themes. From "urban ecology" and "adaptive management" to "sustainability" and "food systems", the discourse of human ecology is ever more focused and practical. In short, human ecology has found its feet. By moving from a diverse intellectual inquiry toward a constellation of applications, human ecology has more clearly defined itself professionally—as colleagues gather and share their respective approaches "within" the unifying perspective of human ecology.

Finally, looking ahead, there is little doubt the advent of electronic connectivity is a timely development. From the outset, the complexity of human ecology's interdisciplinary and international mandate was hampered by the limits of communication. The tools—so necessary for collegiality at a distance—perhaps at last are at hand. The challenge now is to keep up with these opportunities, to infuse them into education, and to prepare the next generation of young men and women to discover what it might be to live and do well in a humane, sustainable and worthwhile world.

Endnotes:

[1]In the United States a "land grant" university is a state-supported institution of higher education and research.
[2]Stephen Boyden's contributions to this section on the origins of the program at the ANU are gratefully acknowledged.
[3]The term 'transdisciplinary' that Steffen uses here is for the purpose of this paper taken to be sufficiently close in meaning with Tengström's term 'interdisciplinary', as used in the earlier classification of Human Ecology into distinct phases. For a discussion of the important differences between these terms see Mobbs and Crabb (2002).
[4]The "Bologna Process" is designed to standardize education across Europe, ostensibly to make all degrees offered comparable in standard and compatible between one another. Some see it as having a "normalizing" effect that is inimical to the approaches to learning approaches that characterizes Human Ecology, as discussed here.

Concluding Remarks

PRISCILA LOPES AND ALPINA BEGOSSI

The studies in this book are unified by the challenge to address and achieve sustainability through the interaction of society and the environment in a constantly changing and diverse cultural landscape. Socio-ecological changes and adaptations, and the impact of indigenous behaviour versus industrial societies onto the physical and biological environment associated with potential mitigation actions are examined with theories in human ecology. The core of this volume can be categorised in four domains: 1. human ecology and its current and future developments; 2. sustainability in different cultures; 3.means and alternatives to reach sustainability, and 4. global environmental concerns such as climate change, health, and education.

Human ecology and its current and future developments

Two chapters bring an overview of human ecology with its definition, concepts, and its prospects. Avila-Pires, in Chapter Nine, suggests that Human Ecology is as a discipline of its own, having society mediated by culture within the biophysical environment. The author stresses how traditional universities still have insufficient curricula to answer to the demands of interdisciplinary sciences such as human ecology. Dyball et al. in Chapter Eleven account for the different historical phases of human ecology, having the 20th century being characterized by a moment where human ecology and other ecological concepts extrapolate the limits of sociology and geography, blending urban planning, agriculture, and public health, among others. For a long time, concepts from different disciplines were used together as independent contributions but rarely blended. The ideal interdisciplinary moment necessary to address sustainability is yet to be achieved and has been pursued since the last decade of the 20th century.

Future developments in human ecology are necessarily linked to advances in its formal education as it becomes part of academic curricula. In this context, Dyball et al. highlight the relevance of the use of internet as a means to facilitate information exchange and collaboration between worldwide partners in real time. Such technological advances in the last years are seen as an enormous catalyst to improve learning and educational exchange in human ecology.

Human ecology and sustainability in different cultural contexts

Burns (Chapter Three) analyzes culture within networks of information and values. He approaches how the natural environment is affected by the evolution of culture and how technological innovations, and their cultural perception have dramatically affected human interactions with the environment. The pathways to overcome environmental problems should take into account the institutions embedded into cultures. Nevertheless, due to different temporal turnovers between environmental degradation and cultural adaptations, an increasing lag between ecological problems and cultural solutions are expected. Susan Crate (Chapter Two) explores how culture affects the perception of environmental issues/conflicts. The different ways that people understand and feel the impacts of climate change are argued as fundamental to international climate policy negotiations of solutions, agreements and equal representation. People are expected to change and adapt to new environmental conditions that affect their subsistence mode, but changes at unprecedented rates could lead to social-cultural disorientation with unknown consequences. Morrison and Singh address (Chapter Six) the importance of a deep understanding of indigenous knowledge, as some cultures can have sustainability concepts within them. Even though such concepts can present effective ways to sustainability and alternatives to adaptations by the dominant global consumer culture, they should be carefully approached. The simple adoption of indigenous practices by western societies is not enough. It is necessary to create authentic dialogue between cultures seeking to promote cooperative partnerships towards sustainability.

Human ecology and the roads to sustainability-Methods towards management

New problems require new solutions. The novelty of solutions can be, however, not in the method itself, but in the way such method is applied, as is the case of rediscovering traditional knowledge. Around the world, there are still small groups of people who have been living marginally to industrialized globalization, making a living off of natural resources sometimes in sustainable ways. Whether their practices can lead to global solutions is not known, but they certainly contribute to improve other local initiatives with wider consequences on the conservation of natural resources and on the resilience of social-ecological systems. In this book, six of the chapters address direct or indirectly the importance of understanding, respecting and applying traditional knowledge, both because scientific knowledge benefits from it and because there is a growing concern in keeping the social-ecological sustainability of such traditional groups.

Silvano et al. (Chapter Four) address *local or indigenous knowledge* and the management of aquatic and fishery resources in tropical areas, highlighting how such knowledge can be used where scientific information on natural resources is limited and lack long term databases. Scientists benefit from dialogue with fishers, gathering past information on stocks and on natural history of vulnerable and currently rare fish species. Also, scientists can upwell a process of co-management, involving local fishers in the research, management and conservation of their own fishing resources. Integrating local knowledge into management policies, according to the authors, increases the effectiveness of mitigation and precautionary actions to conserve aquatic resources. Hanazaki et al. (Chapter Five) goes a step further and shows the connections between local knowledge and the use of land and aquatic resources, scrutinizing the dependency that fishers also have on plant resources, having Brazil as a case study. These authors suggest that the use of agrobiodiversity and management of cultivated and extracted plant resources are "ecologically safe" pathways to make the connection of the household to larger scale markets, as fishers' villages are already in a transition process from semi-isolated to semi-urban populations. The diversity of local knowledge systems is viewed as a factor that bridges the household to policies, programs and projects that aims to achieve sustainable agricultural systems. Still within the indigenous knowledge scope, Morrison and Singh argue in Chapter Six that current societies have lost their capabilities to be aware and co-evolve with the ecosystems they live in. Taking into account the tools of indigenous knowledge through interdisciplinary frameworks, truly collaborative approaches can be constructed to facilitate sustainable outcomes.

Adaptive co-management, where people "learn by doing" in a partnership arrangement among local populations, governments and NGO's, is suggested by Seixas et al. (Chapter Seven) as a tool to develop initiatives to manage aquatic resources. The existence of supportive local organizations (institutions) facilitates the dialogue between resource-users and the government. Nevertheless, the authors draw attention to the inexperience of local communities and the governments themselves in co-management processes, demonstrated for example by the disempowerment of local communities and by the tendency of establishing centralized actions. This can result in discouragement and low participation of local people, decreasing the chances of success of such initiatives.

Coastal management has two modelling procedures highlighted in this book. Firstly, Glaeser et al. (Chapter Eight) propose concepts and methods from both social and natural sciences that can be applied to coastal and marine research. According to these authors, integrated coastal zone management (ICZM) holds potential if it considers the complexity of natural, economic, social and cultural factors. It is recommended an explicit use of knowledge systems where local knowledge must be included in addition to scientific and managerial knowledge when making policy decisions. Finally, using the example of the ECOST Project, a multilateral European project, Serge et al. (Chapter Ten) suggest alternatives to model the complexity of ecological, economical and social interactions for marine environment management. This project aims to include multiple costs (ecological, economic and social) usually neglected by mechanisms of market analysis to give a better estimate on the value of fishing activities. For that, a difficult goal should be achieved, the understanding of the values given by society to the sea. By extending and embodying previous more simplistic models and developing integrative methodologies, the ECOST project attempts to contribute to fisheries policy making and management.

Human Ecology and current global environmental concerns

Beginning with a review on slash-and-burn cultivation, a common agricultural practice throughout the tropical regions, Pedroso-Junior et al. (Chapter One) lays out all the pros and cons of this practice. Although mostly done in small-scale, slash-and-burn cultivation contributes to the expansion of secondary tropical forests, soil impoverishment and is an agent of deforestation and climate change. Changes in environmental and economic policies, market-driven forces and access to technologies can accelerate the destructive scale of this agricultural practice, especially by reducing fallow periods. There are alternatives, however, such as the strengthening of market chains in a way that the small producer can have better profits without increasing the cleared areas. The planting of different cultures, such as leguminous trees, can replenish part of the carbon stock lost by the deforestation phase. Such agroforestry systems are viewed by the authors as sound alternatives, especially considering that millions of people depend upon slash-and-burn cultivation to make a living. Besides, if done in small scale with adequate fallow periods, it can maintain variability and even increase biodiversity.

Susan Crate's anthropological review on climate change (Chapter Two) stresses the importance of the engagement of anthropologists in issues of climate change. Since climate changes have cultural implications on lives of different human groups including the reorganization of these groups in extreme situations, anthropologists can act as practitioners and educators, by translating concepts such as sustainability and helping to put them into practice. Crate uses the Hurricane Katrina as an example of how anthropologists can use participatory research to increase a group's capacity for adaptation. By using this example, Crate also demystifies the idea that climate change effects are yet felt only by the ones who live in extreme environments: Katrina brought the issue of climate change to the American domestic agenda. Finally Crate relates consumption in industrialized countries and climate change, as the first is one of the main drivers of the last. Anthropologists should play a role in transforming the consumer culture into a culture of sustainability, by bringing new models and using local knowledge and perceptions in collaborative researches.

Current human ecology has the challenge of understanding the relationship between human and nature and human behavioural patterns across various scales of simultaneous change. Under the human ecology umbrella, global concerns such as large scale deforestation and climate change can be addressed under multiple perspectives, where ecologists, anthropologists, economists and social scientists, among others, work together to find solutions that could mitigate the fast changes witnessed in these times and the unforeseen impacts on Earth and its biota.

Appendix A

2nd LOICZ Symposium on "Social-Ecological Systems Analysis" - List of Presentations at Rio Conference

Conference held in Rio de Janeiro, at the International Conference of the Society for Human Ecology (SHE), 5 October 2007

Symposium Organizers: Bernhard Glaeser, Marion Glaser, Michael K. Orbach

SES Session 1: Planning and understanding SES (October 5, 8:00-9:30 AM)
Session chair: Bernhard Glaeser
Co-chairs: Marion Glaser, Laurence Mee

Glaeser, Bernhard and Glaser, Marion:
Understanding complex social-ecological dynamics: Integrating different geographical scales in coastal systems
(Presentation of the symposium input paper, co-authored by B. Glaeser and M. Glaser)
Wang, Rusong:
Hainan Island: An Integrative and Adaptive Eco-province Planning and Management
Xue, Julia:
A Social-ecological analysis on the Eco-Province Approach of Shandong and the Jiaolai Marine Canal
Bruckmeier, Karl:
Knowledge Management in Social-ecological Systems. Experiences from ICZM in European Countries
Mee, Laurence:
European Lifestyles and Marine Ecosystems: Innovations in predictive modelling of social-ecological systems

SES Session 2: Inter- and transdisciplinarity (October 5, 10:00-11:30 AM)
Session chair: Michael K. Orbach
Co-chair: Bernhard Glaeser

Orbach, Michael K.:
The "Total Ecology" of Environmental Issues
Santoro, Francesca M.:
A transdisciplinary approach to coastal planning
Yanurita, Dewi, M. Yusran, and M. Nell:
Patron-Client Institution as a Pathway to Strengthen Socio-Ecological Resilience of Coastal Community in Spermonde Islands, Indonesia
Ramesh, R.:
Socio-Ecological Impacts and Integrated Coastal Management: A case Study from North Chennai Coast, India
Das, Saudamini:
Storm Protection Services of Mangroves and Super Cyclone of October 1999

SES Session 3: People in rural communities (October 5, 1:30-3:00 PM)
Session chair: Laurence Mee
Co-chairs: Marion Glaser, Bernhard Glaeser

Blandtt, Lucinaldo:
A systemic resilience in rural communities of the Brazilian Amazon:
An analysis for sustainablilty and the participative intervention of social programs
Glaser, Marion:
Mangroves and People: A Social-Ecological System on the Caeté peninsula, Pará, North Brazil
Krause, Gesche:
Options for resilience in a co-evolutionary social-ecological framework: Employing scenarios for a sustainable future
Maneschy, Maria Cristina:
Becoming a "pescadora": A gender-based view of vulnerabilities and resilience in North-Brazilian coastal fisheries
Huang, Jinlou:
The dynamics and adaptive management of costal mudflat wetland social-ecological system in Yanchen, China

Bibliography

Abizaid, Christian, and Oliver T. Coomes. 2004. Land use and forest fallowing dynamics in seasonally dry tropical forests of the southern Yucatan Peninsula, Mexico. *Land Use Policy* 21: 71-84.

Abbott, Isabella A. 1996. Ethnobotany of seaweeds: Clues to uses of seaweeds. *Hydrobiologia* 326-327: 15-20.

Abreu, Casimiro de. 1930. *Dialogos das grandezas do Brasil.* Rio de Janeiro: Academia Brazileira de Letras.

Acheson, James M. 2006. Institutional failure in resource management. *Annual Review of Anthropology* 35: 117-134.

Adams, Cristina. 1994. As florestas virgens manejadas. *Boletim do Museu Paraense Emílio Goeldi, Série Anthropologia* 10: 3-20.

Adams, Cristina. 2000a. *Caiçaras na Mata Atlântica: pesquisa científica versus planejamento e gestão ambiental.* São Paulo: Annablume.

Adams, Cristina. 2000b. As roças e o manejo da Mata Atlântica pelos caiçaras: uma revisão. *Interciência* 25: 143-150.

Adams, Cristina. 2003. The pitfalls of synchronicity: a case study of the Caiçaras in the Atlantic rainforest of south-eastern Brazil. In *Ethnographies of environmental under-privilege*, org. David G. Anderson, and Eva Berglund, 19-31. Oxford: Berghahn.

Adams, Cristina, and Rui Sérgio Sereni Murrieta. 2008. Agricultura de corte e queima e florestas tropicais em um mundo em mudança. *Boletim do Museu Paraense Emílio Goeldi, Ciências Humanas* 3, no.2: Foreword.

Adger, W. Neil, Nigel Arnell, and Emma Thompkins. 2005. Adapting to climate change: perspectives across scales. *Global Environmental Change* 15: 75-76.

Adriaanse, A., Bringezu, S., Hammond, A., Moriguchi, Y., Rodenberg, E., Rogich, D. and Schütz, H. 1997. *Resource flows: the material basis of industrial economies.* Washington, DC: World Resources Institute.

Agenda 21. 1993. *Agenda 21: Earth Summit–The United Nations programme of action from Rio.* New York: United Nations Publications.

Agrawal, Arun. 1999. On power and indigenous knowledge. In *Cultural and spiritual values of biodiversity*, ed. Paul Sillitoe. London: Intermediate Technologies Publications.

Agrawal, Arun. 2002. Indigenous knowledge and the politics of classification. *International Social Science Journal* 173: 287-297.

Ahmed, Mahfuzuddin, Ana D. Capistrano, and Mokammel Hossain. 1997. Experience of partnership models for the co-management of Bangladesh fisheries. *Fisheries Management and Ecology* 4, no. 3: 233–248.

Aiono-Le Tagaloa, Fanaafi. 2003. *Tapuai: Samoan worship*. Samoa: Malua Printing Press.

Alarcon, Daniela T., and Alexandre Schiavetti. 2005. O conhecimento dos pescadores artesanais de Itacaré sobre a fauna de vertebrados (não peixes) associados às atividades pesqueiras. *Gestão Costeira Integrada* 4: 1-4.

Albuquerque, Ulysses P. de, and Reinaldo Farias de P. Lucena. 2005. Can appearance affect the use of plants by local people in tropical forests? *Interciencia*. 30, no. 8: 506-511.

Alegre, Julio C., and D. K. Cassel. 1996. Dynamics of soil physical properties under alternative systems to slash-and-burn. *Agriculture, Ecosystems & Environment* 58: 39-48.

Alexander, Jeffrey C. 2003. *The meaning of social life: a cultural sociology*. New York: Oxford University Press.

Alford, Donald. 1992. Streamflow and sediment transport from mountain watersheds of the Chao Phraya Basin, northern Thailand: A reconnaissance study. *Mountain Research and Development* 12: 257-268.

Almeida, Oriana T., and Christopher Uhl. 1995. Developing a quantitative framework for sustainable resource-use planning in the Brazilian Amazon. *World Development* 23: 1745-1764.

Altieri, Miguel A. 1999. The ecological role of biodiversity in agroecosystems. *Agriculture, Ecosystems & Environment* 74: 19-31.

Altieri, Miguel A., M. Kat Anderson, and Laura C. Merrick. 1987. Peasant agriculture and the conservation of crop and wild plant resources. *Conservation Biology* 1: 49-58.

Amundsen, Randi. 2005. *The changing roles of the embryo in evolutionary thought: structure and synthesis*. Cambridge: Cambridge University Press.

Anderies, Jon Marty, Brian Harrison Walker, and Ann P. Kinzig. 2006. Fifteen weddings and a funeral. *Ecology and Society* 11(1): 21, http://www.ecologyandsociety.org/vol11/iss1/art21/ (accessed November 20, 2008).

Anderson Jr., Eugene N. 1969. Sacred fish. *Man, New Series* 4, no.3: 443-449.

Anderson, Lee. G. 1986. *The economics of fisheries management.* Caldwell, New Jersey: Blackburn Press.

Andrade, German I., and Heidi Rubio-Torgler. 1994. Sustainable use of the Tropical Rain Forest: evidence from the avifauna in a shifting-cultivation habitat mosaic in the Colombian Amazon. *Conservation Biology* 8: 545-554.

Andreae, Meinrat O. 1991. Biomass burning: its history, use and distribution, and its impact on environmental quality and global climate. In *Global burning: atmospheric, climatic and biospheric implications,* ed. Joel S. Levine, 3-21. Cambridge, MA: MIT Press.

Andrew, Alex M. 2000. Self-organisation in artificial neural nets. *Kybernetes* 29, no. 5/6: 638

Andriesse, J. P., and R. M. Schelhaas. 1987a. A monitoring study on nutrient cycles in soils used for shifting cultivation under various climatic conditions in tropical Asia. ll. Nutrient stores in biomass and soil. *Agriculture. Ecosystems and Environment* 19: 285-310.

Andriesse, J. P., and R. M. Schelhaas. 1987b. A monitoring study on nutrient cycles in soils used for shifting cultivation under various climatic conditions in tropical Asia. Ill. The effects of land clearing through burning on fertility level. *Agriculture. Ecosystems and Environment* 19: 311-332.

Angelsen, Arild. 1995. Shifting cultivation and "deforestation": a study from Indonesia. *World Development* 23: 1713-1729.

ANU (The Australian National University). 1973. *Undergraduate handbook.* Canberra: Australian National University.

Archer, Margaret S. 1988. *Culture and agency: the place of culture in social theory*. New York: Cambridge University Press.

Argyris, Chris, and D. Schon. 1978. *Organizational learning: a theory of action perspective.* Reading, MA: Addison-Wesley.

Armitage, Derek, Fikret Berkes, and Nancy Doubleday, ed. 2007. *Adaptive co-management: collaboration, learning, and multi-level governance*. Vancouver: University of British Columbia Press.

Arnason, Ragnar. 1999. Cost of fisheries management: theoretical and practical implications. Paper presented at the XIth annual conference for the European Association of Fishery Economists, April 6-10, in University of Iceland, Iceland.

Arrow, Kenneth, Gretchen Daily, Partha Dasgupta, Simon Levin, Karl-Göran Mäler, Eric Maskin, David Starrett, Thomas Sterner, and Thomas Tietenberg 2000. Managing Ecosystem Resources. *Environmental Science and Technology* 34: 1401-6.

Ashby, W. Ross. 1956. An introduction to cybernetics. London: Chapman and Hall.

Aswani, Shankar, and Richard J.Hamilton. 2004. Integrating indigenous ecological knowledge and customary sea tenure with marine and social science for conservation of bumphead parrotfish (*Bolpometodon muricatum*) in the Roviana Lagoon, Solomon Islands. *Environmental Conservation* 31: 1-15.

Atkinson, Giles, Simon Dietz, and Eric Neumayer, ed. 2007. *Handbook of sustainable development*. Cheltenham, GL: Edward Elgar.

Attiwill, Peter M. 1994. The disturbance of forest ecosystems: the ecological basis for conservative management. *Forest Ecology and Management* 63: 247–300.

Aulin, Arvid. 1987. Cybernetic causality 111: the qualitative theory of self-steering and social development. *Mathematical Social Sciences* 14: 101-140

Ávila-da-Silva, Antônio O. 2005. Produção pesqueira marinha do Estado de São Paulo no ano 2004. *Série Relatórios Técnicos do Instituto de Pesca* 20: 1-40.

Avila-Pires, Fernando D., Luiz C.Mior, Vilena P. Aguiar, and Suzana R.M.Schlemper. 2000. The concept of sustainable development revisited. *Founations of Scence* 5, no. 3:261-268.

Aweto, Albert O. 1981. Secondary succession and soil fertility restoration in Southwestern Nigeria. *Journal of Ecology* 69: 601-607.

Ayres, Robert U., and Udo Ernst Simonis. 1994. *Industrial metabolism: restructuring for austainable development.* Tokyo, New York, Paris: United Nations University Press.

Balée, William. 1989. Cultura na vegetação da Amazônia Brasileira. In *Biologia e ecologia humana na Amazônia: avaliação e perspectivas,* ed. Walter A. Neves, 95-109. Belém: Museu Paraense Emílio Goeldi – CNPq.

Balée, William C., and David G. Campbell. 1990. Evidence of successional status of liana forest (Xingu river basin, Amazonian Brazil). *Biotropica* 22: 36-47.

Barab, Sasha A, Miriam Cherkes-Julkowski, Rod Swenson, Steve Garret, Robert E. Shaw, and Michael Young. 1999. Principles of self-Organisation: Learning as Participation in Autocatakinetic Systems. *The Journal of the Learning Sciences* 8, no. 3/4: 349-390.

Barlett, Peggy, and Benjamin Stewart. 2009. Shifting the university: faculty engagement and curriculum change. In *Anthropology and climate change: from encounters to actions*, ed. Susan A. Crate and Mark Nuttall, 356-369. Walnut Creek, Ca: Left Coast Press.

Barnett, Anthony, and Valerie Brown. 1983. The theory of biology and the education of biologists: a case study. *Studies in Higher Education (Society for Research in Higher Education)* 8, no. 1: 23-32.

Barrera-Bassols, Narciso, and Joseph Alfred Zinck. 2003. Ethnopedology: a worldwide view on the soil knowledge of local people. *Geoderma* 111: 171-195.

Barros, Fábio, Margarida M. R. F. Melo, and Silvia A. C. Chiea. 1991. *Flora fanerogâmica da Ilha do Cardoso: caracterização geral da vegetação e listagem das espécies ocorrentes v. 1.* São Paulo: Hucitec.

Barrow, Colin J., and H. Hicham. 2000. Two complimentary and integrated land uses of the western High Atlas Mountains, Morocco: the potential for sustainable rural livelihoods. *Applied Geography* 20: 369–394.

Barrows, Harlan. 1923. Geography as human ecology. *Annals of the Association of American Geographers* 13, no. 1: 1-14.

Bartlett, Frederick C. 1932. *Remembering*. Cambridge: Cambridge University Press.

Barton, John, and Tim Haslett. 2007. Analysis, synthesis, systems thinking and the scientific method: rediscovering the importance of open systems. *Systems Research and Behavioral Science* 24, no. 2: 143-155.

Basso, Keith. 1996. Wisdom sits in places: notes on a western Apache landscape. In *Senses of Place*, ed. Keith Basso and Steven Feld. Santa Fe, 105-150. Santa Fe: School of American Research Press.

Bateson, Gregory. 1972. *Steps to an ecology of mind: collected essays in anthropology, psychiatry, evolution and epistemology.* Chicago: University of Chicago Press.

Bayley, Peter B., and Miguel Petrere Jr. 1989. Amazon fisheries: assessment methods, current status and management options. *Canadian Special Publications on Fisheries and Aquatic Sciences* 106: 385-398.

Bebbington, Anthony. 1993. Modernization from below: an alternative indigenous development? *Economic Geography* 69, no. 3: 274-292.

Beck, Ulrich, Anthony Giddens, and Scott Lash. 1994. *Reflexive modernization: politics, tradition and aesthetics in the modern social order*. Stanford: Stanford University Press.

Becker, Ernest. 1973. *The denial of death.* New York: Free Press Paperbacks.

Beckerman, Stephen. 1983a. Does the swidden ape the jungle? *Human Ecology* 11: 1-12.

Beckerman, Stephen. 1983b. Barí swiddens gardens: crop segregation patterns. *Human Ecology* 11: 85-101.

Begossi, Alpina. 1993. Ecologia humana: um enfoque das relações homem-ambiente. *Interciência* 18, no. 3: 121-132.

Begossi, Alpina. 1995. Fishing spots and sea tenure: incipient forms of local management in Atlantic Forest coastal communities. *Human Ecology* 23: 387-406.

Begossi, Alpina. 1998. Cultural and ecological resilience among caiçaras of the Atlantic Forest and caboclos of the Amazon, Brazil. In *Linking social and cultural systems for resilience*, ed. Carl Folke and Fikret Berkes, 129-157. Cambridge: Cambridge University Press.

Begossi, Alpina. 2004. Introdução: Ecologia Humana. *In Ecologia de Pescadores*, org. Alpina Begossi, 13-36. São Paulo: Hucitec.

Begossi, Alpina, and José L. Figueiredo. 1995. Ethnoichthyology of southern coastal fishermen: cases from Búzios Island and Sepetiba Bay (Brazil). *Bulletin of Marine Sciences* 56: 710-717.

Begossi, Alpina, Andrea L. Silva, Cristiana S. Seixas, Fábio de Castro, Juarez Pezzuti, Natália Hanazaki, Nivaldo Peroni, and Renato A. M. Silvano. 2004. *Ecologia de pescadores da Mata Atlântica e da Amazônia*. São Paulo: Hucitec.

Begossi, Alpina, Natália Hanazaki, and Nivaldo Peroni. 2000. Knowledge and use of biodiversity in Brazilian hot spots. *Environment, Development and Sustainability* 2: 177–193.

Bekoff, Marc. 2002. *Minding animals: awareness, emotions, and heart.* Oxford: Oxford University Press.

Bell, Daniel. 1973. *The coming of post-industrial society*. New York: Basic Books.

Bell, Daniel. 1976. *The cultural contradictions of capitalism*. New York: Basic Books.

Bennett, Andrew, and Colin Elman. 2006. Qualitative research: recent developments in case study methods. *Annual Review of Political Science* 9: 455-476.

Berger, Peter, and Thomas Luckmann. 1967. *The social construction of reality*. Garden City, NY: Anchor.

Berkes, Fikret 1989. *Common property resource*. London: Belhaven Pres.

Berkes, Fikret. 1994. Co-management: bridging the two solitudes. *Northern Perspectives* 22, no. 2-3: 18-20.

Berkes, Fikret. 1999/2008. *Sacred ecology: traditional ecological knowledge and resource management*. Philadelphia: Taylor and Francis.

Berkes, Fikret, and Carl Folke. 1998. *Linking social and ecological systems: management practices and social mechanisms for building resilience.* Cambridge: Cambridge University Press.

Berkes, Fikret, and Carl Folke. 2000. *Linking social and ecological systems.* Cambridge: Cambridge University Press.

Berkes, Fikret, Peter George, and Richard J. Preston. 1991. Co-management: the evolution in theory and practice of the joint administration of living resources. *Alternatives* 18, no. 2: 12-18.

Berkes, Fikret., Johan Colding, and Carl Folke. 2000. Rediscovery of traditional ecological knowledge as adaptive management. *Ecological Applications* 10: 1251-1262.

Berkes, Fikret, and Carl Folke. 2002. Back to the future: ecosystem dynamics and local knowledge. In: *Panarchy: understanding transformations in human and natural systems*, ed. Lance H. Gunderson and Crawford S. Holling, 121-146. Washington DC: Island Press.

Berkes, Fikret, Johan Colding, and Carl Folke, ed. 2003. *Navigating social-ecological systems: building resilience for complexity and change.* Cambridge: Cambridge University Press.

Berkes, Fikret, and Nancy J. Turner. 2006. Knowledge, learning and the evolution of conservation practice for social-ecological system resilience. *Human Ecology* 34: 479-494.

Berkes, Fikret, Johan Colding, and Carl Folke. 2000. Rediscovery of traditional ecological knowledge as adaptive management. *Ecological Applications* 10, no.5: 1251-1262.

Berlin, Brent. 1992. *Ethnobiological classification: principles of categorization of plants and animals in traditional societies.* New Jersey: Princeton University Press.

Berrill, N J. 1955. *The origins of vertebrates.* London: Oxford University Press.

Bertalanffy, Ludwig von. 1950a. An outline of general systems theory. *British Journal of the Philosophy of Science* 1: 134-165.

Bertalanffy, Ludwig von. 1950b. The theory of open systems in Physics and Biology. *Science* 111: 23-29.

Bertalanffy, Ludwig von. 1968. *General system theory.* New York: George Braziller.

Bews, John. 1935. *Human ecology.* London: Oxford University Press.

Bilharza, Michael, Sylvia Lorekb, and Katharina Schmittc. 2008. Key points of sustainable consumption. In *Proceedings of the 2nd Conference of the Sustainable Consumption Research Exchange (SCORE), 2008*, 287-306. Belgium: Halles des Tanneurs.

Binswanger, Hans P. 1991. Brazilian policies that encourage deforestation in the Amazon. *World Development* 19: 821-829.

Birminghan, Denise M. 2003. Local knowledge of soils: the case of contract in Côte d'Ivoire. *Geoderma* 111: 481-502.

Blaikie, Piers. 1985. *The political economy of soil erosion in developing countries.* London: Longman Development Studies.

Blum, Benjamin. 2007. Strategies for the dissemination of non-mechanical and fire-free land preparation on small-scale farms at the Transamazon Highway in the eastern Amazon, Brazil. Paper presented at the XV International Conference of the Society for Human Ecology, October 4-7, in Rio de Janeiro, Brazil.

Boischio, Ana A.P., and Diane Henshel. 2000. Fish consumption, fish lore, and mercury pollution: risk communication for the Madeira River people. *Environmental Research, Section A* 84: 108-126.

Bolin, Inge. 2009. The glaciers of the Andes are melting: indigenous and anthropological knowledge merge in restoring water resources. In *Anthropology and climate change: from encounters to actions*, ed. Susan A. Crate and Mark Nuttall, 228-239. Walnut Creek, Ca: Left Coast Press.

Boncoeur, Jean, and Guyader Olivier. 1996. *Programme Amure*. Miméo Brest: Centre d'économie et de droit maritime (CEDEM).

Borden, Richard J. 1989. An international overview of the origins of human ecology and the restructuring of higher education: on defining an evolving process. In *Integration of environmental education into general university teaching in Europe,* Charles Susanne, Luc Hens and Dimitri Devuyst, ed., 297-309. Brussels: VUB Press.

Borden, Richard J. 1991. Human Ecology in the United States. In *Human Ecology—coming of age: an international overview,* ed. Shosuke Suzuki, Richard J. Borden and Luc Hens, 201-223. Brussels: Vrije Universiteit Brussel (VUB) Press.

Borden, Richard J. 2008. A brief history of SHE: reflections on the founding and first twenty five years of the Society for Human Ecology. *Human Ecology Review* 15, no. 1: 95—108.

Borden, Richard J., and Jamien Jacobs. 1989. *International directory of human ecologists*. Maine, USA: Society for Human Ecology, College of Atlantic.

Borgenhoff Mulder, Monique. 1992. Reproductive decisions. In *Evolutionary ecology and human behavior*, ed. Eric Alden Smith and Bruce Winterhalder, 339-374. New York: Aldine de Gruyter

Borrini-Feyerabend, Grazia. 1996. Collaborative management of protected areas: tailoring the approach to the context. Gland, Switzerland: Union Internationale pour la Conservation de la Nature et de ses Ressources.

Boserup, Esther. 1965. *The conditions of agricultural growth: the economics of agrarian change under population pressure.* London: G. Allen and Unwin.

Bossel, Hartmut. 1998. *Earth at a crossroad–paths to a sustainable future.* Cambridge UK: Cambridge University Press.

Bossel, Hartmut. 2001. Assessing viability and sustainability: a systems-based approach for deriving comprehensive indicator sets. *Ecology and Society* 5, no. 2: 12, http://www.ecologyandsociety.org/vol5/iss2/art12/ (accessed October 23, 2008).

Boster, James. 1983. A comparison of the diversity of Jivaroan gardens with that of the tropical forest. *Human Ecology* 11: 47-68.

Boudon, Raymond. 1999. *Le sens des valeurs.* Paris: Presses Universitaires de France.

Bourdieu, Pierre. 1977. *Outline of a theory of practice*. London: Cambridge University Press.

Bourdieu, Pierre. 1984. *Distinction: a social critique of the judgement of taste*. Trans. Richard Nice. Cambridge: Harvard University Press.

Bourdieu, Pierre. 1985. The forms of capital. In *Handbook of theory and research for the sociology of education*, ed. John G. Richardson, 241-258. New York: Greenwood.

Bourdieu, Pierre. 1990. *The logic of practice.* Trans. Richard Nice. Cambridge: Polity Press.

Bowler, Peter. 1992. *The Environmental Sciences.* London: Fontana Press.

Box, John, and Charles Harrisson. 1993. Natural spaces in urban places. *Town and Country Planning* 62, no. 9:231-235.

Boyd, Robert, and Peter J. Richerson. 1985. *Culture and the evolutionary process.* Chicago: University of Chicago Press.

Boyden, Stephen. 1995. *Human ecology? 21 years, past and future*. CRES Symposium 21st Birthday, Canberra: Australian National University.

Boyden, Stephen. 2003 *Interviews with Australian Scientists.* Australia: Australian Academy of Science. www.science.org.au/scientists/sb.htm (accessed November 5, 2008).

Boyden, Stephen. 1987. *Western civilization in biological perspective.* Oxford: Clarendon Press.

Boyden, Stephen. 2004. *The biology of civilisation.* Sydney: University NSW Press.

Boyden, Stephen, Sheelagh Millar, Ken Newcombe, and Beverley O'Neill. 1981. *The ecology of a city and its people: the case of Hong Kong.* Canberra: Australian National University Press.

Brady, Nyle C. 1996. Alternatives to slash-and-burn: a global imperative. *Agriculture, Ecosystems & Environment* 58: 3-11.

Brand, Jürg, and Jean Laurent Pfund. 1998. Site-and watershed-level assessment of nutrient dynamics under shifting cultivation in eastern Madagascar. *Agriculture, Ecosystems & Environment* 71, no. 1: 169-183.

Brauns, Hans-Jochen. 2003. *Private Hochschulen in Deutschland: Eine Bestandsaufnahme. Gutachten im Auftrag der OTA Stiftung für berufliche Bildung*. Berlin: WISO Institut für Wirtschaft & Soziales GmbH.

Brearley, Francis Q., Sukaesih Prajadinata, Petra S. Kidd, and John P. Suriantata. 2004. Structure and floristics of an old secondary rain forest in Central Kalimantan, Indonesia, and a comparison with adjacent primary forest. *Forest Ecology and Management* 195: 385–397.

Brinkmann, W. L. F., and J. C. de Nascimento. 1973. The effect of slash and burn agriculture on plant nutrients in the tertiary region of Central Amazonia. *Turrialba* 23, no. 3: 284-290.

Broadbent, Noel, and Patrik Lantto. 2009. Terms of engagement: an Arctic perspective on the narratives and politics of global climate change. In *Anthropology and climate change: from encounters to actions*, ed. Susan A. Crate and Mark Nuttall, 341-355. Walnut Creek, Ca: Left Coast Press.

Brondízio, Eduardo. 2005. Intraregional analysis of land-use change in the Amazon. In *Seeing the forest and the trees. Human environment interactions in forest ecosystems,* ed. Emilio F. Moran and Elinor Ostrom, 223-252. Cambridge, Massachussets: MIT Press.

Brondízio, Eduardo. 2006. Intensificação Agrícola, Identidade Econômica e Invisibilidade entre Pequenos Produtores rurais amazônicos: caboclos e colonos numa perspectiva comparada. In *Sociedades Caboclas Amazônicas: modernidade e invisibilidade,* ed. Cristina Adams, Rui S. S. Murrieta and Walter A. Neves, 191-232. São Paulo: Annablume/FAPESP.

Brondízio, Eduardo S., Emilio F. Moran, Paul Mausel, and You Wu. 1994. Land use change in the Amazon estuary: patterns of caboclo settlement and landscape management. *Human Ecology* 22: 249-278.

Brondízio, Eduardo. S., Stephen D. McCracken, Emilio F. Moran, Andrea D. Siqueira, Donald R. Nelson, and Carlos Rodriguez-Pedraza. 2002. The colonist footprint: toward a conceptual framework of deforestation trajectories among small farmers in frontier amazônia. In *Deforestation and land use in the Amazon,* ed. Charles Wood and Roberto Porro, 133-166. Gainesville, Florida: University Press of Florida.

Brookfield, Harold. 2001. *Exploring agrobiodiversity*. New York: Columbia University Press.

Brookfield, Harold, and Christine Padoch. 2007. Managing biodiversity in spatially and temporally complex agricultural landscapes. In *Managing biodivesity in agricultural ecosystems*, org. Devra I. Jarvis, Christine

Padoch and H. David Cooper, 338-361. New York: Biodiversity International and Columbia University Press.

Brown, David, and Kathrin Schreckenberg. 1998. Shifting Cultivators as agents of deforestation: assessing the evidence. *Natural Resource Perspectives* 29: 1-14.

Brown, Richard Harvey. 1987. *Society as Text*. Chicago: University of Chicago Press.

Brown, Sandra, and Ariel E. Lugo. 1990. Tropical secondary forests. *Journal of Tropical Ecology* 6: 1-32.

Brown, Valerie. 1978. *Holism and the university curriculum: promise or performance? An investigation into ways of teaching synthesis, with a case study of the Human Sciences Program, ANU 1974*. PhD diss., Australian National University, Canberra.

Brubacher, D., John T. Arnason, and J. D. Lambert. 1989. Woody species and nutrient accumulation during the fallow period of Milpa farming in Belize, C.A. *Plant and Soil* 114: 165-17.

Bruckmeier, Karl. 2005. Interdisciplinary conflict analysis and conflict mitigation in local resource management. *Ambio* 34: 65-73.

Brücher, Heinz. 1992. Useful plants of Neotropical origin (and their wild relatives). Berlin: Springer-Verlag.

Brush, Stephen B. 1995. In situ conservation of landraces in centers of crop diversity. *Crop Science* 35: 346–354.

Brush, Stephen B. 2000. The issues of in situ conservation of crop genetic resources. In *Genes in the field: on-farm conservation of crop diversity*, ed. Stephen B. Brush, 3-26. Rome: IPGRI/IDRC/Lewis Publishers.

Buber, Martin. 1970. *I and Thou*. New York: Simon and Shuster.

Bundy, Alida, Ratana Chuenpagdee, Svein Jentoft, and Robin Mahon. 2008. If science is not the answer, what is? An alternative governance model for the world's fisheries. *Frontiers in Ecology and the Environment* 6, no. 3: 152-155.

Burke, Kenneth. 1966. *Language as symbolic action: essays of life, literature, and method*. Berkeley: University of California Press.

Burke, Kenneth. 1969a. *A grammar of motives*. Berkeley: University of California Press.

Burke, Kenneth. 1969b. *A rhetoric of motives*. Berkeley: University of California Press.

Burns, Thomas J. 1992. Class dimensions, individualism and political orientation. *Sociological Spectrum* 12, no. 4: 349-362.

Burns, Thomas J. 1999. Rhetoric as a framework for analyzing cultural constraint and change. *Current Perspectives in Social Theory* 19: 165-185.

Burns, Thomas J., and Terri LeMoyne. 2001. How environmental movements can be more effective: prioritizing environmental themes in political discourse. *Human Ecology Review* 8: 26-38.

Burns, Thomas J., and Terri LeMoyne. 2003. Epistemology, culture and rhetoric: some social implications of human cognition. *Current Perspectives in Social Theory* 22: 71-97.

Burns, Thomas J., Byron L. Davis, and Edward L. Kick. 1997. Position in the world-system and national emissions of greenhouse gases. *Journal of World-Systems Research* 3, no. 3: 432-466.

Burns, Thomas J., Edward L. Kick, and Byron Davis. 2003. Theorizing and rethinking linkages between the natural environment and the modern world-system: deforestation in the late 20th century. *Journal of World-Systems Research* 9, no. 2: 357-390.

Burt, Ronald. 1982. *Toward a structural theory of action: network models of social structure, perception, and action*. New York: Academic Press.

Buss, Leo W. 1987. *The evolution of individuality*. Princeton: Princeton University Press

Butler, Kiera. 2007. Earning a degree in green. *Plenty Magazine,* December 11th. http://www.plentymag.com/features/2007/12/college_of_the_atlantic.php (accessed November 5, 2008)

Buttel, Frederick H. 2000. World society, the nation-state, and environmental protection: comment on Frank, Hironaka, and Schofer. *American Sociological Review* 65: 117-121.

Button, Gregory, and Kristina Peterson. 2009. Participatory action research: community partnership with social and physical scientists. In *Anthropology and climate change: from encounters to actions*, ed. Susan A. Crate and Mark Nuttall, 327-340. Walnut Creek, Ca: Left Coast Press.

Byron, Neil, and Michael Arnold. 1999. What futures for the people of the tropical forests? *World Development* 27, no 5: 789-805.

Cairns, Malcom, and Dennis P. Garrity. 1999. Improving shifting cultivation in Southeast Asia by building on indigenous fallow management strategies. *Agroforestry Systems* 47: 37–48.

Callicott, Baird. 1986. The preservation of species: the value of biological diversity. In *On the Intrinsic Value of Non-Human Species*, ed. Bryan G. Norton, 138–72. Princeton: Princeton University Press.

Calvente, Maria Del Carmen M. H., M. T. B. Martinez, Wanda Maldonado, and Wladimir C. Fuscaldo. 2004. Caiçaras, mestres, professores e turistas: a resistência da territorialidade em um processo de transformação do território. In *Enciclopédia caiçara: o olhar do pesquisador*, org. Antonio C. Diegues, 263-273. São Paulo: Hucitec.

Caras, Roger. 1997. *A perfect harmony: the interwining lives of animals and humans throughout history*. New York: Fireside.

Carlsson, Lars, and Fikret Berkes. 2005. Co-management: concepts and methodological implications. *Journal of Environmental Management* 75: 65-76.

Carneiro, Robert L. 1988. Indians of the Amazonian forest. In *People of the Tropical Rain Forest,* ed. Julie S. Denslow and Christine Padoch. London/Berkeley: University of California Press.

Carpenter, David, and Robert Dyball. 2006. "Outside In"—experiential education for sustainability. In *Innovation, education and communication for sustainable development*, ed. Walter Leal Filho, 379-394. Frankfurt: Peter Lang.

Carrithers, Michael. 1992. *Why humans have cultures: explaining anthropology and social diversity*. Oxford: Oxford University Press.

Carson, Rachel. 1962. *The Silent Spring*. Boston: Houghton Mifflin.

Castella, Jean Christophe., Tran Ngoc Trung, and Boissau, Stanislau. 2005. Participatory simulation of land-use changes in the northern mountains of Vietnam: the combined use of an agent-based model, a role-playing game, and a geographic information system. *Ecology and Society* 10, no. 1: 2 7, http://www.ecologyandsociety.org/vol10/iss1/art27/ (accessed October 23, 2008)

Castile, George Pierce, and Robert E. Blee, ed. 1992. *State and reservation: new perspectives on federal Indian policy*. Tucson: University of Arizona Press.

Castilla, Juan C., and Miriam Fernandez. 1998. Small-scale benthic fisheries in Chile: on co-management and sustainable use of benthic invertebrates. *Ecological Applications* 8, no. 1: S124-S132.

Castro, F. 2000. Fishing accords: the political ecology of fishing intensification in the Amazon. Ph.D. Diss. *CIPEC dissertation series* no. 4. Indiana University, USA.

Catton, William R. Jr., and Riley E. Dunlap. 1978. Environmental sociology: a new paradigm? *The American Sociologist* 13: 41-49.

Chagas, Carlos. 1912. Sobre um trypanosomo do tatú, *Tatusia novemcincto*, transmitido pela *Triatoma geniculata* Latr. (1811). Possibilidade de ser o tatú um depositario do *Trypanosoma Cruzi* no mundo exterior. *Brazil-Medico* 26, no. 80: 305-306.

Chagas, Carlos. 1924. Informações prestadas pelo Dr. Carlos Chagas a Academia de medicina. *Arquivos Brazileiros de Medicina* 14:69-88.

Chambers, Robert. 1999. *Relaxed and participatory appraisal: notes on practical approaches and methods. Notes for participants in PRA familiarisation workshops in the second half of 1999*. Sussex: IDS.

Chase-Dunn, Christopher, and Thomas D. Hall. 1997. *Rise and demise: comparing world-systems*. Boulder, CO: Westview.

Chase-Dunn, Christopher, Yukio Kawano, and Benjamin D. Brewer. 2000. Trade globalization since 1795: waves of integration in the world-system. *American Sociological Review* 65: 77-95.

Checkland, Peter. 1981. *Systems thinking, systems practice*. Chicester UK: John Wiley.

Checkland, Peter, and Jim Scholes. 1990. *Soft systems methodology in action*. Chicester UK: John Wiley.

Chew, Sing. 2001. *World ecological degradation: accumulation, urbanization, and deforestation, 3000 B.C.-A.D. 2000*. Walnut Creek, CA: AltaMira.

Chidumayo, Emmanuel N. 1987. A shifting cultivation land use system under population pressure in Zambia. *Agroforestry Systems* 5: 15-25.

Chidumayo, Emmanuel N., and L. Kwibisa. 2003. Effects of deforestation on grass biomass and soil nutrient status in Miombo woodland, Zambia. *Agriculture, Ecosystems & Environment* 96, no. 1: 97-105.

Christensen, Villy, and Daniel Pauly. 1992. ECOPATH II-A software for balancing steady-state ecosystem models and calculating network characteristics. *Ecological Modelling* 61: 169-85.

Christensen, Villy, and Pauly, Daniel, ed. 1993. *Proceedings of the ICLARM (International Center for Living Aquatic Resources Management) Conference no.* 26: *Trophic Models of Aquatic Ecosystem.* Manille, Philippines: ICLARM.

Christensen, Villy, Karl A. Aiken, and Maria C. Villanueva. 2007. Threats to the ocean: on the role of ecosystem approaches to fisheries. *Social Science Information* 46, no. 1: 67-86.

Clammer, John. 2002. Beyond the cognitive paradigm. In *Participating in development: approaches to indigenous knowledge*, ed. Paul Sillitoe, Alan Bicker and Johan Pottier, 43-63. London: Routledge.

Clark, Brett. 2002. The indigenous environmental movement in the United States: transcending borders in struggles against mining, manufacturing, and the capitalist state. *Organisation and Environment* 15, no.4: 410-442.

Clark, Brett, John B. Foster, and Richard York. 2007. The critique of Intelligent Design: Epicurus, Marx, Darwin, and Freud and the materialist defense of science. *Theory and Society* 36: 514-546.

Clark, Jeff. 1997. Learning to ask the right questions. *Down East* 43, no. 10: 36-39, 55-56.

Clarke, William. 1977. The structure of permanence: the relevance of self-subsistence communities for world ecosystem management. In *Subsistence and survival: rural ecology in the Pacific,* ed. Tim Bayliss-Smith and Richard Feacham, 365-384. London: Academic Press.

Clausen, Rebecca, and Brett Clark. 2005. The metabolic rift and marine ecology: an analysis of the ocean crisis within capitalist production. *Organization and Environment* 18, no 4: 422-444.

Clement, Charles. 1999. 1492 and the loss of Amazonian crop genetic resources. I. The relation between domestication and human population decline. *Economic Botany* 53: 188-202.

Clements, Frederic E., and Victor E. Shelford. 1939. *Bio-ecology.* New York: John Wiley.

Cleveland, David. A., Daniela Soleri, and Steven E. Smith. 1994. Do folk crop varieties have a role in sustainable agriculture? *Bioscience* 44, no. 11: 740–751.

Coase, Ronald H. 1960. The problem of social cost. *Journal of Law and Economics* 3: 1-44.

Cockburn, Aidan. 1977. Where did our infectious diseases come from? The evolution of infectious disease. In *Proceedings of the Ciba Foundation Symposium, Health and disease in tribal societies, 1977, 103-113.* Amsterdam: Elsevier.

Cohen, Joel E. 1995. *How many people can the earth support?* New York: W.W. Norton.

Coleman, James S. 1986. *Individual interests and collective action.* Cambridge: Cambridge University Press.

Coleman, James S. 1988. Social capital in the creation of human capital. *American Journal of Sociology* 94: 95-121.

Collet, Serge. 1979. Sujets, état, civilta. Propositions théoriques pour une anthropologie critique des modes de vie. PhD diss., École des Hautes Études en Sciences Sociales, Paris.

Collet, Serge. 1995. Haleutica phoenicia I. Contribution à l´etude de la place des activités halieutiques dans la culture phénicienne: point de vue d'un non–archéologue. *Information sur les Sciences Sociales* 34, no. 1: 107–73.

Collet, Serge. 2001. De l'usage possible de la rémotion dans la réinvention de nouveaux rapports entre les sociétés halieutiques et leurs natures. In *Droits de propriété, économie et environnement: les ressources marines,* ed. Max Falque and Henri Lamotte, 285-314. Paris: Dalloz.

Collet, Serge. 2002. Appropriation of marine resources: from management to an ethical approach to fisheries governance. *Social Science Information* 41: 531-53.

Collet, Serge. 2007a. Values at sea, value of the sea: mapping issues and divides. *Social Science Information* 46, no. 1: 35-66.

Collet, Serge. 2007b. Marine values in a great state: a complex and committing stake. Poster presented at Sustainable Day Conference, November 22-23, in Versailles: University of St. Quentin.

Colombi, Benjamin. 2009. Salmon nation: climate change and tribal sovereignty. In *Anthropology and climate change: from encounters to actions*, ed. Susan A. Crate and Mark Nuttall, 186-196. Walnut Creek, Ca: Left Coast Press.

Commoner, Barry. 1992. *Making Peace with the Planet*. New York: New Press.

Conklin, Harold C. 1961. The study of shifting cultivation. *Current Anthropology* 2: 27-61.

Constanza, Robert, and Herman Daly. 1992. Natural capital and Sustainable Development. *Conservation Biology* 6, no.1: 37-46

Cooke, Bill and, Uma Kothari. 2001. *Participation: the new tyranny?* London: Zed Books.

Cooper, Joel. 2007. *Cognitive dissonance: fifty years of a classic theory*. London: Sage

Cordeiro, Alexandre Z., and Isabela B. Curado. 2007. *Reservas Extrativistas: desafios de gestão por parte do Estado*. Teresópolis: III SAPIS.

Cordell, John. 1984. Defending customary sea rights. In *Maritime Institutions of the Western Pacific,* Senri Ethnological Studies 17, ed. Keneth Ruddle and Tomoya Akymichi, pp. 302–26. Osȧka: National Museum of Ethnology.

Cordell, John. 2001. Vers une reconfiguration des espaces marins, rôle et importance des zones protégées par les tenures marines coutumières. In *Droits de propriété, économie et environnement. Les Ressources marines*, ed. Max Falque and Henri Lamotte, 315–42. Paris: Dalloz.

Coura, José R. 2007. Human ecology and Chagas disease: the case of the Amazon. Paper presented at the XV International Conference of the Society for Human Ecology, October 4-7, in Rio de Janeiro, Brazil.

Cramb, Robert A. 1993. Shifting cultivation and sustainable agriculture in east malaysia: a longitudinal case study. *Agricultural Systems* 42: 209-226.

Crate, Susan. 2003. Viliui Sakha post-Soviet adaptation: A subarctic test of Netting's smallholder theory. *Human Ecology* 31, no.4: 499-528.

Crate, Susan. 2006. *Cows, kin and globalization: an ethnography of sustainability*. Walnut Creek: AltaMira Press.

Crate, Susan. 2008. Gone the bull of winter? Grappling with the cultural implications of and anthropology's role(s) in global climate change. *Current Anthropology* 49, no. 4: 569-596.

Crate, Susan A. and Mark Nuttall, ed. 2009. *Anthropology and climate change: from encounters to action*. Walnut Creek, Ca: Left Coast Press.

Cruikshank, Julie. 2001. Glaciers and climate change: perspectives from oral tradition. *Arctic* 54, no. 4: 377-393.

Cruikshank, Julie. 2005. *Do glaciers listen? Local knowledge, colonial encounters, and social imagination*. Vancouver: UBC Press.

Crumley, Carole, ed. 1994. *Historical ecology: cultural knowledge and changing landscapes*. Santa Fe: School of American Research Press.

Crumley, Carole. 2001. *New directions in anthropology and the environment: intersections*. Walnut Creek, CA: AltaMira Press.

Cuc, Le Trong. 1996. *Swidden Agriculture in Vietnam. Montane mainland southeast Asia in transition*. Chiang Mai, Thailand: Chiang Mai University Consortium.

Cury, Philippe, Lyne Shannon, and Yunne-Jai Shin. 2001. The functioning of marine ecosystems. Paper presented at the Reykjavic Conference on Responsible Fisheries in the Marine Ecosystem, October 1-4, 2001, in Reykjavik, Iceland.

D'Andrade, Roy. 1995. *The development of cognitive anthropology*. Cambridge: Cambridge University Press.

Damasio, Antonio. 2003. *Looking for Spinoza, joy, sorrow and the feeling brain*. Orlando FL, New York, London: Harcourt, Inc.

Darwin, Charles. 1859. *On the origin of species*. John Murray, London.

Dash, S. S., and M. K. Misra. 2001. Studies on hill agro-ecosystems of three tribal villages on the Eastern Ghats of Orissa, India. *Agriculture, Ecosystems, and Environment* 86, no. 1: 287-302.

Davidson, Eric A., Cláudio J. R. de Carvalho, Adelaine M. Figueira, Françoise Y. Ishida, Jean Pierre H. B. Ometto, Gabriela B. Nardoto, Renata T. Saba, Sanae N. Hayashi, Eliane C. Leal, Ima Célia G. Vieira, and Luiz A. Martinelli. 2007. Recuperation of nitrogen cycling in Amazonian forests following agricultural abandonment. *Nature* 447: 995-997.

Davidson, Eric A., Tatiana Deane A. Sá, Claudio J. R. Carvalho, Ricardo de O. Figueiredo, Maria do Socorro A. Kato, Osvaldo R. Kato and Françoise Yoko Ishida. 2008. An integrated greenhouse gas assessment of an alternative to slash-and-burn agriculture in eastern Amazônia. *Global Change Biology* 14: 998–1007.

Davidson-Hunt, Iain J. 2006. Adaptive learning networks: developing resource management knowledge through social learning forums. *Human Ecology* 34 no. 4: 593-614

Davis, Anthony, and John Wagner. 2003. Who knows? On the importance of identifying "experts" when researching local ecological knowledge. *Human Ecology* 31, no. 3: 463-489.

Davis, Wade. 2001. *Light at the edge of the world.* Washington DC: National Geographic Society.

Dawkins, Richard. 1979. *O gene egoísta.* São Paulo: Editora da Universidade de São Paulo.

Dawkins, Richard. 2006. *God delusion.* Boston: Houghton Mifflin Company.

De Beer, Gavin R. 1954. *Embryos and ancestors.* Oxford: Oxford University Press.

de Boer, Willem.F., Annemieke M.P.van Schie, Domingos F. Jocene, Alzira B.P. Mabote, and Almeida Guissamulo. 2001 The impact of artisanal fishery on a tropical intertidal benthic fish community. *Environmental Biology of Fishes* 61: 213 – 229.

De Jong, Wil. 1997. Developing swidden agriculture and the threat of biodiversity loss. *Agriculture, Ecosystems & Environment* 62: 187-197.

De Walt, Billie R. 1994. Using indigenous knowledge to improve agriculture and resource management. *Human Organization* 53, no. 2: 123-131.

Dean, Warren. 1996. *A Ferro e fogo: a história e a devastação da Mata Atlântica brasileira.* São Paulo: Companhia das Letras.

Denevan, William M. 1991. The Americas before and after 1492. *Current Geographical Research* 369-385.

Denevan, William M. 1992. The pristine myth: the landscape of the Americas in 1492. *Annals of the Association of American Geographers* 82: 369-385.

Denevan, William M. 1996. A bluff model of riverine settlement in prehistoric Amazonia. *Annals of the Association of American Geographers* 86: 654-681.

Denich, Manfred, Paul L. G. Vlek, Tatiana D. Sá, Konrad Vielhauer, and Wolfgang Lücke. 2005. A concept for the development of fire-free fallow management in the Eastern Amazon, Brazil. *Agriculture, Ecosystems & Environment* 110: 43–58.

Detwiler, R. P., and Charles A. Hall. 1988. Tropical forests and the global carbon cycle. *Science* 239: 42-49.

Diamond, Jared. 2005. *Collapse: how societies choose to fail or succeed.* New York: Viking.

Diaw, Mariteuw Chimère. 1997. Si, Nda bot and Ayong: shifting cultivation, land use and property rights in southern Cameroon. *Rural Development Forestry Network* 21e: 1-28.

Di Beneditto, Ana Paula, Renata M. A. Ramos, and Neusa R. W. Lima. 2001. *Os golfinhos: origem, classificação, captura acidental, hábito alimentar*. Porto Alegre: Editora Cinco Continentes.

Diegues, Antonio C. 1999. Human populations and coastal wetlands: conservation and management in Brazil. *Ocean & Coastal Management* 42: 187-210.

Diegues, Antonio Carlos. 2004. A mudança como modelo cultural: o caso da cultura caiçara e a urbanização. In *Enciclopédia caiçara: o olhar do pesquisador*, org. Antonio Carlos Diegues, 21-48. São Paulo: Hucitec.

Dietz, Thomas, and Eugene A. Rosa. 1994. Rethinking the environmental impacts of population, affluence and technology. *Human Ecology Review* 1: 277-300.

Dilthey, Wilhelm. 1976. The construction of the historical world in the human studies. In *Wilhelm Dilthey: Selected Writings,* ed. H.P. Rickman, 168- 245. London: Cambridge University Press.

DiMaggio, Paul J. 1997. Culture and Cognition. *Annual Review of Sociology* 23: 263-287.

DiMaggio, Paul J., and Walter W. Powell. 1991. The iron cage revisited: institutional isomorphism and collective rationality in organizational fields. In *The new institutionalism in organizational analysis,* ed. Walter W. Powell and Paul J. DiMaggio, 63-82. Chicago: The University of Chicago Press

Djelic Marie-Laure, and Sigrid Quack. 2007. Overcoming path dependency: path generation in open systems. *Theory and Society* 36: 161-186.

Dombrowski, Daniel A. 1997. *Kazantakis and God.* SUNY Series in Constructive Postmodern Thought. New York: SUNY.

Doolittle, William E. 1992. Agriculture in North America on the eve of contact: a reassessment. *Annals of the Association of American Geographers* 82: 386-401.

Douglas, Mary. 1970. *Natural symbols*. London: Barrie & Rockliff.

Doulman, David. 2007. Coping with the extended vulnerability of marine ecosystems: implementing the 1995 FAO code of conduct for responsible fisheries. *Social Science Information* 46, no. 1: 189–237.

Dove, Michael R. 1993. Smallholder rubber and swidden agriculture in Borneo: A sustainable adaptation to the ecology and economy of the tropical forest. *Economic Botany* 47: 136-147.

Dove, Michael R. 2002. Hybrid histories and indigenous knowledge among Asian rubber smallholders. *International Social Science Journal* 173: 349-359.

Dove, Michael R., and D. M. Kammen. 1997. The epistemology of sustainable resource use: managing forest products, swiddens, and high-yielding variety crops. *Human Organization* 56: 91-101.

Drew, Joshua A. 2005.Use of traditional ecological knowledge in marine conservation. *Conservation Biology* 19: 1286-1293.

Ducourtieux, Olivier, Phoui Visonnavong, and Julien Rossard. 2006. Introducing cash crops in shifting cultivation regions–the experience with cardamon in Laos. *Agroforestry Systems* 66: 65-76.

Dulvy, N.K., and Nicholas V.C. Polunin. 2004. Using informal knowledge to infer human-induced rarity of a conspicuous reef fish. *Animal Conservation* 7: 365–374.

Dumont, Louis. 1983. *Essais sur l'individualisme. Une perspective anthropologique sur l'idéologie moderne*. Paris: Le Seuil.

Duncan, Otis Dudley. 1953. Human ecology and population studies. In *The study of populations*, ed. Philip M. Hauser and Otis Dudley Duncan, 678-713. Chicago: University of Chicago Press.

Duncan, Otis Dudley. 1964. From social system to ecosystem. *Sociological Inquiry* 31: 140-149.

Duncan, Otis Dudley. 1973. From social system to ecosystem. In *Population, environment and social organization: current issues in human ecology*, ed. Michael Micklin, 107-177. Hinsdale, IL: Dryden Press.

Dunlap, Riley, and William R. Catton Jr. 1994. Struggling with human exemptionalism: the rise, decline and revitalization of environmental sociology. *American Sociologist* 25: 5-30.

Dunlap, Riley, and William R. Catton Jr. 2002. Which functions of the environment do we study? A comparison of environmental and natural resource sociology. *Society and Natural Resources* 14: 239-249.

Durkheim, Emile. 1964 (1893). *The division of labor in society*. New York: Free Press.

Durkheim, Emile, and Marcel Mauss. 1961 (1902). Social structure and the structure of thought. In *Theories of society: foundations of modern sociological theory*, ed. Talcott Parsons, Edward Shils, Kaspar D. Naegele and Jesse R. Pitts, 1065-1068. New York: Free Press.

Dyball, Robert, and David Carpenter. 2006. Human ecology and education for sustainability. In *Sustainability in the Australasian university context,* ed. David Carpenter and Walter Leal Filho, 45-67. Frankfurt: Peter Lang.

Dyer, Cristopher.L., and James R. McGoodwin, ed. 1994. Folk *management in the world's fisheries*. Niwot: University Press of Colorado.

Eastmond, Amarella, and Betty Faust. 2006. Farmers, fires, and forests: a green alternative to shifting cultivation for conservation of the Maya forest? *Landscape and Urban Planning* 74: 267-284.

Eden, Michael J. 1987. Traditional shifting cultivation and the tropical forest system. *Tree* 2: 340-343.

Eden, Michael J., and Angela Andrade. 1987. Ecological aspects of swidden cultivation among the Andoke and Witoto Indians of the Colombian Amazon. *Human Ecology* 15: 339-359.

Egan, Dave and, Evelyn Howell, ed. 2001. *The historical ecology handbook*. Washington, DC: Island Press.

Ehrhardt-Martinez, Karen. 1998. Social determinants of deforestation in developing countries: a cross-national study. *Social Forces* 77: 567-586.

Ehrlich, Paul R., and Anne H. Ehrlich. 1990. *The population explosion*. New York: Simon & Schuster.

Eisenberg, Evan. 1998. *The ecology of Eden: an inquiry into the dream of paradise and a new vision of our role in nature*. New York: Vintage.

Ekins, Paul, and Manfred Max-Neef, ed. 1992. *Real-life economics: understanding wealth creation*. London: London.

Eldredge, Niles. 2000. *The triumph of evolution and the failure of creationism*. New York: W.H. Freeman.

Elias, Marianne, Laurent Penet, Prune Vindry, Doyle McKey, and Thierry Robert. 2001. Unmanaged sexual reproduction and the dynamics of genetic diversity of a vegetatively propagated crop plant, cassava (*Manihot esculenta* Crantz), in a traditional farming system. *Molecular Ecology* 10: 1895-1907.

Ellen, Roy F. 1970. Trade, environment, and the reproduction of local systems in the Moluccas. In *The ecosystem approach in anthropology: from concept to practice*, ed. Emilio F. Moran, 191-227. Ann Arbor: The University of Michigan Press.

Ellul, Jacques. 1980. *The technological system*. New York: Continuum.

Embratur. 2007. *Estatísticas básicas do turismo*. http://200.189.169.141/site/arquivos/dados_fatos/estatisticas_basicas_do_turismo/estatisticas_basicas_do_turismo___brasil_2002_a_2006__22nov07_.pdf (accessed October 30, 2008).

Emperaire, Laure, and Nivaldo Peroni. 2007. Traditional management of agrobiodiversity in Brazil: the case of cassava. *Human Ecology* 36: 761-768.

European Commission. 1999. *Lessons from the European commission's demonstration programme on Integrated Coastal Zone Management*

(ICZM), April 1999. http://ec.europa.eu/environment/iczm/discdoc2.htm (accessed November 26, 2008).

Evans, Francis. 1956. Ecosystem as the basic unit in ecology. *Science* 123: 1127-1128.

Evans-Pritchard, E. E. 1940. *The Nuer*. Oxford: Clarendon Press.

Ewel, John, Cory Berish, Becky Brown, Norman Price, and James Raich. 1981. Slash and burn impacts on a Costa Rican wet forest site. *Ecology* 62, no. 3: 816-829.

Fagerström, Minh Ha Hoang, S. I. Nilsson, M. van Noordwilk, Thai Phien, M. Olsson, A. Hansson, and C. Svensson. 2002. Does Tephrosia candida as fallow species, hedgerow or mulch improve nutrient cycling and prevent nutrient losses by erosion on slopes in northern Viet Nam? *Agriculture, Ecosystems & Environment* 90, no. 3: 291-304.

Failler, Pierre, and Sophie des Clers. 2002. Perspectives d'extension économique et sociale du modèle ECOPATH. In *La recherche halieutique et le développement durable des ressources naturelles marines en Afrique de l'Ouest, quels enjeux?,* Rapport Recherche Halieutique ACP/UE, n° 11, ed. Pierre Failler, Alkaly Doumbouya, Moctar Bâ, Nicolas Lecrivain.,133-143, Brussels: Office des publications officielles des Communautés Européennes.

Fain, Heidi D., Thomas J. Burns, and Mindy Sartor. 1994. Group and individual selection in the human social environment: from behavioral ecology to social institutions. *Human Ecology Review* 1, no. 2: 335-350.

FAO (Food and Agricultural Organization). 1985. *The tropical forestry action plan.* Rome: UN Food and Agricultural Organization.

Farella, Nicolina, Marc Lucotte, Robert Davidson, and S. Daigle. 2006. Mercury release from deforested soils triggered by base cation enrichment. *Science of the Total Environment* 368: 19–29.

Fearnside, Philip M. 1996. Amazonian deforestation and global warming: carbon stocks in vegetation replacing Brazil's Amazon forest. *Forest Ecology and Management* 80: 21-34.

Fearnside, Philip M. 2001. Land-tenure issues as factors in environmental destruction in Brazilian Amazonia: the case of Southern Paraná. *World Development* 29: 1361-1372.

Fearnside, Philip M. 2005. Deforestation in Brazilian Amazonia: history, rates, and consequences. *Conservation Biology* 19: 680–688.

Fearnside, Philip M., and Walba M. Guimarães. 1996. Carbon uptake by secondary forests in Brazilian Amazon. *Forest Ecology and Management* 80: 35-46.

Fenner, Frank, and David Curtis. 2001. *The John Curtin School of Medical Research: the first fifty years, 1948-1998.* Gundaroo: Brolga Press.

Ferguson, Bruce G., John Vandermeer, Helda Morales, and Daniel M. Griffith. 2003. Post-Agricultural Succession in El Petén, Guatemala. *Conservation Biology* 17, no.3: 818–828.

Fernandez-Gimenez, Maria E., Heidi L. Ballard, and Victoria E. R. Sturtevant. 2008. Adaptive management and Social learning in Collaboratiove and Community-Based Monitoring: a Study of Five Community-Based Forestry Organisations in the western USA. *Ecology and Society* 13, no. 2: 4, http://www.ecologyandsociety.org/vol13/iss2/art4/ (acessed October 23, 2008).

Ferreira, Renata G., Érico Santos Jr., Kelly Pansard, and Sandra Ananias. 2005. Reserva de fauna costeira de Tibau do Sul - Em defesa de um turismo sustentável no município. Report presented to Federal, State and Municipal Secretariats for the Environment. Rio Grande do Norte, Brasil.

Festinger, Leon. 1962 (1957). *A theory of cognitive dissonance*. Stanford, CA: Stanford University Press.

Figueiredo, Gisela M., Hermógenes F. Leitao-Filho, and Alpina Begossi. 1993. Ethnobotany of Atlantic forest coastal communities: Diversity of plant uses in Gamboa (Itacuruca Island, Brazil). *Human ecology* 21: 419-430

Figueiredo, Gisela M., Hermógenes F. Leitao-Filho, and Alpina Begossi. 1997. Ethnobotany of Atlantic forest coastal communities: II. Diversity of plant uses at Sepetiba Bay (SE Brazil). *Human Ecology* 25: 353-360.

Finan, Tim. 2009. Storm warnings: the role of anthropology in adapting to sea-level rise in southwestern Bangladesh. In *Anthropology and climate change: from encounters to actions*, ed. Susan A. Crate and Mark Nuttall, 175-185. Walnut Creek, Ca: Left Coast Press.

Firth, Raymond. 1964. *Essays on social organisation and values.* London: Althon Press.

Fish, Suzane, and Paul Fish. 1970. An archeological assessment of ecosystems in the Tucson Basin of Southern Arizona. In *The ecosystem approach in anthropology: from concept to practice*, ed. Emilio F. Moran, 159-187. Ann Arbor: The University of Michigan Press.

Fischer, Claude S. 1978. Urban-to-rural diffusion of opinions in contemporary America. *American Journal of Sociology* 84, no. 1: 151-159.

Fischer, Michael. 2004. Powerful knowledge: applications in a cultural context. In *Development and local knowledge*, eds Alan Bicker, Paul Sillitoe and Johan Pottier. London: Routledge.

Fischer-Kowalski, Marina. 1997a. Society's metabolism: on the childhood and adolescence of a rising conceptual star. In: *The international handbook of environmental sociology,* ed. Michael Redclift and Graham Woodgate, 119-137. Cheltenham, Northhampton: Edward Elgar.

Fischer-Kowalski, Marina 1997b. Dynamik und Selbststeuerung industrieller Gesellschaften. In *Gesellschaftlicher Stoffwechsel und Kolonisierung von Natur*, ed. Marina Fischer-Kowalski, Helmut Haberl, Walter Hüttler, Harald Payer, Heinz Schandl, Verena Winiwarter and Helga Zangerl-Weisz, 203-222. Amsterdam: Gordon&Breach.

Fischer-Kowalski, Marina. 1998. Society's metabolism: the intellectual history of material flow analysis, Part I: 1860-1970. *Journal of Industrial Ecology* 2: 61-78.

Fischer-Kowalski, Marina, and Helga Weisz. 1999. Society as hybrid between material and symbolic realms: toward a theoretical framework of society-nature interaction. *Advances in Human Ecology* 8: 215-251.

Fischer-Kowalski, Marina, and Helmut Haberl. 1997. Tons, joules and money: modes of production and their sustainability problems. *Society and Natural Resources* 10: 61-85.

Fischer-Kowalski, Marina, and Walter Hüttler. 1998. Society's metabolism: the intellectual history of material flow analysis, Part II: 1970-1998. *Journal of Industrial Ecology* 2: 107-137.

Fischer-Kowalski, Marina, and Helmut Haberl, ed. 2007. *Socioecological transitions and global change*. Cheltenham UK, Northamption USA: Edward Elgar.

Fiske, Shirley. 2009g. Global change policymaking from inside the beltway: engaging anthropology. In *Anthropology and climate change: from encounters to actions*, ed. Susan A. Crate and Mark Nuttall, 277-291. Walnut Creek, Ca: Left Coast Press.

Folke, Carl, Stephen Carpenter, Thomas Elmqvist, Lance Gunderson, Crawford S. Holling, and Brian Walker. 2002. Resilience and sustainable development: building adaptive capacity in a world of transformations. *Ambio* 31: 437-440.

Folke, Carl, Thomas Hahn, Per Olson, and Jon Norberg. 2005. Adaptive governance of social-ecological systems. *Annual review of Environmental Resources* 30: 441-473

Fonseca-Kruel, Viviane S., and Ariane L. Peixoto. 2004. Etnobotânica na Reserva Extrativista Marinha de Arraial do Cabo, RJ, Brasil. *Acta Botanica Brasilica* 18: 177-190.

Fontalvo, Martha, Marion Glaser, and Adagenor Ribeiro. 2007. A method for the participatory design of an indicator system as a tool for local coastal management. *Ocean and Coastal Management* 50: 779-795.

Ford, Edward David. 2000. *Scientific method for ecological research.* Cambridge: Cambridge University Press.

Forde, Daryll. 1934. *Habitat, economy and society: a geographical introduction to ethnology.* London: Methuen.

Forsyth, Tim. 1996. Science, myth and knowledge: testing Himalaian environmental degradation in northern Thailand. *Geoforum* 27: 375-392.

Foster, John Bellamy. 1999. Marx's theory of metabolic rift: classical foundations for environmental sociology. *American Journal of Sociology* 105: 366-402.

Fox, Jefferson, Dao Minh Truong, A. Terry Rambo, Nghim P. Tuyen, Le Trong Cuc, and Stephen Leisz. 2000. Shifting cultivation: a new old paradigm for managing tropical forest. *BioScience* 50, no. 6: 521-528.

Foxon, Timothy J., Mark S. Reed, and Lindsay C. Stringer. 2008. Governing long-term social-ecological change: what can the adaptive management and transition management approaches learn from each other? *Sustainability Research Institute: Working Papers* 13, http://www.see.leeds.ac.uk/research/sri/working_papers/SRIPs-13.pdf (accessed October 24, 2008).

Francini-Filho, Ronaldo Bastos, and Rodrigo Leão de Moura. 2008. Dynamics of fish assemblages on coral reefs subjected to different management regimes in the Abrolhos Bank, eastern Brazil. *Aquatic Conservation: Marine and Freshwater Ecosystems* 18, no. 7: 1166-1179.

Frank, David John, Ann Hironaka, and Evan Schofer. 2000a. The nation-state and the natural environment over the twentieth century. *American Sociological Review* 65: 96-116.

Frank, David John, Ann Hironaka, and Evan Schofer. 2000b. Environmentalism as a global institution: Reply to Buttel. *American Sociological Review* 65: 122-127.

Freudenburg, William R., and Scott Frickel. 1995. Beyond the nature/society divide: learning to think about a mountain. *Sociological Forum* 10: 361-392.

Frizano, Jacqueline, D. R. Vann, Arthur H. Johnson, Christine M. Johnson, Ima C. G. Vieira, and Daniel J. Zarin. 2003. Labile phosphorus in soils of forest fallows and primary forest in the Bragantina Region, Brazil. *Biotropica* 35: 2-11.

Froese, Rainier, and Daniel Pauly, ed. 2000. *Fishbase 2000: concepts, design and data sources*. Los Baños, Laguna, Philippines: ICLARM.

Fromm, Erich. 1947. *Man for himself: an inquiry into the psychology of ethics*. London: Routledge.

Fromm, Erich. 1962. *Beyond the chains of illusion: my encounter with Marx and Freud*. London: Continuum.

Fromm, Erich. 1991. *The fear of freedom*. London: Routledge.

Fujisaka, Sam. 1991. A diagnostic survey of shifting cultivation in northem Laos: targeting research to improve sustainability and productivity. *Agroforestry Systems* 13: 95-109.

Fujisaka, Sam, G. Escobar, and E. J. Veneklaas. 2000. Weedy fields and forests: interactions between land use and the composition of plant communities in the Peruvian Amazon. *Agriculture, Ecosystems & Environment* 78: 175–186.

Fujisaka, Sam, William Bell, Nick Thomas, Liliana Hurtado, and Euan Crawford. 1996. Slash-and-burn agriculture, conversion to pasture, and deforestation in two Brazilian Amazon colonies. *Agriculture, Ecosystems & Environment* 59: 115-130.

Fuller, Steve. 2000. The coming biological challenge to social theory and practice. In *For sociology: legacies and prospects*, ed. John Eldridge, John MacInnes, Sue Scott, Chris Washurst and Anne Witz, 174-190. Durham: The Sociology Press.

Gaedke, Ursula. 1995. A comparison of whole-community and ecosystem approaches (biomass size distributions, food web analysis, network analysis, simulation models) to study the structure, function and regulation of pelagic food webs. *Journal of Plankton Research* 17: 1273-305.

Gafur, Abdul J., Jens R. Jensen, Ole K. Borggaard, and Leif Petersen. 2003. Runoff and losses of soil and nutrients from small watersheds under shifting cultivation (Jhum) in the Chittagong Hill: tracts of Bangladesh. *Journal of Hydrology* 279: 293–309.

Garcia-Alte, Anna, Josep Ma Ollé, Fernando Antoñanzas, and Joan Colom. 2002. The social cost of illegal drug consumption in Spain. *Addiction* 97: 1145-1153.

García-Oliva, Felipe, Robert L. Sanford Jr., and Eugene Kelly. 1999. Effects of slash-and-burn management on soil aggregate organic C and N in a tropical deciduous forest. *Geoderma* 88: 1-12.

Gardner, Katy, and David Lewis. 1996. *Anthropology, development and the post-modern challenge.* London: Pluto Press.

Garstang, Walter. 1951. *Larval forms and other zoological verses*. Chicago: University of Chicago Press.

Gasalla, Maria A. 2003a. Linking fishers' knowledge with models: past fisheries and species interactions in the South Brazil Bight. In *Abstracts of the 21st Lowell Wakefield Fisheries Symposium, Assessment and management of new and developed fisheries in data-limited situations, Anchorage, October 22-25, 2003*, 15-16. Anchorage: Alaska Sea Grant.

Gasalla, Maria A. 2003b. Ethnoecological models of marine ecosystems: "fishing for fishermen" to address local knowledge in Southeastern Brazil industrial fisheries. *Fisheries Centre Research Reports* 11: i-ii.

Gasalla, Maria A. 2004a. Impactos da pesca industrial no ecossistema da plataforma continental interna do Sudeste do Brasil: a abordagem ecossistêmica e a integração do conhecimento. Ph.D diss., University of São Paulo, São Paulo, Brazil.

Gasalla, Maria A. 2004b. Modelling the state of fisheries before the expansion of industrial fisheries. In Abstract Volume of the IV World Fisheries Congress, May 2-6, 2004, 48-49. Vancouver, Canada.

Gasalla, Maria A. 2007. Beyond historical records: how industrial fishers perceive changes in marine communities and contribute to detect long-term alterations in the South Brazil Bight. Paper presented at the XV International Conference of the Society for Human Ecology, October 4-7, in Rio de Janeiro, Brazil.

Gash, John H. C., Carlos A. Nobre, John M. Roberts, and Reynaldo L. Victoria. 1996. *Amazonian deforestation and climate*. Chichester: John Wiley and Sons.

Gaskins, Richard H. 1992. *Burdens of proof in modern discourse*. New Haven, CT: Yale University Press.

Gaziano, Emanuel. 1996. Ecological metaphors as scientific boundary work: innovation and authority in interwar sociology and biology. *The American Journal of Sociology* 101, no. 4: 874-907.

Geels, Frank W. 2002. Technological transitions as evolutionary reconfiguration processes: a multi-level perspective and a case study. *Research Policy* 31: 1257-1274

Geer Ken, Theo, Arnold Tukker, Carlo Vezzoli, and Fabrizio Ceschin ed. 2008. *Proceedings of the Second Conference of the Sustainable Consumption Research Exchange (SCORE), March 10-11, 2008: Sustainable consumption and production: framework for action*. Belgium: Halles des Tanneurs.

Geertz, C. 1963. *Agricultural involution: the process of ecological change in Indonesia*. Berkeley: University of California Press.

Geertz, Clifford. 1973. *Interpretation of cultures*. New York: Basic Books.

Geist, Helmut J., and Eric F. Lambin. 2002. Proximate causes and underlying driving forces of tropical deforestation. *BioScience* 52, no. 2: 143-150.

Gerhardinger, Leopoldo C., Athila, A. Bertoncini, and Mauricio Hostim-Silva. 2006. Local ecological knowledge and Goliath grouper spawning aggregations in the South Atlantic Ocean: Goliath grouper spawning aggregations in Brazil. *SPC Traditional Marine Resource Management and Knowledge Information Bulletin* 20: 33-34.

German, Laura A. 2003. Historical contingencies in the coevolution of environment and livelihood: contributions to the debate on Amazonian Black Earth. *Geoderma* 111: 307–331.

Gilmour, John L., and J. W. Gregor, 1939. Demes: a suggested new terminology. *Nature* 144:333.

Glaser, Barney G. 2002. Conceptualization: on theory and theorizing using grounded theory. *International Journal of Qualitative Methods* 1, no. 2: 1-31.

Glaser, Marion. 2006. The social dimension in ecosystem management: strengths and weaknesses of human-nature mind maps. *Human Ecology Review* 13, no. 2: 122-142.

Glaser, Marion, and Karen Diele. 2004. Asymmetric outcomes: assessing the biological, economic and social sustainability of a mangrove crab fishery, *Ucides cordatus* (Ocypodidae), in North Brazil. *Ecological Economics* 49, no.3: 361–373.

Glaser, Marion, and Rosete da Silva Oliveira. 2004. Prospects for the co-management of mangrove ecosystems on the North Brazilian coast: whose rights, whose duties and whose priorities? *Natural Resources Forum* 28: 224-233.

Glaser, Marion, Gesche Krause, Ulrich Saint-Paul, Joachim Harms, and Gabriele Boehme. 2006. Fachübergreifende Nachhaltigkeitsforschung-Das brasilianisch-deutsche Mangroven Projekt MADAM. In *Fachübergreifende Nachhaltigkeitsforschung. Stand und Visionen am Beispiel nationaler und internationaler Forscherverbünde*, ed. Bernhard Glaeser, 265-297. Munich: Oekom, Edition Humanökologie.

Glaser, Marion, and Gesche Krause. In Preparation. Mangroves and people: a social-ecological system: sustainable ecosystem (co)management. In *Mangroves dynamics and management in North Brazil. Ecological Studies Series*, ed. Ulrich Saint Paul and Horacio Schneider. Heidelberg: Springer.

Glendinning, Anthony, Ajay Mahapatra, and C. Paul Mitchell. 2001. Modes of communication and effectiveness of agroforestry extension in eastern India. *Human Ecology* 29, no.3: 283-305.

Godfrey-Smith, Peter. 2006. Local interaction, multilevel selection, and evolutionary transitions. *Biological Theory* 1, no.4: 372-380

Goffman, Erving. 1974. *Frame analysis: an essay on the organization of experience*. New York: Harper.

Goldsmith, Edward. 1970. The stable society—can we achieve it? *The Ecologist* 1, no. 6

Goleman, Daniel. 1996. *Emotional intelligence: why it can mater more than IQ*. London: Bloomsbury.

Goleman, Daniel. 2006. *Social intelligence: the new science of human relationships*. New York: Bantam Books.

Goleman, Daniel, Richard E. Boyatzis, and Annie McKee. 2002. *The new leaders: transforming the art of leadership into the science of results*. London: Little Brown.

Gómez-Pompa, Arturo, José Salvador Flores, and Victoria Sosa. 1987. The "pet kot": a man made tropical forest of the Maya. *Interciencia* 12: 10-15.

Gore, Al. 1993. *Earth in the balance: ecology and the human spirit*. New York: Plume/Penguin.

Gorsky, Yu M, A. M. Stepanov and A. G. Tesilinov. 2000. Homeostatics: science of the twenty-first century. *Kybernetes* 29, no. 9/10: 1150

Gottschall, Jonathan, and David S. Wilson. 2005. *The literary animal, evolution and the nature of narrative*. Evanston, Illinois: Northwestern University Press.

Gould, Stephen J. 1977. *Ontogeny and phylogeny*. Cambridge: Belknap.

Gould, Stephen J. 1992. *Ever since Darwin*. New York: Norton.

Gould, Stephen J. 1999. *Rocks of ages: science and religion in the fullness of life*. New York: Ballantine.

Goulding, Michael. 1980. *The fishes and the forest: Explorations in Amazonian Natural History*. Berkeley: University of California Press.

Granovetter, Mark. 1985. Economic action and social structure: the problem of embeddedness. *American Journal of Sociology* 91: 481-510.

Gray, Leslie C., and Philippe Morant. 2003. Reconciling indigenous knowledge with scientific assessment of soil fertility changes in south western Burkina Faso. *Geoderma* 111: 425-437.

Green, Donna. 2009. Opal waters, rising seas: how sociocultural inequality reduces resilience to climate change among indigenous Australians. In *Anthropology and climate change: from encounters to actions*, ed. Susan A. Crate and Mark Nuttall, 218-227. Walnut Creek, Ca: Left Coast Press.

Grubler, Arnulf. 1991. Diffusion: long-term patterns and discontinuities. *Technological Forecasting and Social Change* 39: 159-180.

Grusin, R. 1993. Thoreau, extravagance, and the economy of nature. *American Literary History* 5, no. 1: 30-50.

Guariguata, Manuel R., and Rebecca Ostertag. 2001. Neotropical secondary forest succession: changes in structural and functional characteristics. *Forest Ecology and Management* 148: 185-206.

Gunderson, Lance H. 2000. Ecological resilience: in theory and application. *Annual Review of Ecology and Systematics* 31: 425-439.

Gunderson Lance, and Holling, Crawford S., ed. 2002. *Panarchy: understanding transformations in human and natural systems.* Washington: Island Press.

Gunderson, Lance H., Crawford S. Holling, and Stephen S. Light, ed. 1995. *Barriers and bridges to the renewal of ecosystems and institutions*. New York: Columbia University Press.

Gupta, Atul K. 2000. Shifting cultivation and conservation of biological diversity in Tripura, Northeast Índia. *Human Ecology* 28, no. 4: 605-629.

Gupta, Atul K., and Ajith Kumar. 1994. Feeding ecology and conservation of the phayre's leaf monkey *Presbytis phayrei* in northeast India. *Biological Conservation* 69: 301-306.

Gupta, Prabhat K., V. Krishna Prasad, C. Sharma, A. K. Sarkar, Yogesh Kant, K. V. S. Badarinath, and A. P. Mitra. 2001. CH_4 emissions from biomass burning of shifting cultivation areas of tropical deciduous forests–experimental results from ground-based measurements. *Chemosphere-Global Change Science* 3, no. 2: 133-143.

Gustavson, Kent R., Stephen C. Longeran, and H. Jack Ruitenbec K. 1999. Selection and modelling of sustainable development indicators; a case study of the Fraser River Basin, British Columbia. *Ecological Economics*. 28: 117-132

Guttenberg, Albert Z. 1975. The urban system. In *Human ecology,* ed. Norman D. Levine, 296-324. Belmont: Duxbury Press.

Haami, Brad and Mere Roberts. 2002. Genealogy as taxonomy. *International Social Science Journal* 173: 403-412.

Haggan, Nigel, Claire Brignall, and Louisa Wood, ed. 2003. *Proceedings of the Putting fisher's knowledge to work Conference, August 27-30, 2001.* Vancouver: Fisheries Centre Research Reports 11.

Haggan, Nigel, Barbara Neis, and Ian Baird, ed. 2007. *Fisher's knowledge in fisheries science and management*. Coastal Management Sourcebooks 4, Paris: UNESCO Publishing.

Hall, Max. 1994. A distaste for walls. *Harvard Magazine* 97: 2.

Halvorsen, Kathleen E., Stuart M. Kramer, Smriti Dahal, and Barry S. Solmon. 2007. Cultural models, climate change and biofuels. Paper presented at the XV International Conference of the Society for Human Ecology, October 4-7, in Rio de Janeiro, Brazil.

Hames, Raymond. 1983. Monoculture, polyculture, and polyvariety in tropical forest swidden cultivation. *Human Ecology* 11: 13-34.

Hamilton, Henning. 1997. Slash-and-burn in the history of the Swedish forests. *Rural Development Forestry Network* 21f: 1-24.

Hanazaki, Natalia. 1997. Conhecimento e uso de plantas, pesca e dieta em comunidades caiçaras do município de Ubatuba (SP). Master diss., University of São Paulo, São Paulo, Brazil.

Hanazaki, Natalia. 2004. Etnobotânica. In *Ecologia de pescadores da Mata Atlântica e da Amazônia,* ed. Alpina Begossi, 37-57. São Paulo: FAPESP/Hucitec.

Hanazaki, Natalia, Fabio de Castro, Vivian G. Oliveira, and Nivaldo Peroni. 2007. Between the sea and the land: the livelihood of estuarine people in southeastern Brazil. *Ambiente e Sociedade* 10: 121-136.

Hanazaki, Natalia, Jorge Y. Tamashiro, Hermógenes F. Leitão-Filho, and Alpina Begossi. 2000. Diversity of plant uses in two Caicara communities from the Atlantic Forest coast, Brazil. *Biodiversity and Conservation* 9: 597-615.

Handel, Warren. 1982. *Ethnomethodology: how people make sense.* Englewood Cliffs, NJ: Prentice Hall.

Hardin, Garrett. 1968. The Tragedy of the Commons. *Science* 162: 1243-1248.

Hardin, Garrett, 1993. Second thoughts on the Tragedy of the Commons. In *Valuing the Earth: economics, ecology, and ethics,* ed. Herman E. Daly and Kenneth N. Townsend. 145-151. Cambridge, MA: MIT Press.

Hardter, Rolf, Woo Yin Chow, and Ooi Soo Hock. 1997. Intensive plantation cropping, a source of sustainable food and energy production in the tropical rain forest areas in southeast Asia. *Forest Ecology and Management* 91: 93-102.

Hardy, Alister C. 1954. Escape from especialization. In *Evolution as a process*, ed. Julian Huxley, Alister C. Hardy and E. B. Ford, 122-142. London: George Allen and Unwin.

Harries-Jones, Peter. 1995. *A recursive vision: ecological understanding and Gregory Bateson.* Toronto: University of Toronto Press.

Harris, David R. 1971. The ecology of swidden cultivation in the Upper Orinoco Rain Forest, Venezuela. *The Geographical Review* 61: 475-495.

Harris, David R. 1972. The origins of agriculture in the tropics. *American Scientist* 60: 180-193.

Harris, Marvin. 1977. *Cannibals and kings*. New York: Random House.

Harris, Marvin. 1979. *Cultural materialism: the struggle for a science of culture.* New York: Random House.

Hassel, Lucas B. 2006. Conhecimentos e práticas de comunidades pesqueiras sobre a conservação de mamíferos marinhos na costa leste do Estado do Rio de Janeiro, Brasil: um estudo de caso das comunidades de Barra de São João e Armação dos Búzios. Rio de Janeiro. Master's diss., Instituto Oswaldo Cruz, Rio de Janeiro, Brazil.

Hauck, Maria, and Merle Sowman, ed. 2003. *Waves of changes: coastal and fisheries co-management in South Africa*. Cape Town: University of Cape Town Press.

Hauser, Marc D. 2000. *The evolution of communication*. Cambridge, MA: The MIT Press.

Hawley, Amos H. 1968. Human ecology. In *International Encyclopedia of the Social Sciences*, Vol. 4, ed. David L. Sills, 328-337. New York: The Free Press, New York.

Hawley, Amos H. 1982. *Ecología humana*. Madrid: Editorial Technos.

Hawley, Amos. 1986. *Human Ecology: a theoretical essay.* Chicago: The University of Chicago Press.

Heckenberger, Michael J., Afukaka Kuikuro, Urissapá T. Kuikuru, J. Christian Russell, Morgan Schmidt, Carlos Fausto, and Bruna Franchetto. 2003. Amazonia 1492: pristine forest or cultural parkland? *Science* 301: 1710-1714.

Hediger, Werner. 2000. Sustainable development and social welfare. *Ecological Economics* 32: 481-492

Heidegger, Martin. 1956. *The question of being.* London: Vision Press.

Heidegger, Martin. 1966 (1932). *Being and time.* Albany, NY: SUNY Press.

Heidegger, Martin 1992. *The metaphysical foundations of logic*. Bloomington: Indiana University Press.

Heidegger, Martin. 1999. *The hermeneutics of facticity*. Bloomington, IN: Indiana University Press.

Heidegger, Martin. 2006. *Mindfulness*. London: Continuum.

Heizer, Robert F. 1987. Venenos de pesca. In *Suma etnológica brasileira*, Ed. Berta Ribeiro and Darcy Ribeiro, 95-99. Petrópolis: Vozes.

Hens, Luc, Richard J. Borden, Shosuke Suzuki, and Gianumberto Caravello, ed. 1998. *Research in human ecology: an interdisciplinary overview*. Brussels: Vrije Universiteit Brussel (VUB) Press.

Hens, Luc, and Alpina Begossi. 2008. Diversity and Management: from extractive to farming systems. *Environment, Development, and Sustainability* 10: 559-563.

Henshaw, Anne. 2003. Climate and culture in the North: The interface of archaeology, paleoenvironmental science and oral history. In *Weather, climate, culture,* ed. Sarah Strauss and Benjamin Orlove, 217-23. New York: Berg Publishers.

Henshaw, Anne. 2009. Sea ice: the socio-cultural dimensions of a melting environment. In *Anthropology and climate change: from encounters to actions*, ed. Susan A. Crate and Mark Nuttall, 153-165. Walnut Creek, Ca: Left Coast Press.

Hernani, Luiz Carlos, Emilio Sakai, Francisco Lombardi Neto, and Igo F. Lepsch. 1987. Effects of clearing of secondary forest on a yellow Latosol of the Ribeira Valley, São Paulo, Brasil. 2. *Revista Brasileira de Ciência do Solo* 11: 215-219.

Hetzel, Bia, and Liliane Lodi. 1993. *Baleias, botos e golfinhos: guia de identificação para o Brasil.* Rio de Janeiro: Editora Nova Fronteira.

Hiraoka, Mario S., and Shozo Yamamoto. 1980. Agricultural development in the upper Amazon of Ecuador. *American Geographical Society* 70, no.4: 423-445.

Hohnwald, Stefan, Barbara Rischkowsky, Ari Pinheiro Camarão, Rainer Schultze-Kraft, José Adérito Rodrigues Filho, and John M. King. 2006. Integrating cattle into the slash-and-burn cycle on smallholdings in the Eastern Amazon, using grass-capoeira or grass-legume pastures. *Agriculture, Ecosystems & Environment* 117: 266–276.

Holling, Crawford S. 2001. Understanding the complexity of economic, ecological and social systems. *Ecosystems* 4: 390-405

Holling, Crawford S., Lance H. Gunderson, and Donald H. Ludwig. 2002. In quest of a theory of adaptive change. In *Panarchy: understanding transformations in human and natural systems,* ed. Lance H. Gunderson and Crawford S. Holling, 3-22. Washington D.C.: Island Press.

Hölscher, Dirk H., Bernard Ludwig, Maria Regina Freire Möller, and Horst Fölster. 1997. Dynamic of soil chemical parameters in shifting agriculture in the Eastern Amazon. *Agriculture, Ecosystems & Environment* 66: 153-163.

Homma, Alfredo K. O., Robert T. Walker, Frederick N. Scatena, Arnaldo José de Conto, Rui de A. Carvalho, Antonio Carlos P. N. da Rocha, Célio Armando P. Ferreira, and Antonio I. M. dos Santos. 1993. Dynamics of deforestation and burning in Amazonia: a microeconomic analysis. *Rural Development Forestry Network* 16c: 1-19.

Homma, Alfredo, Kingo Oyama, Robert T. Walker, Frederick N. Scatena, Arnaldo J. Conto, Rui A. Carvalho, Célio A. P. Ferreira, and Antonio I. M. Santos. 1998. Redução do desmatamento na Amazônia: política agrícola ou ambiental. In *Amazônia: meio ambiente e desenvolvimento agrícola*, ed. Alfredo K. O. Homma, 119-141. Belém: Embrapa-SPI.

Hornborg, Alf. 1998. Towards an ecological theory of unequal exchange: articulating world systems theory and ecological economics. *Ecological Economics* 25: 127-136.

Hornborg, Alf. 2001. *The Power of the machine. Global inequalities of economy, technology, and environment*. Walnut Creek/Lanham/New York/Oxford: AltaMira Press.

Hornborg, Alf, John Robert McNeill, and Juan Martinez Alier. 2007. *Rethinking environmental history: world-system history and global environmental change*. Lanham, MD: AltaMira.

Houghton, Richard A. 1993. The role of the world's forest in global warming. In *The world forests for the future,* ed. Kilaparti Ramakrishna and George M. Woodwell, 21-58. New Haven, CT: Yale University Press.

Houghton, Richard A., Daniel S. Lefkowitz, and David L. Skole. 1991. Changes in the landscape of Latin America between 1850 and 1985 I. Progressive loss of forests. *Forest Ecology and Management* 38: 143-172.

House, Paul. 1997. Forest Farmers: a case study of traditional shifting cultivation in Honduras. *Rural Development Forestry Network Paper* 21a: 1-36.

Howes, Frank N. 1930. Fish-poison plants. *Bulletin of Miscellaneous Information (Royal Gardens, Kew)* 1930: 129-153.

Hunn, Eugene. 1982. The utilitarian factor in folk biological classification. *American Anthropologist* 84, no. 4: 830-847.

Huntington, Henry P. 1998. Observations on the utility of the semi-directive interview for documenting traditional ecological knowledge. *Arctic* 51: 237-242.

Huntington, Henry P. 2000a. Using traditional ecological knowledge in science: methods and applications. *Ecological Applications* 10: 1270-1274.

Huntington, Henry P. 2000b. Traditional knowledge of the ecology of Belugas *Delphinapterus leucas*, in Cook Inlet, Alaska. *Marine Fisheries Review* 62: 134-140.

Huntington, Henry P., and the communities of Buckland, Elim, Koyuk, Point Lay and Shaktoolik. 1999. Traditional ecological knowledge of beluga whales (*Delphinapterus leucas*) in the eastern Chukchi and northern Bering Seas, Alaska. *Arctic* 52: 49-61.

Huntington, Henry P., Robert S. Suydam, and Daniel H. Rosenberg. 2004. Traditional knowledge and satellite tracking as complementary approaches to ecological understanding. *Environmental Conservation* 31: 177-180.

Husserl, Edmund. 1965. *Phenomenology and the crisis of philosophy*. New York: Harper and Row.

Huxley, Aldous. 1962. *Island.* London: Chatto and Windus.

Huxley, Julian. 1942. *Evolution: the modern synthesis.* London: George Allen and Unwin.

Ibarra, Peter R., and John I. Kitsuse. 1993. Vernacular constituents of moral discourse: an interactionist proposal for the study of social problems. In *Reconsidering social constructionism*, ed. Gale Miller and James A. Holstein, 25-58. New York: Aldine de Gruyter.

ICRAF (International Center for Research in Agroforestry). 1996. *Alternatives to slash and burn programme: strategy and funding requirements 1997–2000.* Nairobi: ICRAF.

Igbozurike, Matthias U. 1971. Ecological balance in tropical agriculture. *Geographical Review* 61, no. 4: 519-529.

Intergovernmental Panel on Climate Change (IPCC). 2007. Working Group II Summary for Policymakers. In *Climate change 2007: impacts, adaptation and vulnerability..* Geneva: IPCC Secretariat. http://www.ipcc-wg2.org/ (accessed September 3, 2008).

Imbernon, Jacques. 1999. Changes in agricultural practice and landscape over a 60-year period in North Lampung, Sumatra. *Agriculture, Ecosystems & Environment* 76: 61-66.

Inayatullah, Sohail. 2005. Spirituality as the fourth line? *Futures* 37: 573-579.

Inglehart, Ronald. 1990. *Culture Shift in Advanced Industrial Society.* Princeton, NJ: Princeton University Press.

Inglehart, Ronald. 1997. *Modernization and postmodernization: cultural, economic, and political change in 43 societies.* Princeton, NJ: Princeton University Press.

Inglehart, Ronald, and Wayne E. Baker. 2000. Modernization, Cultural Change, and the Persistence of Traditional Values. *American Sociological Review* 65: 19-51.

Ingold, Timothy. 2000. *The perception of the environment. Essays in livelihood, dwelling and skill.* London and New York: Routledge.

Instituto Brasileiro do Meio Ambiente e dos Recursos Naturais e Renováveis (IBAMA). 2001. *Mamíferos aquáticos do Brasil. Plano de ação.* Brasília: MMA/IBAMA.

IUCN (The International Unit for Conservation of Nature). 1970. *The ecology of man in the tropical environment.* Morges: IUCN Publ., New Series.

Iversen, Johannes. 1956. Forest clearance in the Stone Age. *Scientific American* 194: 36-41.

Jacka, Jerry. 2009. Global averages, local extremes: the subtleties and complexities of climate change in Papua New Guinea. In *Anthropology and climate change: from encounters to actions*, ed. Susan

A. Crate and Mark Nuttall, 197-208. Walnut Creek, Ca: Left Coast Press.

Jakobsen, Jens. 2006. The role of NTFPs in a shifting cultivation system in transition: a village case study from the uplands of North Central Vietnam. *Danish Journal of Geography* 106, no. 2: 103-114.

Jansen, Marco A., and John M Anderies. 2007. Robustness trade-offs in social-ecological systems. *International Journal of the Commons* 1: 43-65.

Jägersten, G. 1972. *Evolution of the metazoan cycle*. London: Academic Press.

Jayatissa, Loku P., Sanath Hettiarachi, and Farid Dahdouh-Guebas. 2006. An attempt to recover economic losses from decadal changes in two lagoon systems of Sri Lanka through a newly patented mangrove product. *Environment, Development and Sustainability* 8: 585-595.

Jefferson, Thomas A., Stephen Leatherwood, and Marc A. Webber. 1993. *Marine mammals of the World: FAO Species identification guide*. Rome: UNEP.

Jentoft, Svein. 1985. Models of fishery management. *Marine Policy* 9: 322-331.

Jentoft, Svein. 1989. Fisheries co-management. *Marine Policy* 13: 137-154.

Jentoft, Svein. 2000. Co-managing the coastal zone: is the task too complex? *Ocean and Coastal Management* 43: 527-535.

Jentoft, Svein. 2003. Co-management: the way forward. In *The fisheries co-management experience: accomplishments, challenges and prospects*, ed. Douglas C. Wilson, Jesper R. Nielsen, and Poul Degnbol, 1-14. Dordrecht: Kluwer Academic Publisher.

Johannes, Robert E. 1981. Working with fishermen to improve coastal tropical fisheries and resource management. *Bulletin of Marine Sciences* 31: 673-680.

Johannes, Robert E. 1998. The case for data-less marine resource management: examples from tropical nearshore finfisheries. *Trends in Ecology and Evolution* 13: 243-246.

Johannes, Robert E. 2002. The renaissance of community-based marine resource management in Oceania. *Annual Review of Ecology and Systematics* 33: 317-340.

Johannes, Robert E. 2003. Use and misuse of traditional ecological knowledge and management practices. In *Values at sea: ethics for the marine environment*, ed. Dorinda G. Dallmeyer, 111-126. Athens: University of Georgia Press.

Johannes, Robert E., Milton R. Freeman, and Richard J. Hamilton. 2000. Ignore fishers' knowledge and miss the boat. *Fish and Fisheries* 1: 257-271.

Johannes, Robert E., Milton R. Freeman, and Richard J. Hamilton. 2001. Marine biologists miss the boat by ignoring fishers' knowledge. *Fish and Fisheries* 1: 257-265.

Johnson, Christine M., Ima C. G. Vieira, Daniel J. Zarin, Jacqueline Frizano, and Arthur H. Johnson. 2001. Carbon and nutrient storage in primary and secondary forests in eastern Amazonia. *Forest Ecology and Management* 147, no. 2: 245-252.

Jonas, Hans. 1984 (1979). *The Imperative of responsibility. In search of an ethics for the technological age*. Chicago: University of Chicago Press.

Jorgenson, Andrew K. 2003. Consumption and environmental degradation: a cross-national analysis of the ecological footprint. *Social Problems* 50: 374-394.

Jorgenson, Andrew K. 2004. Uneven processes and environmental degradation in the world economy. *Human Ecology Review* 11: 103-113.

Jorgenson, Andrew K., and Thomas J. Burns. 2007. The political-economic causes of change in the ecological footprints of nations, 1991-2001: a quantitative investigation. *Social Science Research* 36: 834-853.

Judge, Anthony. 1994. *Configuring conceptual polarities in question: metaphoric pointers to self-reflexive coherence*. http://www.laetusinpraesens.org/docs00s/writpol2.php (accessed October 24, 2008).

Juo, Anthony S. R., and Andrew Manu. 1996. Chemical dynamics in slash-and-burn agriculture. *Agriculture, Ecosystems & Environment* 58: 49-60.

Kalikoski, Daniela C., Marcelo Vasconcellos, Les Lavkulich. 2002. Fitting institutions and ecosystems: the case of artisanal fisheries management in the Patos lagoon. *Marine Policy* 26, no. 3: 179-196.

Kalikoski, Daniela C., and Theresa T. Satterfield. 2004. On crafting a fisheries co-management arrangement in the estuary of Patos Lagoon (Brazil): opportunities and challenges faced through implementation. *Marine Policy* 28: 503-522.

Kalikoski, Daniela C., Cristiana S. Seixas, and Tiago Almudi. In Press. Gestão compartilhada e comunitária da pesca no Brasil: avanços e desafios. *Ambiente & Sociedade*.

Kant, Immanuel. 1950 (1783). *Prolegomena to any future metaphysics*, ed. Lewis W. Beck. New York: Bobbs-Merrill.

Kant, Immanuel. 1958 (1781). *Critique of pure reason*, trans. Norman K. Smith. New York: Random House.

Karlen, A. 1995. *Plague's progress. A social history of man and disease*. London: Phoenix.

Kartawinata, Kuswata. 1994. The use of secondary forest species in rehabilitation of degraded forest lands. *Journal of Tropical Forest Science* 7: 76–86.

Kates, Robert W., William C. Clark, Robert Corell, J. Michael Hall, Carlo C. Jaeger, Ian Lowe, James J. McCarthy, Hans Joachim Schellnhuber, Bert Bolin, Nancy M. Dickson, Sylvie Faucheux, Gilberto C. Gallopin, Arnulf Gruebler, Brian Huntley, Jill Jäger, Narpat S. Jodha, Roger E. Kasperson, Akin Mabogunje, Pamela Matson, Harold Mooney, Berrien Moore III, Timothy O'Riordan, and Uno Svedin. 2000. *Sustainability science*. Research and assessment systems for sustainability program. Discussion Paper 2000-2033. Cambridge, MA: Environment and Natural Resources Program, Belfer Center for Science and International Affairs, Kennedy School of Government, Harvard University.

Kato, Maria S., Osvaldo R. Kato, Manfred Denich, and Paul L. G. Vlek. 1999. Fire-free alternatives to slash-and-burn for shifting cultivation in the eastern Amazon region: the role of fertilizers. *Field Crops Research* 62: 225-237.

Kay, James, Henry Regier, Michelle Boyle, and George Francis. 1999. An ecosystem approach for sustainability: addressing the challenge of complexity. *Futures* 31: 721-742.

Kearney, Richard. 1995. *Poetics of modernity: towards a hermeneutic imagination*. Atlantic Highlands, NJ: Humanities Press.

Keen, Meg, and Sango Mahanty. 2006. Learning in sustainable natural resource management: challenges and opportunities in the Pacific. *Society and Natural Resources* 19: 497-513

Kempton, Willett. 2001. Cognitive anthropology and the environment. In *New directions in anthropology and the environment*, ed. Carole Crumley, 49-71. Walnut Creek, CA: AltaMira Press.

Kempton, Willett, James S.Boster, and Jennifer A Hartley. 1996. *Environmental values in American culture*. Cambridge: MIT Press.

Kendrick, Anne. 2003. Caribou co-management in northern Canada: fostering multiple ways of knowing. In *Navigating social-ecological systems: building resilience for complexity and change,* ed. Fikret Berkes, Johan Colding, and Carl Folke, 241-267. Cambridge: Cambridge University Press.

Kerr, Warwick E., and Charles R. Clement. 1980. Práticas agrícolas de conseqüências genéticas que possibilitaram aos índios da Amazônia uma melhor adaptação às condições ecológicas da região. *Acta Amazônica* 10: 251-261.

Ketterings, Quirine M., Titus Tri Wibowo, Meine van Noordwijk, and Eric Penot. 1999. Farmers' perspectives on slash-and-burn as a land clearing method for small-scale rubber producers in Sepunggur, Jambi Province, Sumatra, Indonésia. *Forest Ecology and Management* 120: 157-169.

Kinzig, Ann P.; Stephen Carpenter, Michael Dove, Geoff Heal, Simon Levin, Jane Lubchenco, Stephen J. Schneider, and David Starrett. 2000. *Executive Summary of the National Science Foundation Meeting, June 5-8, 2000: Nature and Society: an imperative for integrated environmental research*. Tempe, Arizona: NSF. http://stephenschneider.stanford.edu/Publications/PDF_Papers/NatureAndSociety.pdf (accessed October 24, 2008).

Kitcher, Philip. 1983. *Abusing science: the case against creationism*. Cambridge, MA: MIT Press.

Kitcher, Philip. 2007. *Living with Darwin: evolution, design, and the future of faith*. New York: Oxford University Press.

Klein,Carlos H., Nelson A. Silva, Armando R. Nogueira; Katia V. Bloch, and Lúcia H.Campos. 1995. Hipertensão arterial na Ilha do Governador, Rio de Janeiro, Brasil. I. Metodologia. *Cadernos de Saúde Pública* 11, no. 2:187-201.

Kleinman, Peter J., David Pimentel, and Ray B. Bryant. 1995. The ecological sustainability of slash-and-burn agriculture. *Agriculture, Ecosystems & Environment* 52, no. 2/3: 235-249.

Kluckhohn, Clyde. 1951. Value and values orientations in the theory of action: an exploration in the definition and classification. In *Towards a general theory of action*, ed. Talcott Parsons and Edward A. Shills, 388-433. Cambridge, MA: Harvard University Press.

Knight, Dennis H. 1975. A phytosociological analysis of species-rich tropical forest on Barro Colorado Island, Panama. *Ecological Monographs* 45: 259–284.

Koffman, Theodore and Richard J. Borden. 1998. Building unlikely partnerships: human ecology models for collaborative leadership and the management of complexity.1998. In *Research in human ecology: an interdisciplinary overview*, ed. Luc Hens, Richard J. Borden, Shosuke Suzuki and Gianumberto Caravello, 11-30. Brussels: Vrije Universiteit Brussel (VUB) Press.

Kohn, Marek. 2004. *A reason for everything: natural selection and the English imagination*. London: Faber & Faber.

Kormandy, Edward. 1978. Ecology/economy of nature—synonyms? *Ecology* 59, no. 6: 1292-1294.

Kotto-Same, Jean, Paul L. Woomer, Moukam Appolinaire, and Zapfack Louis. 1997. Carbon dynamics in slash-and-bum agriculture and land use alternatives of the humid forest zone in Cameroon. *Agriculture, Ecosystems & Environment* 65: 245-256.

Krause, Gesche, Dirk Schories, Marion Glaser, and Karen Diele. 2001. The spatial patterns of the Bragantinian mangrove ecosystem of North-Brazil. *Ecotropica* 7: 93-107.

Krause, Gesche, Jon Norberg, Marion Glaser, Patrik Rönnbäck, and Nils Kautsky. In Press. Options and resilience in a social-ecological mangrove system in Pará (North Brazil). *Ecosytems.*

Krause, Gesche, and Marion Glaser. In Preparation. Mangroves and people: a social-ecological system–ecosystem use, driving forces and co-evolution. In *Mangroves dynamics and management in North Brazil. Ecological Studies Series,* ed. Ulrich Saint Paul and Horacio Schneider. Heidelberg: Springer.

Krupnik, Igor, and Dyanna Jolly, ed. 2002. *The earth is faster now: indigenous observations of Arctic environment change.* Fairbanks, Alaska, USA: Arctic Research Consortium of the United States.

Kurien, John. 2003. People and the sea: a "Tropical-Majority" world perspective. *Maritime Studies* 12: 9-26.

Kyuma, Kazutake, Theparitt Tulaphitak, and Chaitat Parintra. 1985. Changes in soil fertility and tilth under shifting cultivation. I. General description of soil and effect of burning on the soil characteristics. *Soil Science and Plant Nutrition* 31, no. 2: 227-238.

Lahsen, Myanna. 2007. Anthropology and the trouble of risk society. *Anthropology News* December: 10–11.

Laland, Kevin N., and Gillian R. Brown. 2002. *Sense & Nonsense: evolutionary perspective on human behavior.* Oxford: Oxford University Press.

Lanly, Jean Paul. 1982. *Tropical forest resources.* Rome: UN Food and Agriculture Organization.

Larrère, Catherine. 1997. *Les philosophies de l'environnement.* Paris: Presses Universitaires de France.

Larrère, Raphaël. 1994. L'art de produire de la nature; une leçon de Rousseau. *Le Courrier de l'Environnement* 34: 1-9.

Lassiter, Luke Eric. 2005. Collaborative ethnography and public anthropology. *Current Anthropology* 6, no. 1: 83-107.

Latour, Bruno 2005. *Reassembling the social: an introduction to actor-network-theory.* Oxford: Oxford University Press.

Lawrence, Deborah, David R. Peart, and Mark Leighton. 1998. The impact of shifting cultivation on a rainforest landscape in West Kalimantan: spatial and temporal dynamics. *Landscape Ecology* 13: 135–148.

Lazrus, Heather. 2009. The governance of vulnerability: climate change and agency in Tuvalu, South Pacific. In *Anthropology and climate change: from encounters to actions*, ed. Susan A. Crate and Mark Nuttall, 240-249. Walnut Creek, Ca: Left Coast Press.

Leakey, Louis. 1934. *Adam's ancestors*. London: Methuen and Co..

Leakey, Louis. 1960. *Adam's ancestors*. New York, Harper and Row, 4th edition.

Leduc, Timothy. 2007. Approaching a climatic research etiquette. *Ethics & The Environment* 12, no. 2: 45-69.

Leopold, Aldo. 1949. *A sand county almanac and sketches here and there*. Oxford: Oxford University Press.

Levi-Strauss, Claude. 1966. *The savage mind*. Nature of Human Society Series. London: Weidenfeld & Nicolson.

Lewin, Tamar. 2007. Eco-education. *The New York Times*. November 4, http://www.nytimes.com/2007/11/04/education/edlife/lewin-atlantic.html (accessed November 5, 2008).

Lianzela.1997. Effects of shifting cultivation on the environment: with special reference to Mizoram. *International Journal of Social Economics* 24, no.7-9: 785-791.

Liebezeit, Gerd, and Monica T. Rau. 2006. New Guinean mangroves - Traditional usage and chemistry of natural products. *Senckenbergiana Maritima* 36: 1-10.

Liebovitch, Larry S., and Daniela Scheurle. 2000. Two lessons from fractals and chaos: changes in the way we see the world. *Complexity* 5, no. 4: 34-43.

Lima, Hedinaldo N., Carlos E. R. Schaefer, Jaime W. V. Mello, Robert J. Gilkes, and João C. Ker. 2002. Pedogenesis and pre-Colombian land use of ''Terra Preta Anthrosols'' (''Indian black earth'') of Western Amazônia. *Geoderma* 110: 1–17.

Lindbladh, Matts, and Richard Bradshaw. 1998. The origin of present forest composition and pattern in Southern Sweden. *Journal of Biogeography* 25: 463-477.

Lloyd, Christopher. 1989. Realism, structurism, and history. *Theory and Society* 18: 451-494.

Lobão, Ronaldo J.S. 2000. Reservas extrativistas marinhas: uma reforma agrária do mar? Master's diss., Universidade Federal Fluminense, Rio de Janeiro, Brazil.

Lofdahl, Corey. 2002. *Environmental impacts of globalization and trade*. Cambridge, MA: MIT Press.

Loker, William M. 1994. Where's the beef? Incorporating cattle into sustainable agroforestry systems in the Amazon Basin. *Agroforestry Systems* 25: 227-241.

Lorenzi, Harri, and Abreu F.J. Matos. 2002. *Plantas medicinais no Brasil: nativas e exóticas*. Nova Odessa: Instituto Plantarum.

Lu, Dengsheng, Paul Mausel, Eduardo S. Brondízio, and Emilio Moran. 2003. Classification of successional forest stages in the Brazilian Amazon basin. *Forest Ecology and Management* 181: 301-312.

Lugo, Ariel E., and Mary J. Sanchez. 1986. Land use and organic carbon content of some subtropical soils. *Plant and Soil* 96: 185-196.

Lumsden, Charles J., and Edward O. Wilson. 1981. *Genes, mind and culture*. Cambridge: Harvard University Press.

Macchi, Mirjam. 2008. *Indigenous and traditional peoples and climate change: Issues Paper*. Switzerland: IUCN.

MacMynowski, Dena P. 2007. Pausing at the brink of interdisciplinarity: power and knowledge at the meeting of social and biophysical science. *Ecology and Society* 12: 1-20, http://www.ecologyandsociety.org/vol12/iss1/art20/ (accessed October 23, 2008).

Maldague, Michel. 1977. L'étude de lénvironnement dans la formation des spécialités. In *Tendances de l'éducation relative à l'environnement,* ed. UNESCO, 177-197. Paris: UNESCO.

Malhi, Yadvinder, J. Timmons Roberts, Richard A. Betts, Timothy J. Killeen, Wenhong Li, and Carlos A. Nobre. 2008. Climate change, deforestation, and the fate of the Amazon. *Science* 319: 169-172.

Mann, Kenneth H. 1988. Towards predictive models for coastal marine ecosystems. In *Concepts of ecosystem ecology: a comparative view. Ecological studies* 67, ed. Lawerence R. Pomeroy and James J. Alberts, 291-316. New York: Springer.

Mansperger, Mark C. 1995. Tourism and cultural change in small-scale societies. *Human Organization* 54: 87-94.

Marcuse, Herbert. 1964. *One-dimensional man*. Boston: Beacon Press.

Margalef, Ramon. 1981. Stress in ecosystems: a future approach. In *Stress effects on natural ecosystems,* ed. Garry W. Barrett and Rutger Rosenberg, 281-289. New York: Wiley.

Marino, Elizabeth, and Peter Schweitzer. 2007. When "the hundred-year-flood" comes every other year: Inupiaq coping and adaptation strategies in a climate of change. Paper presented at the XV International Conference of the Society for Human Ecology, October 4-7, in Rio de Janeiro, Brazil.

Marino, Elisabeth, and Peter Schweitzer. 2009. Talking and not talking about climate change in Northwestern Alaska, In *Anthropology and climate change: from encounters to actions*, ed. Susan A. Crate and Mark Nuttall, 209-218. Walnut Creek, Ca: Left Coast Press.

Markewitz, Daniel, Ricardo O. Figueiredo, and Eric A. Davidson. 2006. CO_2-driven cation leaching after tropical forest clearing. *Journal of Geochemical Exploration* 88: 214– 219.

Marques, José G.W. 1991. Aspectos ecológicos na etnoictiologia dos pescadores do complexo estuarino-lagunar de Mundaú Manguaba, Alagoas. Ph.D diss., Universidade Estadual de Campinas, Campinas, SP, Brazil.

Marques, José G.W. 1995. *Pescando pescadores: etnoecologia abrangente no baixo São Francisco alagoano*. São Paulo: Nupaub-USP.

Martinez-Alier, Juan, ed. 1990. *Ecological economics. Energy, environment and society.* Oxford, UK; Cambridge, USA: Blackwell.

Martinez-Alier, Juan. 1999. The Socio-ecological embeddedness of economic activity: the emergence of a transdisciplinary field. In *Sustainability and the social sciences,* ed. Egon Becker and Thomas Jahn, 112-139. London, New York: Zed Books.

Martins, Paulo S. 1994. Biodiversity and agriculture: patterns of domestication of Brazilian native plant species. *Anais da Academia Brasileira de Ciências* 66: 219-226.

Martins, Paulo S. 2001. Dinâmica evolutiva em roças de caboclos amazônicos. In *Diversidade biológica e cultural da Amazônia*, ed. Ima Célia G. Vieira, José Maria C. Silva, David C. Oren and Maria Ângela D´Incao, 369-384. Belém: Museu Paraense Emílio Goeldi.

Martins, Paulo S. 2005. Dinâmica evolutiva em roças de caboclos amazônicos. *Estudos Avançados* 19: 209-220.

Marx, Karl. 1967 (1867). *Capital*, vol. 1. New York: International Publishers.

Matthews, Emily, Christof Amann, Stefan Bringezu, Marina Fischer-Kowalski, Walter Hüttler, Rene Kleijn, Yuichi Moriguchi, Christian Ottke, Eric Rodenburg, Don Rogich, Heinz Schandl, Helmut Schütz, Ester Van der Voet, and Helga Weisz. 2000. *The weight of nations: material outflows from industrial economies.* Washington, D.C.: World Resources Institute.

Maurstad, Anita, Trine Dale, and Pål A. Bjørn. 2007. You wouldn't spawn in a septic tank, would you? *Human Ecology* 35: 601–610.

Max-Neef, Manfred A., Antonio Elizalde, and Martin Hopenhayn. 1991. *Human scale development: conception, application and further reflections*. New York: The Apex Press.

McCarthy, James J., Osvaldo E.Canziani, N. A. Leary, D. J. Dokken, and Kasey S. White, ed. 2001. *Climate change 2001: impacts, adaptation, and vulnerability*. Cambridge: Cambridge University Press.

McCloskey, Deirdre N. 1985. *The rhetoric of economics*. Madison, WI: University of Wisconsin Press.

McClure, John. 1991. *Explanations, accounts, and illusions: a critical analysis*. Cambridge: Cambridge University Press.

McDonald, M. A., J. R. Healey, and P. A. Stevens. 2000. The effects of secondary forest clearance and subsequent land-use on erosion losses and soil properties in the Blue Mountains of Jamaica. *Agriculture, Ecosystems & Environment* 92: 1-19.

McGee, Michael C. 1975. In search of "the people": a rhetorical alternative. *Quarterly Journal of Speech* 61, no.3: 235-249.

McGee, Michael C. 1980. The "Ideograph": a link between rhetoric and ideology. *Quarterly Journal of Speech* 66, no. 1: 1-16.

McGowan, John. 1991. *Postmodernism and its critics*. Ithica: Cornell University Press.

McGrath, David G. 1987. The role of biomass in shifting cultivation. *Human Ecology* 15: 221-242.

McGrath, David G., Fábio de Castro, Celia Futemma, Benedito D. Amaral, and Juliana Calabria. 1993. Fisheries and the evolution of resource management on the lower Amazon floodplain. *Human Ecology* 21: 167-196.

McHarg, Ian. 1969. *Design with nature.* New York: American Museum of Natural History/Natural History Press.

McHarg, Ian. 1981. Human ecology planning at Pennsylvania. *Landscape Planning* 18, no. 2: 109-120.

McIntosh, Alastair. 2004. *Soil and soul: people versus corporate power.* London: Aurum Press.

McIntosh, Roderick, Joseph Tainter, and Susan Keech McIntosh, ed. 2000. *The way the wind blows: climate, history, and human action*. New York: Columbia University Press.

McKenzie, Roderick. 1926. *The scope of human ecology.* Pennsylvania: Hutchinson Ross Publishing.

McKeown, Thomas. 1988. *The origins of human disease*. Oxford: Blackwell Publishing.

McMillin, Jennifer and Robert Dyball. 2008. Developing a whole-of-university approach to educating for sustainability: linking curriculum, research and sustainable campus operations. *Journal of Education for Sustainable Development,* forthcoming

McNamara, Kenneth. 1997. *Shapes of time*. Baltimore: Johns Hopkins.

McNeill, John Robert. 2000. *Something new under the sun: an environmental history of the twentieth-century world*. New York: W.W. Norton.

Meggers, Betty. 1971. *Amazonia: man and culture in a counterfeit paradise*. Chicago: Aldine.

Mehan, Hugh, and Houston Wood. 1975. The *reality of ethnomethodology*. New York: Wiley.

Mellor, Mary. 1997. *Feminism and ecology*. Cambridge: Polity Press.

Mendonça, Ana Lucia F. 2000. A Ilha do Cardoso – o parque estadual e os moradores. Master diss., University of São Paulo, São Paulo, Brazil.

Mendonça, Jocemar T., and Mario Katsuragawa, M. 2001. Caracterização da pesca artesanal no complexo estuarino-lagunar da Cananéia-Iguape, Estado de São Paulo, Brasil (1995-1996). *Acta Scientiarum* 23: 535-547.

Mendonça, Mário J. C., Maria del Carmen V. Diaz, Daniel Nepstad, Ronaldo S. da Motta, Ane Alencar, João Carlos Gomes, and Ramon A. Ortiz. 2004. The economic cost of the use of fire in the Amazon. *Ecological Economics* 49: 89– 105.

Mendoza-Vega, Jorge, Erik Karltun, and Mats Olsson. 2003. Estimations of amounts of soil organic and fine root carbon in land use and land cover classes, and soil types of Chiapas highlands, Mexico. *Forest Ecology and Management* 177: 1-16.

Mertz, Ole. 2002. The relationship between fallow length and crop yields in shifting cultivation: a rethinking. *Agroforestry Systems* 55: 149-159.

Mertz, Ole, Reed L. Wadley, and Andreas E. Christensen. 2005. Local land use strategies in a globalizing world: Subsistence farming, cash crops and income diversification. *Agricultural Systems* 85: 209–215.

Metzger, Jean Paul. 2002. Lanscape dynamics and equilibrium in areas of slash-and-burn agriculture with short and long fallow period (Bragantina region, NE Brazilian Amazon). *Landscape Ecology* 17: 419-431.

Meuriot, Éric. 1987. *Les modèles bio-économiques d'exploitation des pêcheries: démarches et enseignements*. Rapports économiques et juridiques de 1' IFREMER 4, Paris: Institut français d'exploitation de la mer, http://www.ifremer.fr/docelec/doc/1987/rapport-1462.pdf (accessed October 24, 2008).

Meyer, John. 1977. The effects of education as an institution. *American Journal of Sociology* 83: 55-77.

Meyer, John. 1980. The world polity and the authority of the nation-state. In *Studies of the modern world-system*, ed. Albert J. Bergesen, 109-137. New York: Academic Press.

Meyer, John, David John Frank, Ann Hironaka, Evan Schofer, and Nancy Brandon Tuma. 1997. The structuring of the world environmental regime, 1870-1990. *International Organization* 51: 623-651.

Michod, Richard E. 1998. Evolution of individuality. *Journal of Evolutionary Biology* 11: 225-227.

Millennium Ecosystem Assessment (MEA). 2005. *Ecosystems and human well-being*. Washington DC: World Resources Institute.

Miller, Patricia M., and J. Boone Kauffman. 1998. Effects of slash and burn agriculture on species abundance and composition of a tropical deciduous forest. *Forest Ecology and Management* 103: 191-201.

Ministério do Meio Ambiente. 2002. *Avaliação e ações prioritárias para a conservação da biodiversidade das zonas costeira e marinha.* Brasília: MMA/SBF.

Miranda, Tatiana M., and Natalia Hanazaki. 2008. Conhecimento e uso de recursos vegetais de restinga por comunidades das ilhas do Cardoso (SP) e de Santa Catarina (SC), Brasil. *Acta Botanica Brasílica*. 22: 203-215.

Mithen, Steven. 2003. *After the ice: a global human history, 20.000-5.000 BC*. London: Weidenfeld & Nicolson.

Mobbs, Catherine, and Peter Crabb. 2002. *Building strengths in interdisciplinary research on resource and environmental issues: an appraisal of CRES experience 1973 – 2000.* Canberra: Centre for Resource and Environmental Studies, Australian National University.

Monastersky, Richard. 2007. Colleges strain to reach climate-friendly future. *Chronicle of Higher Education* 13, http://chronicle.com/free/v54/i16/16a00101.htm (accessed November 5, 2008).

Moran, Emilio F. 1990. Ecosystem ecology in biology and anthropology: a critical assessment. In *The ecosystem approach in anthropology: from concept to practice*, ed. Emilio F. Moran, 3-40. Ann Arbor: The University of Michigan Press.

Moran, Emilio F. 2000. *Human adaptability: an introduction to ecological anthropology.* Boulder: Westview Press.

Morán, Emilio F. 2007. Human adaptability: an introduction to cultural anthropology. Boulder: Westview Press.

Moran, Emilio F., Eduardo S. Brondizio, Joanna M. Tucker, Maria Clara Silva-Forsberg, Stephen; McCracken, and Italo Falesi. 2000. Effects of soil fertility and land use on forest succession in Amazônia. *Forest Ecology and Management* 139: 93-108.

Moretti, Christian, and Pierre Grenand. 1982. Les nivrées ou plantes ichtyotoxiques de la Guyane Française. *Journal of Ethnopharmacology* 6: 139-160.

Morril, Warren. T. 1967. Ethnoichthyology of the Cha-Cha. *Ethnology* 6: 405-417.

Morris, David. 2005. Animals and human, thinking and nature. *Phenomenology and the Cognitive Sciences* 4: 49-72.

Morris, David. 2007. Faces and the invisible of the visible: toward an animal ontology. *PhaenEx* 2, no.2: 124-169.

Morrison, Keith. 2002. Wilderness as the kingdom of God. *Ecotheology* 6: 23-41

Morrison, Keith. 2008. Indigenous entrepreneurship in Samoa in the face of neo-colonialism and globalisation. *Journal of Enterprising Communities: People and Places in the Global Economy* 2, no.3: 240-253

Moser, Susanne and Lisa Dilling, ed. 2007. *Creating a climate for change: communicating climate change and facilitating social change*. Cambridge: Cambridge University Press.

Mott, Kenneth, Philip. Desjeux, Alvaro Moncayo, P. Ranque, and P. de Raadt. 1990. Parasitic diseases and urban development. *Buletin of the World Health Organization* 68, no. 6:691-698.

Moura, Rodrigo L., Guilherme F. Dutra, Ronaldo Francini-Filho, Carolina Minte-Vera, Isabel B. Curado, Fernanda J. Guimarães; Ronaldo F. Oliveira, and Diego C. Alves. 2007. Fisheries management in the Extractive Reserve of Corumbau, Bahia. In *Aquatic protected areas as fisheries management tools*, org. Ministério do Meio Ambiente, 175-187. Brasília: Protected Areas of Brazil Series, 4.

Müller, Daniel, and Manfred Zeller. 2002. Land use dynamics in the central highlands of Vietnam: a spatial model combining village survey data with satellite imagery interpretation. *Agricultural Economics* 27: 333-354.

Munda, Giuseppe 2005. "Measuring Sustainability": a multi-criterion framework. *Environment, Development and Sustainability*. 7: 117-134

Myers, Norman. 1993. Tropical forests: the main deforestation fronts. *Environmental Conservation* 20: 9-16.

Mymrin, Nicolai I., the communities of Novoe Chaplino, Sireniki, Uelen and Yanrakinnot, and Henry P. Huntington. 1999. Traditional knowledge of the ecology of beluga whales (*Delphinapterus leucas*) in the northern Bering Sea, Chukotka, Russia. *Arctic* 52: 62-70.

Nadav, Malin. 2000. The cost of green materials. *Building Research and Information* 28: 408–412.

Nagy, Laszlo, and John Proctor. 1999. Early secondary forest growth after shifting cultivation. In *Early secondary forest growth. Management of secondary and logged-over forest in Indonesia,* ed. Plinio Sist, Cesar

Sabogal and Yvonne Byron, 1-12. Bogor, Indonesia: Centre for International Forestry Research.

Nakagawa, Michiko, Hideo Miguchi, and Tohru Nakashizuka. 2006. The effects of various forest uses on small mammal communities in Sarawak, Malaysia. *Forest Ecology and Management* 231: 55–62.

Nakano, Kazutaka 1978. An ecological study of swidden agriculture at a village in Northern Thailand. *Southeast Asian Studies* 16: 411-446.

Natcher, David. 2007. Notions of time and sentience: methodological considerations for arctic change. Paper presented at the XV International Conference of the Society for Human Ecology, October 4-7, in Rio de Janeiro, Brazil.

National Research Council. 1999. *Human dimensions of global environmental change: research pathways for the next decade*. Washington, DC: National Academy Press.

Negrelle, Raquel B., and Katia R. C. Fornazzari. 2007. Ethnobotanical study in two rural communities (Limeira and Riberião Grande) in Guaratuba (Paraná, Brazil). *Revista Brasileira de Plantas Medicinais* 9: 36-54.

Neis, Barbara, 1997. Fishers' ecological knowledge and stock assessment in Newfoundland. In How *deep is the ocean? Historical essays on Canada's Atlantic Fishery*, eds. James E. Candow and Carol Corbin, pp. 243-260. Sydney: University College of Cape Breton Press and the Louisbourg Institute.

Neis, Barbara, David C. Schneider, Lawrence Felt, Richard L. Haedrich, Johanne Fischer, and Jeffrey A. Hutchings. 1999. Fisheries assessment: what can be learned from interviewing resource users? *Canadian Journal of Fisheries and Aquatic Sciences* 56: 1949–1963.

Nelson, John N., and Deirdre .N. McCloskey, ed. 1987. *The rhetoric of the human sciences: language and argument in scholarship and public affairs*. Madison, WI: University of Wisconsin Press.

Nepstad, Daniel C., Claudia M. Stickler, Britaldo Soares-Filho, and Frank Merry. 2008. Interactions among Amazon land use, forests and climate: prospects for a near-term forest tipping point. *Philosophical Transactions of the Royal Society B*. 363: 1737–1746.

Netting, Robert M. 1968. *Hill farmers of Nigeria: cultural ecology of the Kofyar of the Jos Plateau.* Seattle: University of Washington Press.

Netting, Robert M. 1993 *Smallholders, householders: farm families and the ecology of intensive, sustainable agriculture*. Stanford, CA: Stanford University Press.

Neuwinger, Hans D. 2004. Plants used for poison fishing in tropical Africa. *Toxicon* 44: 417-430. Neves, Walter A. 1995. Sociodiversity and biodiversity, two sides of the same equation. In *Brazilian perspectives on sustainable development of the Amazon region,* ed. Miguel Clüsener-Godt and Ignacy Sachs, 91-124. Man and the Biosphere Series, 15. London/Paris: Parthenon.

Newell, Barry, Carole L. Crumley, Nordin Hassan, Eric F. Lambin, Claudia Pahl-Wostl, Arild Underdal, and Robert Wasson. 2005. A conceptual template for integrative human-environment research. *Global Environmental Change* Part A 15: 299-307.

Newell, Patrick T., Simon Wing, and Frederick J. Rich. 2007. Cusp for high and low merging rates. *Journal of Geophysical Research* 112: A09205, doi: 10.1029/2007JA012353

Nielsen, Jesper R., Poul Degnbol, K. Kuperan Viswanathan, Mahfuzuddin Ahmed, Mafaniso Hara, and Nik M. R. Abdullah. 2004. Fisheries co-management: an institutional innovation? Lessons from South East Asia and Southern Africa. *Marine Policy* 28, no. 2: 151-160.

Ngidang, Dimbab. 2002. Contradictions in land development schemes: the case of joint ventures in Sarawak, Malaysia. *Asia Pacific Viewpoint* 43: 157-180.

Nnochiri, Enyinnaya. 1968. *Parasitic disease and urbanization in a developing community.* London: Oxford University Press.

Noble, Ian R., and Rodolfo Dirzo. 1997. Forests as human-dominated ecosystems. *Science* 277: 522-525.

Norgaard, Richard B. 1984. Coevolutionary development potential. *Land Economics* 60, no. 2: 160-172.

Norgrove, Lindsay, and Stefan Hauser. 2002. Yield of plantain grown under different tree densities and "slash and mulch" versus "slash and burn"' management in an agrisilvicultural system in southern Cameroon. *Field Crops Research* 78: 185-195.

Nuttall, Mark. 2009. Living in a world of movement: human resilience to environmental instability in Greenland. In *Anthropology and climate change: from encounters to actions*, ed. Susan A. Crate and Mark Nuttall, 292-310. Walnut Creek, Ca: Left Coast Press.

Nye, Peter H., and Dennis J. Greenland. 1960. The soil under shifting cultivation. Technical communications 51. Harpenden, UK: Commonwealth Bureau of Soils.

O'Brien, Willian E. 2002. The nature of shifting cultivation: redemption. *Human Ecology* 30: 483-502.

Odum, Eugene. 1953. *Fundamentals of ecology*. Philadelphia: Saunders.

Ogburn, William. 1961 (1932). The hypothesis of cultural lag. In *Theories of Society: foundations of modern sociological theory,* ed. Talcott

Parsons, Edward Shils, Kaspar D. Naegele and Jesse R. Pitts, 1270-1274. New York: Free Press.

Ohtsuka, Ryutaro. 1983. *Oriomo Papuans: ecology of sago-eaters in lowland Papua.* Tokyo: University of Tokyo Press.

Okasha, Samir. 2003. Recent work on the levels of selection problem. *The Human Nature Review* 3: 349-356.

Oldfield, Margery L., and Janis B. Alcorn. 1987. Conservation of traditional agroecosystems. *Bioscience* 37: 199–208.

Oliveira, Flávia C. de. 2007. Etnobotânica da exploração de espécies vegetais para confecção do cerco-fixo na região do Parque Estadual Ilha do Cardoso, SP. Msc. diss., Universidade Federal de Santa Catarina, Santa Catarina, Brazil.

Oliveira, Rogério R. 2008. When the shifting agriculture is gone: functionality of Atlantic Coastal Forest in abandoned farming sites. *Boletim do Museu Paraense Emílio Goeldi, Ciências Humanas* 3, no. 2: 213-226.

Oliveira, Rogério R., Denise F. Lima, Patricia D. Sampaio, Rogério F. Silva, and Daniel G. Toffoli. 1994. Roça Caiçara: um sistema "primitivo" auto-sustentável. *Ciência Hoje* 18: 44-51.

Olsson, Per, Carl Folke, and Fikret Berkes. 2004a. Adaptive comanagement for building resilience in social-ecological systems. *Environmental Management* 43, no. 1: 75-90.

Olsson, Per, Carl Folke, and Thomas Hahn. 2004b. Social-ecological transformation for ecosystem management: the development of adaptive co-management of a wetland landscape in southern Sweden. *Ecology and Society* 9, no. 4: 2, http://www.ecologyandsociety.org/vol9/iss4/art2/ (accessed October 30, 2008).

Organisation for Economic Co-Operation and Development (OECD). 2000. *Transition to responsible fisheries; economic and policy implications.* Paris: OECD.

Organisation for Economic Co-Operation and Development (OECD). 2003. *The Costs of Managing Fisheries*. Paris: OECD.

Orlove, Benjamin S. 1980. Ecological anthropology. *Annual Review of Anthropology* 9:235-73.

Orlove, Ben, Ellen Wiegandt, and Brian Luckman, ed. 2008. *Darkening peaks: glacier retreat, science and society.* Berkeley: University of California Press.

Ostrom, Ellinor. 1990. *Governing the commons. the evolution of institutions for collective action*. Cambridge: Cambridge University Press.

Ostrom, Elinor, Roy Gardner, and James Walker. 1994. *Rules, games, and common-pool resources*. Michigan: University of Michigan Press.

Padoch, Christine, Emily Harwell, and Adi Susanto. 1998. Swidden, sawah, and in-between: agricultural transformation in Borneo. *Human Ecology* 26: 3-21.

Padoch, Christine, Kevin Coffey, Ole Mertz, Stephen J. Leisz, Jefferson Fox, and Reed L. Wadley. 2007. The demise of swidden in southeast Asia? Local realities and regional ambiguities. *Geografisk Tidsskrift-Danish Journal of Geography* 107: 29-41.

Pahl-Wostl, Claudia. 1993. Food webs and ecological networks across temporal and spatial scales. *Oikos* 66: 415-32.

Pahl-Wostl, Claudia. 2006. The importance of social learning in restoring the multifunctionality of rivers and floodplains. *Ecology and Society* 11, no. 1: 10, http://www.ecologyandsociety.org/vol11/iss1/art10/ (accessed October 23, 2008)

Pahl-Wostl, Claudia. 2007. Transitions towards adaptive management of water facing climate and global change. *Water Resource Management* 21: 49-62

Palamas, Gregory. 1983. *The Triads*. The classics of western spirituality series. New Jersey: Paulist Press.

Palm, Cheryl A., Michael J. Swift, and Paul L. Woomer. 1996. Soil biological dynamics in slash-and-burn agriculture. *Agriculture, Ecosystems & Environment* 58: 61-74.

Palm, Cheryl A., Richard A. Houghton, Jerry M. Melillo, and Davy L. Skole. 1986. Atmospheric carbon dioxide from deforestation in southeast Asia. *Biotropica* 18: 177-188.

Palm, Cheryl, Stephen A. Vosti, Pedro A. Sanchez, and Polly J. Ericksen, ed. 2005. *Slash-and-burn agriculture: the search for alternatives.* New York: Columbia University Pres.

Palmer, Craig T., and Reed L. Wadley. 2007. Local environmental knowledge, talk, and skepticism: using 'LES' to distinguish 'LEK' from 'LET' in Newfoundland. *Human Ecology* 35: 749–760.

Palomares, Maria L. D., and Daniel Pauly. 2004. *West African marine ecosystems: models and fisheries impacts*. Fisheries Centre Research Report Vol. 12, no. 7, Vancouver: University of British Colombia Press.

Paniagua, Alejandro, Johann Kammerbauer, Miguel Avedillo, and Ana M. Andrews. 1999. Relationship of soil characteristics to vegetation successions on a sequence of degraded and rehabilitated soils in Honduras. *Agriculture, Ecosystems & Environment* 72: 215-225.

Parada, Isadora L. S. 2004. Mudanças sociambientais de comunidades caiçaras do Parque Estadual da Ilha do Cardoso. In *Enciclopédia caiçara: o olhar do pesquisador*, org. Antonio C. Diegues, 263-273. São Paulo: Hucitec.

Park, Robert Ezra. 1936. Human ecology. *The American Journal of Sociology* 42, no. 1: 1-14.

Park, Robert, and Ernest Burgess 1921. *Introduction to the science of sociology.* Chicago: The University of Chicago Press.

Parmesan, Camille. 2006. Ecological and evolutionary responses to recent climate change. *Annual Review of Ecology and Systematics* 37: 637-669.

Parra, Guido R. 2006. Resource partitioning in sympatric delphinids: space use and habitat preferences of Australian snubfin and Indo-Pacific humpback dolphins. *Journal of Animal Ecology* 75: 862-874.

Parsons, Talcott, and Shills, Edward, ed. 1951. *Towards a general theory of action*. New York: Harper and Row.

Parsons, Talcott. 1937. *The structure of social action.* New York: MacGraw-Hill.

Parsons, Talcott. 1951. *The social system*. Glencoe, IL: Free Press.

Parsons, Talcott. 1966. *Societies: evolutionary and comparative perspectives*. Englewood Cliffs, NJ: Prentice Hall.

Pascual, Unai. 2005. Land use intensification potential in slash-and-burn farming through improvements in technical efficiency. *Ecological Economics* 52: 497–51.

Patterson, John. 1992. *Exploring Maori Values*. Palmerston North NZ: Dunmore Press.

Patterson, John. 1994. Maori environmental ethics. *Environmental Ethics* 16, no.4: 397-409.

Pauly, Daniel 1995. Anecdotes and the shifting baseline syndrome of fisheries. *Trends in Ecology and Evolution* 10: 430.

Pauly, Daniel. 2001. Importance of the historical dimension in policy and management of natural resource systems. In *Proceedings of the INCO-DEV International Workshop on Information Systems for Policy and Technical Support in Fisheries and Aquaculture, Los Baños, Philippines, 2000*, 05-10. Enrico Feoli and Cornelia E. Nauen, ed. The Netherlands: ACP-EU Fisheries Research Report 8.

Pauly, Daniel, and Christensen Villy. 1996. *Mass balance models of north-eastern pacific ecosystems.* Fisheries Centre Research Reports, Vol. 4, Vancouver: University of British Colombia Press:

Pauly, Daniel, Christensen Villy, and Lucilia Coelho, ed. 1999. *Actes de la conférence EXPO'98 sur les réseaux alimentaires des océans et leur productivité économique*. Rapports de Recherche Halieutique ACP-UE, no. 5. Brussels: Office des publications officielles des Communautés européennes.

Pauly, Daniel, Christensen Villy, Sylvie Guénette, Tony J. Pitcher, U. Rashid Sumaila, Carl J. Walters, R. Watson, and Dirk Zeller. 2002. Towards sustainability in world fisheries. *Nature* 418: 689-695.

Paz, Vilma A., and Alpina Begossi. 1996. Ethnoichthyology of Gamboa fishers (Sepetiba Bay, Brazil). *Journal of Ethnobiology* 16: 157-168.

Pearce, David 1996. *Blueprint 4: capturing global environmental value*. New York: Earthscan Publications Ltd.

Pedroso-Jr., Nelson N., Rui S. Murrieta, Carolina S. Taqueda, Natasha D. Navazinas, Aglair P. Ruivo, Danilo V. Bernardo, and Walter A. Neves. 2008. A casa e a roça: demografia e agricultura em populações quilombolas do Vale do Ribeira, São Paulo, Brasil. *Boletim do Museu Paraense Emilio Goeldi, Ciências Humanas,* 3, no. 2: 227-252.

Pelling, Mark, Chris High, John Dearing, and Denis Smith. 2008. Shadow spaces for social learning: a relational understanding of adaptive capacity to climate change within organisation. *Environment and Planning A* 40: 867-884.

Peluso, Nancy L. 1996. Fruit trees and family trees in an anthropogenic forest: ethics of access, property zones, and environmental change in Indonesia. *Comparative Studies in Society and History* 38: 510-548.

Pennesi, Karen. 2007. Beyond the forecast: narrative in prediction performances. Paper presented at SHE XV, October 4-7, in Rio de Janeiro, Brazil.

Pereira, Cassio A., and Ima Célia G. Vieira. 2001. A importância das florestas secundárias e os impactos de sua substituição por plantios mecanizados de grãos na Amazônia. *Interciencia* 26: 337-341.

Peroni, Nivaldo. 1998. Taxonomia folk e diversidade intraespecífica de mandioca (Manihot esculenta Crantz) em roças de agricultura tradicional em áreas de Mata Atlântica do Sul do Estado de São Paulo. Master's diss., ESALQ/USP, Piracicaba, SP, Brazil.

Peroni, Nivaldo. 2004. Ecologia e genética da mandioca na agricultura itinerante do litoral sul paulista: uma análise espacial e temporal. PhD diss., Universidade de Campinas, São Paulo, Brazil.

Peroni, Nivaldo, and Natalia Hanazaki. 2002. Current and lost diversity of cultivated varieties, especially cassava, under swidden cultivation system in the Brazilian Forest. *Agriculture, Ecosystems & Environment* 92: 171-202.

Peroni, Nivaldo, and Paulo S. Martins. 2000. Influência da dinâmica agrícola itinerante na geração de diversidade de etnovariedades cultivadas vegetativamente. *Interciência* 25: 22-29.

Peroni, Nivaldo, Alpina Begossi, and Natalia Hanazaki. 2008. Artisanal fishers' ethnobotany: from plant diversity use to agrobiodiversity management. *Environment, development and sustainability* 10, no. 5: 623-637.

Perrin, William F., Gregory P. Donovan, and Jay Barlow. 1994. *Gillnets and Cetaceans.* Special Issue 15, Cambridge: International Whaling Commission.

Perz, Stephen G., and Robert T. Walker. 2002. Household life cycles and secondary forest cover among small farm colonists in the Amazon. *World Development* 30: 1009-1027.

Peterson, Débora. 2005. Etnobiologia de botos (*Tursiops truncatus*) e a pesca cooperativa em Laguna, Santa Catarina.Bachelor's monography, Universidade Federal de Santa Catarina, Florianópolis, Brazil.

Peyrusaubes, Daniel. 2007. Climatologists and Merina farmers read the sky in Madagascar: crossed glances. Paper presented at the XV International Conference of the Society for Human Ecology, October 4-7, in Rio de Janeiro, Brazil.

Pezzey Jack and Toman, Michael 2005. Sustainability and its economic interpretations. In *Scarcity and growth: natural resources and the environment in the new millennium*, ed. R. David Simpson, Michael. Toman and Robert Ayers. Washington D.C.: RFF Press.

Phillips, Oliver, and Alwyn H. Gentry. 1993a. The useful plants of Tambopata, Peru: I. Statistical hypothesis tests with a new quantitative technique. *Economic Botany*. 47: 15-32.

Phillips, Oliver, and Alwyn H. Gentry. 1993b. The useful plants of Tambopata, Peru: II. Additional hypothesis testing in quantitative ethnobotany. *Economic Botany*. 47: 33-43.

Piaget, Jean. 1951. *The child's conception of the world.* New York: Humanities Press.

Piaget, Jean. 1954. *The construction of reality in the child.* New York: Basic Books.

Pigou, Arthur Cecil. 1920. *The economics of welfare.* London: MacMillan.

Pimenta, Eduardo, and Paulo Hargreaves. 1999. *Relatório de avaliação técnica da viabilidade de zoneamento costeiro e oceânico para bioprodução e atividades complementares.* Rio de Janeiro: IBAMA-Cabo Frio e COPPE/UFRJ.

Pinheiro, Luciane, and Marta J. Cremer. 2003. Etnoecologia e captura acidental de golfinhos (Cetacea: Pontoporiidae e Delphinidae) na Baía da Babitonga, Santa Catarina. *Desenvolvimento e Meio Ambiente* 8: 69-75.

Pinker, Stephen. 1994. *The language instinct.* New York: William Morrow Co.

Pinkerton, Evelyn W., ed. 1989. *Co-operative management of local fisheries: new directions for improved management and community development*. Vancouver: University of British Columbia Press.

Pinnegar, John K., and Georg H. Engelhard. 2008. The 'shifting baseline' phenomenon: a global perspective. *Reviews in Fish Biology and Fisheries* 18: 1–16.

Pinton, Florence, and Laure Emperaire. 2001. Le manioc en Amazonie brésilienne: diversité variétale et marché. *Genetic Selection and Evolution* 33, suppl 1: 491-512.

Pitcher, Tony J. 2001. Fisheries managed to rebuild ecosystems? Reconstructing the past to salvage the future. *Ecological Applications* 11: 601–617.

Plano de Manejo do Parque Estadual da Ilha do Cardoso. 2000. Plano de Manejo do PEIC - Parque Estadual Ilha do Cardoso. Deliberação Consema – 30, 24/10/2000. Diário Oficial do Estado de São Paulo.

Point, Patrick. 1998. La place de l'évaluation des biens environnementaux dans la décision publique. *Economie Publique* 1: 13-46.

Poizat, Gilles, and Eric Baran. 1997. Fishermen's knowledge as background information in tropical fish ecology: a quantitative comparison with fish sampling results. *Environmental Biology of Fishes* 50: 435-449.

Polanyi, Karl 1944. *The great transformation*. Boston: Beacon Press.

Polanyi, Karl 1977. *The livelihood of man*. New York: Academic Press.

Polmieni, John M., Kozo Mayumi, Mario Giampietro, and Blake Alcott. 2008. *The jevons paradox and the myth of resource efficiency improvements*. London: Earthscan Publications Ltd.

Polunin, Nicolas V.C. 1996. Trophodynamics of reef fisheries productivity. In *Reef Fisheries,* ed. Nicolas V.C. Polunin and Callum M. Roberts, 113-135. London: Chapman & Hall.

Pomeroy, Robert S., and Fikret Berkes. 1997. Two to tango: the role of government in fisheries co-management. *Marine Policy* 21: 465-480.

Pomeroy, Robert S., Brenda M. Katon, and Ingvild Harkes. 2001. Conditions affecting the success of fisheries co-management: lessons from Asia. *Marine Policy* 25, no 3: 197-208.

Ponting, Clive. 1991. *A green history of the world: the environment and collapse of great civilizations*. New York: Penguin.

Portes, Alejandro. 1998. Social Capital: its origins and applications in modern sociology. *Annual Review of Sociology* 24: 1-24.

Posey, Darrel A. 1984. Os Kayapó e a natureza. *Ciência Hoje* 2: 35-41.

Posey, Darrel A. 2002. Upsetting the sacred balance: can the study of indigenous knowledge reflect cosmic connectedness? In *Participating in development: approaches to indigenous knowledge*, ed. Paul Sillitoe, Alan Bicker and Johan Pottier, 24-42. London: Routledge.

Prance, Ghillean T. 1999. The poisons and narcotics of the Amazonian Indians. *Journal of the Royal College of Physicians of London* 33: 368-376.

Prance, Ghillean T., William Baleé, Brian M. Boom, and Robert Carneiro. 1987. Quantitative ethnobotany and the case for conservation in Amazonia. *Conservation Biology* 1: 296-310.

Prasad, V. Krishna, Yogesh Kant, Prabhat K. Gupta, C. Sharma, A. P. Mitra, and K. V. S. Badarinath. 2001. Biomass and combustion characteristics of secondary mixed deciduous forests in Eastern Ghats of India. *Atmospheric Environment* 35: 3085–3095.

Pretty, Jules, and Hugh Ward. 2001. Social capital and the environment. *World Development* 29: 209-227.

Prucha, Francis Paul. 1985. *The Indians in American society: from the revolutionary war to the present.* Berkeley: University of California Press.

Przbylski, Cristiane B., and Emygdio L. A. Monteiro-Filho. 2001. Interação entre pescadores e mamíferos marinhos no litoral do Estado do Paraná – Brasil. *Biotemas* 14: 141-156.

Puntenney, Pamela. 2009. Where managerial and scientific knowledge meet sociocultural systems: local realities, global responsibilities. In *Anthropology and climate change: from encounters to actions*, ed. Susan A. Crate and Mark Nuttall, 311-326. Walnut Creek, Ca: Left Coast Press.

Purcell, Trevor W., and Elizabeth A. Onjoro. 2002. Indigenous knowledge power and parity. In *Participating in development*, ed. Paul Sillitoe, Alan Bicker and Johan Pottier, 180-181. London: Routledge.

Purser, Ronald, Changkil Park, and Alfonso Montuori. 1995. Limits to anthropocentrism: towards an ecocentric organization paradigm. *The Academy of Management Review* 20, no.4: 1053-1089.

Queiroz, Helder L., and William G.R. Crampton. 1999. *Estratégias para manejo de recursos pesqueiros em Mamirauá.* Brasília: MCT/ CNPq and Sociedade Civil Mamirauá.

Queiroz, Renato S. 2006. *Caipiras negros no Vale do Ribeira: um estudo de antropologia econômica.* São Paulo: EDUSP.

Quinn, Naomi, and Dorothy Holland, ed. 1987. *Cultural models in language and thought*. New York: Cambridge University Press.

Rabineau, Louis and Richard J. Borden 1991. Human ecology and education: the founding, growth and influence of College of the Atlantic. In *Human ecology, environmental education and sustainable development*, ed Zena Daysh, 137-141. London: Commonwealth Human Ecology Council.

Radasewsky, A. 1976. Considerações sobre a captura de peixes por um *cerco-fixo* em Cananéia -SP – Brasil. *Boletim do Instituto Oceanográfico–USP* 25: 1-28.

Raffles, Hugh. 2002. Intimate knowledge. *International Social Science Journal* 173: 325-333.

Ramakrishnan, Palayanoor Sivaswamy 1992. Shifting agriculture and sustainable development. An interdisciplinary study from north-eastern India. *Man and biosphere series.* 10. Paris: UNESCO.

Raman, Shankar. 2001. Effect of slash-and-burn shifting cultivation on rainforest birds in Mizoram, northeast India. *Conservation Biology* 15: 685-698.

Rambo, A. Terry. 1996. The composite Swiddening agroecosystems of the Tay ethnic minority of the northwestern mountains of Vietnam. In *Land Degradation and Agricultural Sustainability: Case Studies from Southeast and East Asia*, ed. Aran Patanothai. 69-89. Khon Kaen: SUAN Regional.

Ranjan, Rajiv, and V. P. Upadhyay. 1999. Ecological problems due to shifting cultivation. *Current Science* 77: 1246-1250.

Rappaport, Roy A. 1967. *Pigs for the ancestors: ritual and ecology of the New Guinea people*. New Haven: Yale University Press.

Rappaport, Roy A. 1971. The flow of energy in an agricultural society. *Scientific American* 225: 117-132.

Rappaport, Roy A. 1979. *Ecology, meaning and religion*. Berkeley: North Atlantic Books.

Rappaport, Roy A. 1984. *Pigs for the ancestors: ritual in the ecology of a New Guinea people*. Prospect eights: Waveland Press.

Rappaport, Roy A. 1990. Ecosystems, populations and people. In *The ecosystem approach in anthropology: from concept to practice*, ed. Emilio F. Moran, 41-72. Ann Arbor: The University of Michigan Press.

Rappaport, Roy A. 1999. *Ritual and religion in the making of humanity*. Cambridge UK: Cambridge University Press.

Rasul, Golam, and Gopal B. Thapa. 2006. Financial and economic suitability of agroforestry as an alternative to shifting cultivation: the case of the Chittagong Hill Tracts, Bangladesh. *Agricultural Systems* 91: 29–50.

Record, Samuel J., and Robert W. Hess. 1972. *Timbers of the new world*. New York: Arno Press.

Reed, Mark S. 2008. Stakeholder participation for environmental management: a literature review. *Biological Conservation* 141, no.10: 2417-2431.

Reed, Mark S., Evan D G. Fraser, Stephen Morse, and Andrew J. Dougill 2005. Integrating methods for developing sustainability indicators to facilitate learning and action. *Ecology and Society* 10, no. 1: r3, http://www.ecologyandsociety.org/vol10/iss1/resp3/ (accessed October 23, 2008).

Reed, Mark S., Evan D.G. Fraser, and Andrew J. Dougill. 2006. An adaptive learning process for developing and applying sustainability indicators with local communities. *Ecological Economics* 59: 406-418.

Reed, Mark S, Andrew J. Dougill, and Timothy R. Baker 2008. Participatory indicator development: what can ecologist and local communities learn from each other? *Ecological Applications* 18, no. 5: 1253-1269.

Reeves, Randall R., Brian D. Smith, Enrique A. Crespo, and Giuseppe N. di Sciara,. 2003. *Dolphins, whales and porpoises: 2002 – 1020 conservation action plan for the world's cetaceans.* Gland, Switzerland and Cambridge, UK: IUCN / SSC Cetacean Specialist Group.

Rêgo, José F. 1999. Amazônia: do extrativismo ao neoestrativismo. *Ciência Hoje* 25, no. 147:62-65.

Renzo. 2007. Decision making, cultural context and the "human dimensions" of climate studies. Paper presented at SHE XV, October 4-7, in Rio de Janeiro, Brazil.

Ribeiro, Berta G. 1995. *Os índios das águas pretas: modo de produção e equipamento produtivo*. São Paulo: Companhia das Letras/EdUSP.

Ricardo, David. 2006 (1817). *On the Principles of Political Economy and Taxation*. New York: Cosimo.

Rice, James. 2007. Ecological unequal exchange: consumption, equity, and unsustainable structural relationships within the global economy. *International Journal of Comparative Sociology* 48: 43-72.

Richard, P. R., and D. G. Pike. 1993. Small whale co-management in the eastern Canadian arctic: a case history and analysis. *Arctic* 46: 138-143.

Richards, Michael. 1996. *Stabilising the Amazon frontier: technology, institutions and policies*. Natural Resource Perspectives 10, London: Overseas Development Institute, http://dlc.dlib.indiana.edu/archive/00003347/01/nrp_10.pdf (accessed October 24, 2008).

Richardson, Kurt A., Wendy J. Gregory, and Gerald Midgley, ed. 2005. *Systems thinking and complexity science: implications for action.* Boston: ISCE Press.

Richerson, Peter J. 1977. Ecology and human ecology: a comparison of theories in the biological and social sciences. *American Ethnologist* 4: 1-26.

Richerson, Peter J., and Robert Boyd. 2005. *Not by genes alone, how culture transformed human evolution*. Chicago: The University of Chicago Press.

Ritzer, George. 1993. *The McDonaldization of society*. Thousand Oaks, CA: Pine Forge.

Robben, Antonius C.G.M. 1982. Tourism and change in a Brazilian fishing village. *Cultural Survival Quaterly* 6: 18-19.

Roberts, J. Timmons, and Bradley Parks. 2007. Fueling injustice: globalization, ecologically unequal exchange and climate change. *Globalizations* 4: 193-210.

Robinson, William. 2004. *A theory of global capitalism*. Baltimore: Johns Hopkins University Press.

Roder, Walter, Bouakham Phouaravanh, Somphet Phengchanh, Bounthanth Keoboualapha, and Soulasith Maniphone. 1994. Upland agriculture. In *Shifting cultivation and rural development in the LAO PDR*, Activities by Lao-IRRI Project. Report of the Nabong technical meeting, Nabong College, Vientiane. pp. 152-169.

Roder, Walter, Somphet Phengchanh, and Soulasith Maniphone. 1997. Dynamics of soil and vegetation during crop and fallow period in slash-and-burn fields of northern Laos. *Geoderma* 76: 131-144.

Rodrigues, Hélio C.L., Isabel B. Curado, Rodrigo L. Moura, and Carolina Minte-Vera. In Press. Revelando identidades: gestão participativa e os atores da Reserva Extrativista Marinha do Corumbau. *Revista Floresta e Ambiente*.

Rogers, Paul F.. 1974. Draft report on a survey of human ecology degree courses in the commonwealth. *Unpublished manuscript*. Huddersfield: Huddersfield Polytechnic.

Rolston, Holmes. 1988. *Environmental ethics: duties to and values in the natural world*. Philadelphia: Temple University Press.

Roncoli, Carla, Todd Crane, and Ben Orlove. 2009. Fielding climate change in cultural anthropology. In *Anthropology and climate change: from encounters to actions*, ed. Susan A. Crate and Mark Nuttall, 87-115. Walnut Creek, Ca: Left Coast Press.

Rosen, Arlene Miller. 2007. *Civilizing climate: social responses to climate change in the ancient near East.* Walnut Creek: AltaMira Press.

Rossato, Silvia C., Hermógenes F. Leitão-Filho, and Alpina Begossi. 1999. Ethnobotany of Caicaras of the Atlantic forest coast (Brazil). *Economic Botany* 53: 387-395

Roue, Marie, and Douglas Nakashima. 2002. Knowledge and foresight: the predictive capacity of traditional knowledge applied to environment assessment. *International Social Science Journal* 173: 337-347.

Roux, Dirk J., Kevin H. Rogers, Harry C. Biggs, Peter J. Ashton, and Anne Sergeant. 2006. Bridging the science-management divide: moving from unidirectional knowledge transfer to knowledge interfacing and sharing. *Ecology and Society* 11, no.1:4, http://www.ecologyandsociety.org/vol11/iss1/art4/ (accessed October 23, 2008).

Ruddle, Kenneth, and Francis R. Hickey. 2008. Accounting for the mismanagement of tropical nearshore fisheries. *Environment, Development, and Sustainability* 10, no. 5: 565-589.

Rudel, Thomas, and Jill Roper. 1997. Paths to rainforest destruction. *World Development* 25: 53-65.

Ruffino, Mauro. 2003. Provárzea - a natural resource management project for the Amazon floodplains. Paper presented at the international symposium on the Management of Large Rivers, February 11-14, in Phnom Penh, Kingdom of Cambodia.

Sá, Tatiana A., Osvaldo R. Kato, Claudio J. R. de Carvalho, and Ricardo de O. Figueiredo. 2006/2007. Queimar ou não queimar? De como produzir na Amazônia sem queimar. *Revista USP São Paulo 72:* 90-97.

Sachs, Ignacy. 2008. The biofuels controversy. Paper presented at the United Nations Conference on Trade and Development. http://www.unctad.org/en/docs/ditcted200712_en.pdf (accessed October 24, 2008).

Sáenz-Arroyo, Andrea, Callum M. Roberts, Jorge Torre, and Micheline M. Cariño-Olivera. 2005a. Fishers' anecdotes, naturalists' observations and grey reports to reassess marine species at risk: the case of the Gulf grouper in Gulf of California, Mexico. *Fish and Fisheries* 6: 121–133.

Sáenz-Arroyo Andrea, Callum M. Roberts, Jorge Torre, Micheline M. Cariño-Olivera, and Roberto R. Enriquez-Andrade. 2005b. Rapidly shifting environmental baselines among fishers of the Gulf of California. *Proceedings of the Royal Society* 272: 1957-1962.

Said, Edward W. 1993. *Culture and imperialism.* New York: Knopf.

Saint-Paul, Ulrich, and Horacio Schneider, ed. In press. *Mangroves dynamics and management in North Brazil.* Ecological Studies Series. Heidelberg : Springer.

Saito, Kazuki, Linquist Bruce, Bounthanh Keobualapha, Tatsuhiko Shiraiwa, and Takeshi Horie. 2006. Farmers' knowledge of soils in relation to cropping practices: a case study of farmers in upland rice based slash-and-burn systems of northern Laos. *Geoderma* 136: 64-74.

Saldarriaga, Juan G., and Christopher Uhl. 1991. Recovery of forest vegetation following slash-and-burn agriculture in the upper rio Negro. In *Tropical rain forest: regeneration and management,* ed. Arturo Gomez-Pompa, Timothy C. Whitmore, and Malcolm Hadley, 303-312. New York: Blackwell.

Saldarriaga, Juan G., Darrel C. West, M L Tharp, and Christopher Uhl. 1988. Long-term chronosequence of forest succession in the upper Rio Negro of Colombia and Venezuela. *Journal of Ecology* 76: 938-958.

Salick, Jan, Nicoletta Cellinese, and Sandra Knapp. 1997. Indigenous diversity of cassava: generation, maintenance, use and loss among the Amuesha, Peruvian Upper Amazon. *Economic Botany* 51: 6–19.

Sall, Aliou 2007. Loss of bio-diversity: representation and valuation processes of fishing communities. *Social Science Information* 46, no. 1: 153-187.

Salthe, Stanely N. 2007. The natural philosophy of work. *Entropy* 9: 83-99.

Sam, Do Dinh. 1994. *Shifting cultivation in Vietnam: its social, economic, and environment values relative to alternative land use.* London: International Institute for Environment and Development. http://www.iied.org/pubs/pdfs/7505IIED.pdf (accessed October 24, 2008).

Sampaio, Daniela, Vinicius C. Souza, Alexandre A. Oliveira, Juliana Souza-Paula, and Ricardo R. Rodrigues. 2005. *Árvores da restinga: guia ilustrado para identificação das espécies da Ilha do Cardoso.* São Paulo: Neotrópica.

Sanches, Rosely A. 2001. Caiçara communities of the southeastern coast of São Paulo state (Brazil): Traditional activities and conservation policy for the Atlantic rain forest. *Human Ecology Review* 8:52-64.

Sanchez, Pedro A. 2000. Linking climate change research with food security and poverty reduction in the tropics. *Agriculture, Ecosystems & Environment* 82: 371-383.

Sanchez, Pedro A., and M. Hailu, ed. 1996. Alternatives to slash-and-burn agriculture. *Agriculture, Ecosystems & Environment* 58, no. 1, (special issue).

Sanchez, Pedro A., and Roger R. B. Leakey. 1997. Land use transformation in Africa: three determinants for balancing food security with natural resource utilization. *European Journal of Agronomy* 7, no. 1: 15–23.

Sanchez, Pedro A., Cheryl. H. Palm, Stephen. A. Vosti, Thomas P. Tomich, and Joyce Kasyoki. 2005. Alternatives to slash-and-burn: challenge and approaches of an International Consortium. In *Slash-and-burn agriculture: the search for alternatives,* ed. Cheryl H. Palm, Stephen A.

Vosti, Pedro A. Sanchez and Polly J. Ericksen, 3-37. New York: Columbia University Press.

Sanchez, Pedro A., Dale E. Bandy, J. Hugo Villachica, and John J. Nicholaides. 1982. Amazon Basin soils: management for continuous crop production. *Science* 216: 821-827.

Sanford, Robert L., Juan Saldarriaga, Kathleen E. Clark, Christopher Uhl, and Raphael Herrera. 1985. Amazon rain forests fires. *Science* 227: 53-55.

Santos, Katia Maria P. dos, and Maria Elisa de P. E. Garavello. 2003. Um levantamento Etnoecológico da pesca artesanal com cerco no Parque Estadual Ilha do Cardoso. In Anais. I Simpósio de Etnobiologia e Etnoecologia da Região Sul: Aspectos humanos da biodiversidade. SBEE: Florianópolis.

Sargent, Frederick, ed. 1965. Special Issue on Human Ecology. *Bioscience* 15, no. 8: 509-543.

Sargent, Frederick, ed. 1974. *Human Ecology.* New York: American Elsevier.

Scheffer, Marten, Steve Carpenter, Jonathan A. Foley, Carl Folke, and Brian Walker. 2001. Catastrophic shifts in ecosystems. *Nature* 413: 591-596.

Schellnhuber, Hans Joachim, Paul J. Crutzen, William C. Clark; Martin Claussen, and Hermann Held, ed. 2005. *Earth System Analysis for Sustainability.* Cambridge: The MIT Press.

Schmemann, Alexander 1974. *Of the Water & the Spirit: a liturgical study of baptism.* Crestwood, New York: SVSP.

Schnaiberg, Allan, and Kenneth Gould. 1994. *Environment and society: the enduring conflict.* New York: St. Martin's.

Schofer, Evan, and Ann Hironaka. 2005. The effects of world society on environmental outcomes. *Social Forces* 84: 25-47.

Schüle, Wilhelm. 1990a. Human evolution, animal behaviour, and quaternary extinctions: a paleo-ecology of hunting. *Homo* 41: 228-250.

Schüle, Wilhelm. 1990b. Landscapes and climate in prehistory: interactions of wildlife, man and fire. In *Fire in tropical biota. Ecosystem processes and global challenges*, ed. Johann G. Goldammer, 273-318. Berlin and London: Springer-Verlag.

Schüle, Wilhelm. 1992a. Antropogenic trigger effects on Pleistocene climate? *Global Ecology and Biogeography Letters* 2: 33-36.

Schüle, Wilhelm. 1992b. Vegetation, megaherbivores, man and climate in the quaternary and the genesis of closed forests. In *Tropical forests in transition,* ed. Johann G. Goldammer, 45-76. Switzerland: Birkhäuser Verlag Basel.

Schultz, Erich. 1953. *Proverbial expression of Samoa.* Trans. Brother Herman. Auckland: The Polynesian Society.

Schutkowski, Holger. 2006. *Human ecology: biocultural adaptation in human communities.* Heidelberg: Springer.

Schutz, Alfred. 1972. *The phenomenology of the social world.* London: Heinemann.

Scoones, Ian. 1999. New ecology and the social sciences: what prospects for a fruitful engagement? *Annual Review of Anthropology* 28: 479-507.

Scott, Bernard. 1996. Second order cybernetics as cognitive methodology. *Systems Research* 13, no.3: 393-406.

Seixas, Cristiana S., and Alpina Begossi. 2001. Ethnozoology of fishing communities from Ilha Grande (Atlantic Forest Coast, Brazil). *Journal of Ethnobiology* 21: 107 – 135.

Seixas, Cristiana S., and Fikret Berkes. 2003. Dynamics of social-ecological changes in a Lagoon fishery in Southern Brazil. In *Navigating social-ecological systems: building resilience for complexity and change,* ed. Fikret Berkes, Johan Colding, and Carl Folke, 271-298. Cambridge: Cambridge University Press.

Seixas, Cristiana S., and Brian Davy. 2008. Self-organization in integrated conservation and development initiatives. *International Journal of the Commons* 2, no. 1: 99-125. http://www.thecommonsjournal.org/index.php/ijc/article/viewFile/24/20 (accessed October 30, 2008).

Seixas, Cristiana S., and Daniela C. Kalikoski. In Preparation.Gestão participativa da pesca no Brasil: propostas, projetos e documentação dos processos.

Seligman, Martin E.P., and Steven F. Maier. 1967. Failure to escape traumatic shock. *Journal of Experimental Psychology* 74: 1-9.

Sen, Amartya K. 1987. On ethics and economics. London: Blackwell.

Sen, Sevaly, and Jesper R. Nielsen. 1996. Fisheries co-management: a comparative analysis. *Marine Policy* 20, no. 5: 405-418.

Senge, Peter M. 1990. *The fifth discipline: the art & practice of the learning organisation.* New York: Doubleday Currency.

Serbser,Wolfgang H., and Jadranka Mrzljak. 2006. A College of the Atlantic for Europe. *GAIA* 15/4: 307-309.

Serbser,Wolfgang H., and Jadranka Mrzljak. 2007. A College of Human Ecology for Europe. *GAIA* 16/4: 304-306.

Serrão, Emmanuel A., Daniel Nepstad, and Robert T. Walker. 1996. Upland agricultural and forestry development in the Amazon: sustainability, criticality and resilience. *Ecological Economics* 18: 3-13.

Shandra, John M. 2007. International non-governmental organizations and deforestation: good, bad, or irrelevant? *Social Science Quarterly* 88: 665-689.

Shepard, Paul 1967. Whatever happened to human ecology? *Bioscience* 17: 891-894.

Sheppard, Charles. 2006. Traditional nonsense. *Marine Pollution Bulletin* 52: 1–2.

Sherratt, Tim, Tom Griffiths, and Libby Robin. 2003. *A change in the weather: climate and culture in Australia*. Canberra: National Museum of Australia Press.

Siciliano, Salvatore, Ignácio B. Moreno, Érico D. Silva, and Vinícius C. Alves. 2006. *Baleias, botos e golfinhos na Bacia de Campos*. Rio de Janeiro: ENSP/FIOCRUZ.

Sillitoe, Paul, and Alan Bicker. 2004. Hunting for theory, gathering ideology. In *Development and local knowledge*, ed. Paul Sillitoe, Alan Bicker and Johan. Pottier, 1-18. London: Routledge.

Silvano, Renato A.M., and Alpina Begossi. 2001. Seasonal dynamics of fishery at the Piracicaba River (Brazil). *Fisheries Research* 51: 69-86.

Silvano, Renato A.M., and Alpina Begossi. 2002. Ethnoichthyology and fish conservation in the Piracicaba River (Brazil). *Journal of Ethnobiology* 22: 285-306.

Silvano, Renato A.M., and Alpina Begossi. 2005. Local knowledge on a cosmopolitan fish, ethnoecology of *Pomatomus saltatrix* (Pomatomidae) in Brazil and Australia. *Fisheries Research* 71: 43-59.

Silvano, Renato A.M., and John Valbo-Jørgensen. 2008. Beyond fishermen's tales: contributions of fishers' local ecological knowledge to fish ecology and fisheries management. *Environment, Development and Sustainability* 10, no. 5: 657-675.

Silvano, Renato A.M., Andrea L. Silva, Marta Cerone, and Alpina Begossi. 2008. Contributions of ethnobiology to the conservation of tropical rivers and streams. *Aquatic Conservation* 18, no. 3: 241-260.

Silvano, Renato A.M., Priscila F.L. MacCord, Rodrigo V. Lima, and Alpina Begossi 2006. When does this fish spawn? Fishermen's Local knowledge of migration and reproduction of Brazilian coastal fishes. *Environmental Biology of Fishes* 76: 371-386.

Simmel, Georg. 1908/1955. *Conflict and the web of group affiliations*. New York: Free Press.

Simões-Lopes, Paulo C., Marta Fabian, and J.O. Menegheti. 1998. Dolphin interactions with the mullet artisanal fishing on Southern Brazil: a qualitative and quantitative approach. *Brazilian Journal of Zoology* 15: 709-726.

Simon, Herbert A. 1990. Invariants of human bhavior. *Annual Review of Psychology* 41: 1-19.

Simon, Julian L. 1983. Life on earth is getting better, not worse. *The Futurist* 17, no. 4: 7-14.

Singh, Gurmel, Ram Babu, Pratap Narain, L. S. Bhushan, and I. P. Abrol. 1992. Soil erosion rates in India. *Journal of Soil and Water Conservation* 47: 97–99.

Singh, Simron Jit. 2003. *In the sea of influence: a world system perspective of the Nicobar Islands*. Lund Studies in Human Ecology 6. Lund, Sweden: Lund University Press.

Singleton, Sara. 1998. *Construction cooperation: the evolution of institutions of co-management*. Ann Arbor: The University of Michigan Press.

Slik, Ferry J. W., Paul J. A. Kebler, and Peter C. van Welzen. 2003. Macaranga and Mallotus species (Euphorbiaceae) as indicators for disturbance in the mixed lowland dipterocarp forest of East Kalimantan (Indonesia). *Ecological Indicators* 2: 311-324.

Smit, Barry, and Olga Pilifosova. 2003. From adaptation to adaptive capacity and vulnerability reduction. In *Climate change, adaptive capacity and development*, ed. Joel Smith, Richard J. T. Klein and Saleemul Huq, 9-28. London: Imperial College Press.

Smith, Adam. 1999 (1776). *Inquiry into the Nature and Causes of the Wealth of Nations*. New York: Penguin.

Smith, Eric A., and Bruce Winterhalder.1992. *Evolutionary ecology and human behavior*. New York: Aldine de Gruyter.

Smith, Jackie, and Dawn Wiest. 2005. The uneven geography of global civil society: national and global influences on transnational association. *Social Forces* 84: 632-652.

Smith, Joyotee, Petra van de Kop, Keneth Reategui, Ignacio Lombarda, Cesar Sabogal, and Armando Diaz. 1999. Dynamics of secondary forests in slash-and-burn farming: interactions among land use types in the Peruvian Amazon. *Agriculture, Ecosystems & Environment* 76: 85-98.

Smith, Nigel J.H. 1979. *Man, Fishes, and the Amazon.* New York: Columbia University Press.

Snow, Charles Percy. 1959. *The two cultures.* Cambridge: Cambridge University Press.

Snow, Charles Percy. 1998. *The two cultures*. Cambridge, U.K.: Cambridge University Press.

Soemarwoto, Otto, 1974. Rural ecology and development in Java. In *Unifying concepts in ecology,* ed. W. H. van Dobben and Rosemary Lowe-McConnell, 275-281. The Hague: D. W. Junk.

Soja, Edward, W. 1996. *Thirdspace: journey to Los Angels and other real-and-imagined places.* Oxford: Blackwell Publishers

Soto, B., R. Basanta, R. Perez, and Francisco Diaz-Fierros. 1995. An experimental study of the influence of traditional slash-and-burn practices on soil erosion. *Catena* 24: 13-23.

Souza, Shirley P., and Alpina Begossi. 2006. Etnobiologia de *Sotalia fluviatilis* (Gervais, 1853) no litoral norte do estado de São Paulo, Brasil. Paper presented at the Workshop on Research and Conservation of the Genus *Sotalia*, June 19-23, in Búzios, Brazil.

Souza, Shirley P., and Monique Winck. 2005. Captura acidental de pequenos cetáceos no litoral norte de São Paulo. Paper presented at the IV Encontro Nacional sobre Conservação e Pesquisa de Mamíferos Aquáticos, November 12-15, in Itajaí, Brazil.

Souza, Shirley P., Monique Winck, and Beatriz B. Costa. 2008. Hábitos alimentares de pequenos cetáceos no litoral norte do estado de São Paulo. Paper presented at the V Encontro Nacional de Pesquisa e Conservação de Mamíferos Aquáticos, April 6-10, in São Vicente, Brazil.

Spaargaren, Gert, and Arthur P. J. Mol. 1992. Sociology, environment and modernity: ecological modernisation as a theory of social change. *Society and Natural Resources* 5, no. 4: 323-344.

Spinoza, Benedictus de. 1994. *A Spinoza reader: the Ethics and other works.* Trans., edit. Edwin Curley. Princeton: Princeton University Press.

Sponsel, Leslie E. 1986. Amazon ecology and adaptation. *Annual Review of Anthropology* 15: 67-97.

Springate-Baginski, O., and Blaikie, Piers, ed. 2007. Forests, people and power: the political ecology of reform in south Asia. London: Earthscan Publications Ltd.

Steffen, Will 2008. *Fenner school of environment and society.* Canberra: Australian National University. http://fennerschool.anu.edu.au/fenner/welcome/ (accessed November 5, 2008).

Steingraber, Sandra. 1998. *Living downstream: a scientist's personal investigation of cancer and the environment.* New York: Vintage/Random House.

Steward, Julian H. 1955. *Theory of culture change.* Urbana: University of Illinois Press.

Steward, Julian H. 1968. Cultural Ecology. *International Encyclopedia of the Social Sciences,* Vol. 4, 337-344. New York, Macmillan.

Stocks, Anthony. 1983. Candoshi and Cocamilla swiddens in eastern Peru. *Human Ecology* 11: 69-84.

Strauss, Anselm L., Shizuko Fagerhaugh, Barbara Suczek, and Carolyn Wiener, ed. 1997. *Social organization of medical work.* Edison, NJ: Transaction Publishers.

Strauss, Sarah. 2009. Global models, local risks: responding to climate change in the Swiss Alps. In *Anthropology and climate change: from encounters to actions*, ed. Susan A. Crate and Mark Nuttall, 166-174. Walnut Creek, Ca: Left Coast Press.

Strauss, Sarah, and Benjamin Orlove, ed. 2003. *Weather, climate, culture.* New York: Berg Pub.Taddei.

Stromgaard, Peter. 1984. The immediate effect of burning and ash fertilization. *Plant Soil* 80: 307-320.

Stromgaard, Peter. 1985. Biomass, growth, and burning of woodland in a shifting cultivation area of south central Africa. *Forest Ecology and Management* 12: 163-178.

Stromgaard, Peter. 1986. Early secondary succession on abandoned shifting cultivator's plots in the Miombo of South Central Africa. *Biotropica* 18: 97-106.

Styger, Erica R., Harivelo M. Rakotondramasy, Max J. Pfeffer, Erick C.M. Fernandes, and David M. Bates. 2007. Influence of slash-and-burn farming practices on fallow succession and land degradation in the rainforest region of Madagascar. *Agriculture, Ecosystems & Environment* 119: 257-269.

Susanne, Charles, Luc Hens, and Dimitri Devuyst, ed. 1989. *Integration of environmental education into general university teaching in Europe.* Brussels: Vrije Universiteit Brussel (VUB) Press.

Suzuki, Shosuke, Richard J. Borden, and Luc Hens, ed. 1991. Human Ecology—coming of age: an international overview. Brussels: Vrije Universiteit Brussel (VUB) Press.

Swaine, Michael D., and John B. Hall. 1983. Early succession on cleared forest land in Ghana. *Journal of Ecology* 71: 601–627.

Swenson, Roy. 1997. Autocatakinetics, evolution, and the law of maximum entropy production: a principled foundation towards the study of human ecology. *Advances in Human Ecology* 6: 1-47.

Swenson, Roy. 2000. Spontaneous Order, autocatakinetic closure, and the development of space-time. *Annals New York Academy of Science* 901: 311-319.

Swenson, Roy, and M.T. Turvey. 1991. Thermodynamic reasons for perception-action cycles. *Ecological Psychology* 3, no.4: 317-348.

Swidler, Ann. 2001. What anchors cultural practices? In *The practice turn in contemporary theory*, ed. Theodore R. Schatzki, Karin K. Cetina and Eike von Savigny, 74-92. New York: Routledge.

Symes, David, Nathalie Steins, and Juan-Luis Alegret. 2003. Experiences with fisheries co-management in Europe. In *The fisheries co-management experience: accomplishments, challenges and prospects*, ed. Douglas C. Wilson, Jesper R. Nielsen, and Poul Degnbol, 119-133. Dordrecht: Kluwer Academic Publisher.

Szostak, Rick. 2003. Classifying natural and social scientific theories. *Current Sociology* 51, no. 1: 27-49.

Tabarelli, Marcelo, and Waldir Mantovani. 1999. A regeneração de uma floresta tropical montana após corte e queima (São Paulo-Brasil). *Revista Brasileira de Biologia* 59: 239-250.

Taddei, Renzo. 2007. Decision making, cultural context and the "human dimensions' of climate studies. Paper presented at the XV International Conference of the Society for Human Ecology, October 4-7, in Rio de Janeiro, Brazil.

Tansley, Arthur 1935. The use and abuse of vegetational terms and concepts. *Ecology* 16, 284-307.

Taylor, Michael. 1998. Governing natural resources. *Society and Natural Resources* 11, no. 3: 251-258.

Taylor, Peter J. 1997. How do we know we have global environmental problems? Undifferentiated science-politics and its potential reconstruction. In *Changing fife: genomes, ecologies, bodies, commodities* , ed. Peter J. Taylor, Saul E. Halfon, and Paul N. Edwards, 149-174. Minneapolis: University of Minnesota Press.

Tengstrom. E. 1985. Human ecology—a new discipline? A short tentative description of the institutional and intellectual history of human ecology. *Humanekologiska Skrifter* 4: 1-42.

Thompson, William Irwin. 1971. *At the edge of history*. New York: Harper and Row.

Thompson, William Irwin. 1981. *The time falling bodies take to light: mythology, sexuality, and the origins of culture*. New York: St. Martin's.

Tinker, Bernard, John S. I. Ingram, and Sten Struwe. 1996. Effects of slash-and-burn agriculture and deforestation on climate change. *Agriculture, Ecosystems & Environment* 58: 13-22.

Tomich, Thomas, P., Meine Van Noordwijk, Suseno Budidarsono, Andrew N. Gillison, Yanti. Kusumanto, Daniel Murdiyarso, Fred Stolle, and A. M. Fagi. 1998. *Alternatives to slash-and-burn in Indonesia: summary report and synthesis of phase II.* ASB-Indonesia Report Number 8, Bogor, Indonesia: ICRAF.

Tosi, Carolina H., and Renata G. Ferreira. Online First. Behavior of estuarine dolphin, *Sotalia guianensis* (Cetacea, Delphinidae), in controlled boat traffic situation at southern coast of Rio Grande do Norte, Brazil. *Biodiversity and Conservation:* DOI 10.1007/s10531-008-9435-z .

Troesch, Karin. 2003. Highland rice paddy development in mountainous regions of northern Lao PDR. Draft Report. Swiss College of Agriculture. 129 pp.

Truswell, Stewart. 1977. Diet and nutrition of hunter-gatherers. In *Proceedings of the Ciba Foundation Symposium, Health and disease in tribal societies*, *1977, 213-225.* Amsterdam: Elsevier.

Tschakert, Petra C., Oliver T. Coomes, and Catherine Potvin. 2007. Indigenous livelihoods, slash-and-burn agriculture, and carbon stocks in Eastern Panama. *Ecological Economics* 60: 807-820.

Tulaphitak, Thepparit, Pairintra Chaitat, and Kyuma Kazutake. 1985. Changes in soil fertility and tilth under shifting cultivation.ll. Changes in soil nutrient status. *Soil Science and Plant Nutrition* 31: 239-249.

Turner, George 1989. *Samoa: a hundred years ago and long before*. Suva, Fiji: Institute of Pacific Studies, University of the South Pacific.

Turner, Nancy J. 2003. The ethnobotany of edible seaweed (Porphyra abbottae and related species; Rhodophyta: Bangiales) and its use by First Nations on the Pacific Coast of Canada. *Canadian Journal of Botany* 81: 283-293.

Tyrrell, David A., 1977. Aspects of infection in isolated communities. In *Proceedings of the Ciba Foundation Symposium, Health and disease in tribal societies*, *1977, 137-153.* Amsterdam: Elsevier.

Uhl, Cristopher. 1987. Factors controlling sucession following slash-and-burn agriculture in Amazonia. *The Journal of Ecology* 75: 377-407.

Uhl, Cristopher, and Carl F. Jordan. 1984. Succession and nutrient dynamics following forest cutting and burning in Amazonia. *Ecology* 65: 1476-1490.

UN Millennium Project. 2005. *Investing in development: a practical plan to achieve the millennium development goals*. New York: EarthScan, http://www.unmillenniumproject.org/documents/MainReportComplete-lowres.pdf (accessed October 24, 2008).

UNESCO (United Nations Educational, Scientific and Cultural Organization). 2008. *Science and technology policies for sustainable development programme*, Division for Science Policy and Sustainable

Development. http://www.unesco.org/science/psd/programme.shtml (accessed October 24, 2008).

United Nations. 1992. *Long-range world population projections: two centuries of population growth, 1950-2150.* New York: United Nations.

Uotila, Anneli, Jari Kouki, Harry Kontkanen, and Päivi Pulkkinen. 2002. Assessing the naturalness of boreal forests in eastern Fennoscandia. *Forest Ecology and Management* 161: 257-277.

Valbo-Jørgensen, John, and Andreas F. Poulsen. 2000. Using local knowledge as a research tool in the study of river fish biology: experiences from the Mekong. *Environment, Development and Sustainability* 2: 253-276.

Van Andel, Tinde. 2000. The diverse uses of fish-poison plants in Northwest Guyana. *Economic Botany* 54: 500-512.

van Ginkel, R. 2007. Coastal cultures. An anthropology of fishing an whaling traditions. Apeldoorn: Ed. Het Spinhuis.

van Reuler, H., and Bert H. Janssen. 1993. Nutrient fluxes in the shifting cultivation system of south-west Côte d'Ivoire. I. Dry matter production, nutrient contents and nutrient release after slash and burn for two fallow vegetations. *Plant and Soil* 154: 169-177.

Vasconcellos, Marcelo, and Gasalla, Maria A. 2001. Fisheries catches and the carrying capacity of marine ecosystems in southern Brazil. *Fisheries Research* 50: 279-295.

Vázquez de E., Antonio. 1948. *Compendio y descripción de las Indias Occidentales.* Washington, D.C: Smithsonian Institution.

Venkatesan, Krishnamoorthy, Veluchamy Balakrishnan, Konganapuram C. Ravindran, and V. Devanathan. 2005. Ethnobotanical report from mangroves of Pichavaram, Tamil Nadu State, India. *SIDA Contributions to Botany* 21: 2243-2248.

Vézina, Alain F., and Trevor Platt. 1988. Food web dynamics in the ocean. i. best-estimates of flow networks using inverse methods. *Marine and Ecological Progress Series* 42: 269-87.

Vickers, William T. 1983. Tropical Forest mimicry in Swiddens: a reassessment of Geertz's model with Amazonian data. *Human Ecology* 11: 35-45.

Vitebsky, Piers. 2006. Reply *in* Letters. *Natural History* 1152: 10.

Walker, Brian, Stephen Carpenter, John Anderies, Nick Abel, Graeme S. Cumming, Marco Janssen, Louis Lebel, Jon Norberg, Garry D. Peterson, and Rusty Pritchard. 2002. Resilience management in social-ecological systems: a working hypothesis for a participatory approach. *Ecology and Society* 6, no.1: 14, http://www.consecol.org/vol6/iss1/art14/ (accessed October 23, 2008.).

Walker, Robert T., and Alfredo K. O. Homma. 1996. Land use and land cover dynamics in the Brazilian Amazon: an overview. *Ecological Economics* 18: 67-80.

Wallace, David R. 1984. *The Klamath Knot*. San Francisco: Sierra Club.

Wallerstein, Immanuel. 1974. *The modern world system*. Vol. 1. New York: Academic Press.

Warner, Katherine. 1991. *Shifting cultivators: local technical knowledge and natural resource management in the humid tropics*. Rome: UN Food and Agricultural Organization.

Warren, Karen J., and Nisvan Erkal. 1997. *Ecofeminism: women, culture, nature*. Indiana: Indiana University Press.

Watt-Cloutier, Sheila. 2004. Climate change and human rights. In *Human rights dialogue: "environmental rights"*. Carnegie Council, http://www.cceia.org/resources/publications/dialogue/2_11/section_1/4445.html (accessed 9/3/2008).

Watts, Duncan J. 2004. The "new" science of networks. *Annual Review of Sociology* 30: 243-270.

WCED (World Commission on Environment and Development). 1987. *Our common future*. Oxford: Oxford University Press.

Weber, Max. 1948. *From Max Weber: essays in sociology*. Trans., edit. Hans H. Gerth and C. Wright Mills. London: Routledge and Kegan Paul.

Weber, Max. 1978 (1921). *Economy and society: an outline of interpretive sociology*. Berkeley: University of California Press.

Weber, Max. 1985 (1904). *The protestant ethic and the spirit of capitalism*. London: Unwin.

Weber, Thomas. 1987. *Hugging the trees: the story of the Chipko movement*. New York: Viking-Penguin.

Weed, Mike. 2005. "Meta Interpretation": a method for the interpretive synthesis of qualitative research. *Forum Qualitative Social Research* 6, no. 1: 37, http://www.qualitative-research.net/index.php/fqs/article/view/508/1096 (accessed October 23, 2008).

Weeks, Brian, Marko A. Rodriguez, and J.H. Blakeslee. 2004. Panarchy: complexity and regime change in Human Societies. In *Proceedings of Complex Systems Summer School, August, Santa Fe, 2004*, CSSS04. Santa Fe: Santa Fe Institute.

Weeratna, C. S. 1984. The effect of shifting cultivation in the tropics on some soil properties. *Beitrage zur Tropischen Landwirtschaft und Veterinarmedizin* 22: 135-139.

Wenger, Etieenne 1998. *Communities of practice: learning, meaning, and identity*. Cambridge: Cambridge University Press

White, Hayden. 1973. *Metahistory: the historical imagination in nineteenth-century Europe*. Baltimore: Johns Hopkins University Press.

White, Leslie. 1943: Energy and the evolution culture. *American Anthropologist* 45: 335-356.

White, Richard. 1983. *The roots of dependency: subsistence, environment, and social change among the Choctaws, Pawnees, and Navajos.* Lincoln: University of Nebraska Press.

Whitehead Hal, and Janet Mann. 2000. Female reproductive strategies of cetaceans. In *Cetacean Societies: field studies of dolphins and whales*, ed. Janet Mann, Richard C. Connor, Peter L. Tyack, and Hal Whitehead, 222-223. Chicago: The University of Chicago Press.

Whitmore, Timothy C. 1991. Tropical forest dynamics and its implications for management. In *Rain forest regeneration and management,* ed. Gomez-Pompa, Arturo, Timothy C. Whitmore, and Malcolm Hadley, 67-89. UNESCO, Paris, France: The Parthenon Publishing Group.

WHO (World Health Organization). 1993. Dengue prevention and control. Report by the Director-General. *46th World Health Assembly* A46/8: 1-7.

Wiium, Vilhjálmur. 1999. *Reducing highgrading of fish: what can be done?* Glaway, Ireland: National University of Ireland.

Wilden, A. 1980. *System and structure: essays in communication and exchange*. London: Tavistock

Wilk, Richard. 2009. Consuming ourselves to death: the anthropology of consumer culture and climate change. In *Anthropology and climate change: from encounters to actions*, ed. Susan A. Crate and Mark Nuttall, 265-276. Walnut Creek, Ca: Left Coast Press.

Wilkie, David, and J. Timothy Finn. 1990. Slash-and-burn cultivation and mammal abundance in the Ituri Forest, Zaire. *Biotropica* 22: 90-99.

Wilkie, David, Gilda Morelli, Fiona Rotberg, and Ellen Shaw. 1999. Wetter isn't better: global warming and food security in the Congo Basin. *Global Environmental Change* 9: 323-328.

Willis, Katherine J., Lindsey Gillson, and T. M. Brncic. 2004. How "virgin" is virgin rainforest? *Science* 304: 402-403.

Willis, Roy G. 1980. Magic and "medicine" in Ufipa. In *Culture and curing*, ed. Peter Morley and Roy Wallis, 139-151. Pittsburgh: University of Pittisburg Press.

Wilson, David S, Mark Van Vugt, and Rick O'Gorman. 2008. Multilevel selection theory and major evolutionary transitions. *Current Directions in Psychological Science* 17, no. 1: 6-9.

Wilson, Douglas C. 1999. Fisheries science collaborations: the critical role of the community. *IFM Research publication no. 45.*

Wilson, Douglas C., Jesper R. Nielsen, and Poul Degnbol, ed. 2003. *The fisheries co-management experience: accomplishments, challenges and prospects.* Dordrecht: Kluwer Academic Publisher.

Wilson, Douglas C., Mahfuzuddin Ahmed, Susanna V. Siar, and Usha Kanagaratnam. 2006. Cross-scale linkages and adaptive management: fisheries co-management in Asia. *Marine Policy* 30, no. 5: 523-533.

Wilson, Edward O. 1999. *Consilience: the unity of knowledge.* London: Abacus.

Winklerprins, Antoinette M. G. 1999. Local soil knowledge: a tool for sustainable land management. *Society and Natural Resources* 12: 151-161.

Winterhalder, Bruce, and Eric A. Smith. 1981. *Hunter-gatherer foraging strategies.* Chicago: The University of Chicago Press.

Wisner, Ben, Maureen Fordham, Ilan Kelman, Barbara Rose Johnston, David Simon, Allan Lavell, Han Gunter Brauch, Ursula Oswald Spring, Gustavo Wilches-Chaux, Marcus Moench, and Daniel Weiner. 2007. *Climate change and human security*, http://www.radixonline.org/cchs.html (accessed October 23, 2008).

Wooster, Donald 1977. *Nature's economy: the roots of ecology.* San Francisco: Sierra Club Books.

Worster, Donald. 2003. Transformações da Terra: para uma perspectiva agroecológica da história. *Ambiente & Sociedade* 5: 23-44.

Wuthnow, Robert. 1987. *Meaning and moral order: explorations in cultural analysis.* Berkeley: University of California Press.

Xu, Jianchu, Erzi T. Ma., Duojie Tashi, Yongshou Fu, Zhi Lu, and David Melick. 2005. Integrating sacred knowledge for conservation: Cultures and landscapes in southwest China. *Ecology and Society* 10, no. 2: 7, http://www.ecologyandsociety.org/vol10/iss2/art7/ (accessed October 23, 2008).

York, Richard, Eugene A. Rosa, and Thomas Dietz. 2003. Footprints on the earth: the environmental consequences of modernity. *American Sociological Review* 68: 279-300.

Young, Gerald 1974. Human ecology as an interdisciplinary concept: A critical inquiry. *Advances in Ecological Research.* 8: 1-105.

Young, Gerald, ed. 1983. *Origins of Human Ecology.* Stroudsburg, Penn.: Hutchinson Ross.

Young, Gerald 1989. A conceptual framework for an interdisciplinary human ecology. *Acta Oecologica Hominis: International Monographs in Human Ecology* 1, no. 1: 1-137.

Yunus, Muhammad. 1999. *Banker to the poor: micro lending and the battle against world poverty.* New York: Perseus/Public Affairs.

Yunus, Muhammad. 2007. *Creating a world without poverty: social business and the future of capitalism*. New York: Perseus/Public Affairs.

Zarin, Daniel J., Eric A. Davidson, Eduardo Brondizio, Ima C. G. Vieira, Tatiana Sá, Ted Feldpausch, Edward A.G. Schuur, Rita Mesquita, Emilio Morán, Patricia Delamonica, Mark J. Ducey, George C. Hurtt, Cleber Salimon, and Manfred Denich. 2005. Legacy of fire slows carbon accumulation in Amazonian forest regrowth. *Frontiers in Ecology and the Environment* 3: 365-369.

Zhang, Quanfa, Christopher O. Justice, and Paul V. Desanker. 2002. Impacts of simulated shifting cultivation on deforestation and the carbon stocks of the forests of Central Africa. *Agriculture, Ecosystems & Environment* 90: 203-209.

Zimmerer, Karl L. 2004. Cultural ecology: placing households in human-environment studies: the cases of tropical forest transitions and agrobiodiversity change. *Progress in Human Geography* 28: 795-806.

Zinke, Paul J., Sanga Sabhasri, and Peter Kunstadter. 1978. Soil fertility *aspects* of the Lua forest fallow system of shifting cultivation. In *Farmers in the forest. economic development and marginal agriculture in northern Thailand*, ed. Kunstadter, Peter, E. C. Chapman, and Sanga Sabhasri, 134-159. Honolulu: The University Press of Hawaii.

Zollett, Erika A., and Andrew J. Read. 2006. Depredation of catch by bottlenose dolphins (*Tursiops truncatus*) in the Florida king mackerel (*Scomberomorus cavalla*) troll fishery. *Fishery Bulletim* 104: 343–349.

Zizioulas, John D. 1985. *Being as communion: studies in personhood and the church*. New York: SVSP

Zizioulas, John D. 2006. *Communion & Otherness.* New York: T & T Clark.

Contributors

Cristina Adams is an Assistant Professor and director of the Laboratory of Human Ecology at the School of Arts, Sciences and Humanity at the University of São Paulo, Brasil. She holds a PhD in Ecology (Department of Ecology, University of São Paulo) and was a Visiting Research Fellow at the Department of Anthropology at the University of Kent at Canterbury (UK, 1999-00). She has been studying the adaptation of indigenous Brazilian populations to the Amazon and the Atlantic Rainforest for the last ten years. Her most recent co-edited book *Amazon Peasant Societies in a Changing Environment - Political Ecology, Invisibility and Modernity in the Rainforest* has just been published by Springer.

Fernando Dias de Avila-Pires is a Researcher at the Department of Tropical Medicine, Oswaldo Cruz Institute, Rio de Janeiro, Brazil, and a visiting Professor at the International Master's Programme of Human Ecology, Vrije Universiteit Brussel, Belgium. He is currently teaching and doing research on zoonoses at the Master's Programme of Public Health at Federal University of Santa Catarina, Brazil. Professor. Avila-Pires is a member of the Brazilian Academy of Sciences.

Alpina Begossi, Ph.D. in Ecology (University of California, Davis, 1989), has been studying the ecology of Amazonian and Atlantic Forest fisheries as one of her main research lines, among other studies in human ecology. She has about 100 published works, which include articles published in periodicals, such as Ecological Applications, Ecology and Society, Fisheries Research, among others, besides four books with other authors. She is currently the Executive Director of the Fisheries and Food Institute (FIFO), an NGO founded by her and collaborators and also works as a Researcher at the State University of Campinas (Capesca, Preac, UNICAMP).

Richard Borden (Ph.D. in Psychology, Kent State University) is the Rachel Carson Chair in Human Ecology at College of the Atlantic, where he also served as Academic Dean for twenty years. He is past president of the international Society for Human Ecology and continues to serve as its Executive Director. Author of numerous publications, Rich has been a USIA academic consultant and an advisor to human ecology programs in China, Russia and elsewhere in Europe and in North and South America. Course areas: environmental psychology, community planning and decision making, personality and social development, contemporary psychology, and philosophy of human ecology

Karl Bruckmeier is a senior lecturer and researcher at the University of Göteborg, School of Global Studies, Sweden. He has single authored two books and co-authored more than 12 books, besides more than 12 papers. His current research interests are sustainable rural and coastal development.

Thomas J. Burns is Professor of Sociology at the University of Oklahoma. He does both quantitative and qualitative research on the interface between human social organization and the natural environment. In ongoing projects with a number of collaborators, he examines how macro-level cultural, social, demographic, political and economic processes affect environmental outcomes including pollution, deforestation and the ecological footprint.

Serge Collet is a maritime anthropologist based at the Institute of Ethnology of the University of Hamburg and also associate professor at the CIES MAJISE (Fallows), University of Calabria. He has participated the last ten years in diverse European research projects and expertise workshops convened by EU Directorates. He is the author of more than 40 articles and co-author of books. He is the guest editor of "Pursuing of the true value of the people and the sea", Social Science Information, 46,1,March 2007, Sage publications.

Susan A. Crate is an anthropologist specializing in human ecology, sustainability and climate change. She has conducted research in Siberia since 1989 and is an assistant professor at George Mason University in Fairfax, Virginia.

Robert Dyball is the convenor of the Human Ecology Program at the Australian National University. His research interests are in Human Ecology especially Urban Ecology, Food Systems, Education for Sustainability (EfS), and the application of dynamic systems thinking to problems of sustainability. Robert is on the board of the international Society for Human Ecology (SHE) and academic advisor to Australasian Campuses Towards Sustainability (ACTS).

Pierre Failler is a Senior Research Fellow at the Centre for the Economic and Management of Aquatic Resources (CEMARE), University of Portsmouth. He currently manages the ECOST project and the POORFISH project (both European projects). Recent publications include: (with M.

Diop and S. M'Bareck) (2006) Les effets de la libéralisation du commerce. Le cas du secteur des pêches de la République de Mauritanie. Genève: PNUE; with Hoaran Pan (2007), Global value, full value and societal costs; capturing the true cost of destroying marine ecosystems, Social Information Journal, Vol. 46, No. 1, pp. 109-134. With Andy Thorpe, David Whitmarsh and P.F. (2007), The Situation in World Fisheries, Encyclopedia of Live Supports Systems (EOLSS), (Contribution 5.5.5.4)

Célia Futemma is an Associate Professor at the Federal University of São Carlos (campus of Sorocaba), Brazil. Biologist with a major in Anthropology and Environmental Sciences. Publications are found in indexed journals at national and international levels and chapters of books. Main research areas are: community-based management of natural resources (forest and fishing), patterns of land use and land cover (deforestation), human ecology, common-pool resources and household analysis. Main areas of study: Amazon and Atlantic Forest.

Maria A. Gasalla is an Associate Professor of Fisheries at the Department of Biological Oceanography, Instituto Oceanográfico, University of São Paulo, Brazil, lecturing both undergrad and graduate courses with international background. She is presently leading the Fisheries Ecosystems Laboratory (Lab Pesq) and implementing research focused on innovative and integrated approaches to fisheries in Brazil, linking different knowledge systems. She wrote more than 30 peer-review publications included in internatioval scientific journals and books, such as Fisheries Research, Journal of Coastal Research, FAO Technical Papers, Ecological Modeling and REVIZEE series-Brazil, among others.

Bernhard Glaeser is a Ph.D. in Philosophy from the University of Hamburg; with a doctoral dissertation in the area of epistemology: "Critique of the Sociology of Knowledge". He is presently the president of the German Society for Human Ecology (DGH) and Professor at Free University Berlin, Department of Political and Social Sciences. His current research interests are Human Ecology, Sustainable Coastal Management and Environment and Development.

Marion Glaser is a Senior Social Scientist at the Center for Tropical Marine Ecology (ZMT) in Bremen (Germany). Her thematic specializations include transdisciplinary sustainability research, participatory resource management, the social dimensions of ecosystem management, livelihood strategies of rural households and social-ecological systems analysis.

Currently she teaches at under and postgraduate level in Germany and Brazil. Her publications include more than 33 articles in peer reviewed journals, two book chapters and two book editions.

Natalia Hanazaki is an Ecologist (Universidade Estadual Paulista "Júlio de Mesquita Filho" UNESP/Rio Claro), holds a master in Ecology (Institute of Biology, Universidade de São Paulo , 1997); and a doctor degree in Ecology (Institute of Biology, Universidade Estadual de Campinas , 2001). Associate professor of the Department of Ecology and Zoology, Center of Biological Sciences, Universidade Federal de Santa Catarina, where coordinates the Laboratory of Human Ecology and Ethnobotany.

Gesche Krause works at the Center for Tropical Marine Ecology (ZMT) in Bremen, Germany, and since 1997 she has been mainly involved in the Mangrove Dynamics and Management (MADAM) project, a 10-year bilateral co-operation between Germany and Brazil (1995-2005). Since 2006 she is involved in a new social science research cluster in the Science for the Protection of Indonesian Coastal Marine Ecosystems (SPICE) program in Indonesia, which is an interdisciplinary German-Indonesian initiative in earth system research. She is also involved as consultant research partner in the joint research project Coastal Futures that deals with ICZM questions at the western coast of Schleswig-Holstein.

Priscila F. M. Lopes holds a Ph.D. in Ecology (State University of Campinas, Brazil) and currently is an associated researcher at Fisheries and Food Institute. Her research has focused on small-scale fisheries, local fisheries management initiatives, and the use of ecological models to understand human foraging behavior. In the last years she has been working as a consultant on Amazonian small-scale fisheries and indigenous use of natural resources to different Brazilian federal institutes. Some of her research also approach ethnoecology, ethnobiology and diet change processes on fishermen's communities. Her recent papers can be found in Hydrobiologia, Environment, Development & Sustainability, Current Anthropology, Ecology and Society, Environmental Biology of Fishes, Journal of Tropical Ecology, among others.

Tatiana Mota Miranda is an Ecologist (Universidade Estadual Paulista "Júlio de Mesquita Filho" UNESP/Rio Claro/Brazil) and holds a master degree in Plant Biology at the Universidade Federal de Santa Catarina (UFSC), Florianópolis, Brazil.

Keith Morrison is a lecturer at Lincoln University, New Zealand. While studying for his PhD at Lincoln University, he also studied Maori language to assist deepening knowledge of Maori culture in New Zealand. Upon completion of his Phd "A postmodern reconstruction of floodplain management methodologies", which trained him to take up an academic post in human ecology, he first worked at the University of the South Pacific and then returned to work in New Zealand. His research has continued in the South Pacific however, in particular in Samoa. He has focused since then on ethnomethodological work to understand how indigenous communities in the South Pacific facilitate learning. Between 2002 and 2005 he focused on establishing the Sustainable Community Development forum, which is presently extending to become a worldwide network and looking towards becoming a formal independent institute. Currently he supervises PhD and Masters thesis students in sustainable community development and ecological engineering.

Rui Sérgio Sereni Murrieta is an Assistant Professor of the Genetics and Evolutionary Biology Department at the University of São Paulo. In 2003, concluded his post-doctoral research in the Laboratory for Human Evolutionary Studies in the same department. He holds a Ph.D. from the Department of Anthropology at the University of Colorado at Boulder in 2000, under the supervision of Dr. Darna Dufour and Dr. Terry McCabe. His main academic interests are the impact of recent social and economic changes on subsistence activities, especially swidden cultivation, and on the nutritional status of indigenous populations in the Amazon and the Atlantic Rainforest environments.

Flávia Camargo Oliveira is a Biologist (Universidade Federal do Paraná-UFPR, Curitiba/PR), holds a master degree in Plant Biology (Universidade Federal de Santa Catarina-UFSC, Florianópolis, 2007). Currently she is a researcher associated at the "Instituto de Pesquisas Cananéia-IPeC, Cananéia/SP" works as a consultant for Wageningen International (Wageningen, The Netherlands) organizing and facilitating a Tailor-Made Programme entitled "Conservation, management and use of agrobiodiversity" which is held in Ecuador and Brazil.

Nelson Novaes Pedroso Junior is a Biologist. He holds a M.Sc. degree in Ecology from the Federal University of São Carlos (2002), where studied the conflicts between traditional fisheries populations and their maintenance in protected natural areas. Currently, he has concluded his PhD in Ecology (Department of Ecology, University of São Paulo), where he studies the political ecology of slash-and-burn agriculture of quilombola populations in Atlantic Rainforest areas. His research interests are human ecology studies with particular emphasis on the livelihood systems of indigenous people in neotropical forests.

Nivaldo Peroni is Agronomist, holds a master degree in Genetics and Plant Evolution at University of São Paulo (ESALQ – USP), and has a Doctor degree in Plant Biology at the State University of Campinas (Unicamp), Campinas, Brazil. Currently he is the vice-president of Fisheries and Food Institute (FIFO) and a researcher at the Department of Cellular Biology, Embriology and Genetics at the Universidade Federal de Santa Catarina.

Aliou Sall is a fisheries socio-anthropologist. Based in Senegal, he is the executive secretary of CREDETIP (Centre de Recherche pour le Développement des Technologies Intermédiaires de Pêche), a Senegalese fisheries NGO. For the last 16 years he has been working with fishing communities and fishworkers' social movements at the international level. During this same period he has been involved in several scientific studies as senior researcher in collaboration with northern universities, Bergen, Namur and Geneva (Graduate Institute of Development Studies), as well as southern universities and other research-related institutions such as the University Cheikh Anta Diop of Dakar, the University Gaston Berger of Saint-Louis and the Fisheries and Aquaculture Graduate Institute of the University Cheikh Anta Diop.

Cristiana Simão Seixas holds a PhD in Natural Resources and Environmental Management from the University of Manitoba, Canada. She has worked as a consultant in the field of Participatory Fisheries Management and Community-based Conservation and Development initiatives. Presently, she is an assistant professor at the State University of Campinas (UNICAMP), Brazil. Her research interests include common-property resource management, co-management, social-ecological resilience, and local/traditional ecological knowledge. She has studied artisanal fisheries in Brazil in several localities inland and on the coastal area.

Renato A.M. Silvano has been doing research dealing with artisanal fisheries, fish ecology and ethnobiology since 1994, participating in about 11 research projects conducted in marine, estuarine and freshwater environments in the southeast and northeast Brazilian coasts, as well as the Brazilian Amazon. Currently, Renato is professor in the ecology department of the Federal University of Rio Grande do Sul in Porto Alegre, Brazil. He has published about 16 peer-reviewed articles in journals such as Environmental Biology of Fishes, Fisheries Research, Ecological Economics, Hydrobiologia and Aquatic Conservation.

Simron Jit Singh (Ph.D. Human Ecology) is engaged in field-based sustainability studies in Asia, primarily among non-industrial societies on their way to integrate fully into the market economy. His main empirical interests are to examine the altering patterns of society-nature interactions and the impact of development aid on the sustainability of local regions. His books are: In the Sea of Influence: A World System Perspective of the Nicobar Islands (2003), The Nicobar Islands: Cultural Choices in the Aftermath of the Tsunami (2006), Complex Disasters: Humanitarian aid and Sustainable Development (forthcoming). He has been awarded an International Fellowship from the Austrian Ministry of Education, Science and Culture, and a START Fellowship within the framework of the IHDP-IT global environmental change research.

Shirley Souza has been working with marine mammals in the southeastern Brazilian coast in the last 13 years, addressing especially the interactions between small cetaceans and fishing gears. In her doctorate project she is studying fishers' knowledge on cetaceans in several fishing communities along the Brazilian coast. She has four publications about cetacean ecology and ethnobiology, including international journals such as International Journal for Ethnobiology and Ethnomedicine.

Index